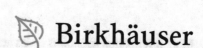 **Birkhäuser**

Static & Dynamic Game Theory: Foundations & Applications

More information about this series at http://www.springer.com/series/10200

Leon A. Petrosyan • Vladimir V. Mazalov •
Nikolay A. Zenkevich

Editors

Frontiers of Dynamic Games

Game Theory and Management,
St. Petersburg, 2019

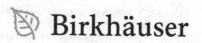 Birkhäuser

Editors
Leon A. Petrosyan
St. Petersburg State University
St. Petersburg, Russia

Vladimir V. Mazalov
Institute of Applied Mathematical Research
Russian Academy of Sciences
Petrozavodsk, Russia

Nikolay A. Zenkevich
Graduate School of Management
St. Petersburg State University
St. Petersburg, Russia

ISSN 2363-8516 ISSN 2363-8524 (electronic)
Static & Dynamic Game Theory: Foundations & Applications
ISBN 978-3-030-51943-8 ISBN 978-3-030-51941-4 (eBook)
https://doi.org/10.1007/978-3-030-51941-4

Mathematics Subject Classification: 90B, 91A, 91B

This book is published under the imprint Birkhäuser, www.birkhauser-science.com, by the registered company Springer Nature Switzerland AG.
The registered company address is: Gewerbestrasse 11, 6330 Cham, Switzerland

Preface

The content of this volume is mainly based on selected talks that were given at the "International Meeting on Game Theory (ISDG12-GTM2019)," as joint meeting of "12th International ISDG Workshop" and "13th International Conference on Game Theory and Management," held in St. Petersburg, Russia on July 03–05, 2019. The meeting was organized by St. Petersburg State University and International Society of Dynamic Games (ISDG).

Every year starting from 2007, an international conference "Game Theory and Management" (GTM) has taken place at the Saint Petersburg State University. Among the plenary speakers of this conference series were the Nobel Prize winners Robert Aumann, John Nash, Reinhard Selten, Roger Myerson, Finn Kidland, Eric Maskin, and many other famous game theorists. The underlying theme of the conferences is the promotion of advanced methods for modeling the behavior that each agent (also called player) has to adopt in order to maximize his or her reward once the reward does not only depend on the individual choices of a player (or a group of players), but also on the decisions of all agents that are involved in the conflict (game). In particular, the emphasis of the ISDG12-GTM2019 conference was on the following topics:

- Cooperative games and applications
- Dynamic games and applications
- Pursuit-evasion games
- Dynamic networks
- Stochastic games and applications
- Market models
- Networking games
- Auctions
- Game theory applications in fields such as strategic management, industrial organization, marketing, public management, financial management, human resources, energy and resource management, and others.

In this volume, two sorts of contributions prevail: chapters that are mainly concerned with the application of game-theoretic methods and chapters where the theoretical background is developed.

In the chapter of Petr Ageev and Svetlana Tarashnina, a three-player hierarchical game with perfect information modeling the competition on telecommunication market is considered. The Nash equilibrium is found and the results are illustrated on an example.

In the chapter of Sadettin Haluk Citci and Kubra Uge, dynamic Bertrand competition in mixed oligopoly where a private firm competes with a social welfare maximizing public firm is considered. The corresponding game-theoretic model is constructed and it is proved that under some conditions the game possesses the unique Bayes equilibrium in dominant strategies.

In the chapter of Denis Fedyanin, the version of Tullock rent-seeking game is considered. The results provide insights into the impact of reflexive analysis on the properties of information control. In the most simple case, the Nash equilibrium is found and its properties are discussed.

In the chapter of Ekaterina Gromova and Anastasiya Malakhova, a special differential 3-person cooperative game on network is considered. For finding solutions in such class of games, the characteristic function must be calculated. There are different approaches for the definition of this function. The authors use classical Neumann–Morgenstern approach and found the values of this function in explicit form.

In the chapter of Igor Konnov, the problem of finding the optimal performance of a composite system taking in account the possible external interference and corresponding protecting measures. In general, the problem can be formulated as a parametric zero-sum game. The author proposes an inexact penalty method for the solution which can be applied also to more general cases.

In the chapter of Alexei Korolev, the network game-theoretic model with production and knowledge externalities and stochastic parameters is considered. The explicit expressions for the dynamics of a single agent and dyad agents in the form of Brownian random processes are derived and the qualitative analysis of the solutions is provided.

In the chapter of David A. Kosian and Leon A. Petrosyan, the new type of characteristic function for the cooperative games with hypergraph communication structure is proposed. The value of the characteristic function for each coalition is computed as sum of the cooperative payoffs of players from this coalition under an additional condition that players outside the coalition are cutting links with coalition members. Is it proved that the corresponding game is convex.

In the chapter of Nikolay A. Krasovskii and Alexander M. Tarasyev, the algorithm is developed for finding the value function of the zero-sum setting for differential dynamic bimatrix games on an infinite time interval. The results are illustrated on nontrivial and interesting example of 2×3 bimatrix game. The construction of the value functions of both associated zero-sum games is provided and a feedback strategy of the first player is found.

In the chapter of Suriya Kumacheva, Elena Gubar, Ekaterina Zhitkova, and Galina Tomilina, the effect of information spreading about future tax audits is considered using an evolutionary game approach. The scenario analysis has shown that the propagation of information about the possible audit may increase the tax collection.

In the chapter of Denis Kuzyutin, Ivan Lipko, Yaroslavna Pankratova, and Igor Tantlevskij, the problem of time consistency of cooperative solutions in dynamic multicriterial games is considered. The variation of a known IDP technique which guarantees the no negativity of stage payments necessary for the realization of time-consistent solutions is proposed. Results are illustrated on interesting examples.

In their chapter, Mario Alberto Garcia Meza and Cesar Gurrola Rios considered conditions under which the cooperation between companies may give better results than competition. While substitute goods hardly find solution for cooperation, in the article the examples are given on how the complementarily can achieve such results. The existing time-consistency problem in cooperation is not considered.

In the chapter of Ekaterina Orlova, the one-shot cooperative game model of Eurasian gas network is considered and different solution concepts such as the Shapley value, core, and nucleolus are compared. Among other topics, the effect of liberalization is studied. Many interesting examples based on real data analysis are presented.

In the chapter of Dmitry Rokhlin and Gennady A. Ougolnitsky, the continuous-time dynamic game in the case of one leader and one follower is considered. The conditions for the existence of Stackelberg solution are derived and the solution is found in explicit form in a game-theoretic model of a nonrenewable resource extraction problem.

In the chapter of Ovanes Petrosian, Maria Nastych, and Yin Li, the differential game model is applied to analyze the world oil market. The looking forward approach is proposed to take into account dynamically updating information. To model the situation on different time intervals, non-cooperative and cooperative approaches are used. The numerical simulations based on open access date are presented.

In the chapter of Simon Rothfuß, Jannik Steinkamp, Michael Flad, and Sören Hohmann, the game-theoretic approach is used to model the interaction of humans and automated assistant systems. The goal is to design emancipated cooperative decision-making systems capable of negotiating with humans. Two game concepts: event-based game and game model based on war of attrition are proposed for describing the negotiation process.

Alexander Sidorov in his chapter tries to show on developed mathematical models that in some cases the typical presumption of the most economic theories that free entry is desirable for social efficiency may not be always true. He considers the one-sector economy with horizontally differentiated good and one production factor-labor with special classes of utility functions.

Igor Shevchenko in his chapter considers a classical game of obstacle tag proposed first by R. Isaaks in his book. The solution of the game in the sense of saddle point is not known although many papers are published on the subject. The author proposes the strategy for the pursuer which allows him to choose the geodesic lines and evaluates the guaranteed results for this strategy.

The ISDG12-GTM2019 program committee thanks all the authors for their active cooperation and participation during the preparation of this volume. Also, the organizers of the conference gratefully acknowledge the financial support given by the Saint Petersburg State University. Last but not least, we thank the reviewers for their outstanding contribution and the science editor.

Saint Petersburg, Russia   L. A. Petrosyan
Petrozavodsk, Russia   V. V. Mazalov
Saint Petersburg, Russia   N. A. Zenkevich

Contents

Contributors

Petr Ageev Saint Petersburg State University, St. Petersburg, Russia

Sadettin Haluk Citci Gebze Technical University, Kocaeli, Turkey

Denis Fedyanin V.A.Trapeznikov Institute of Control Sciences, Moscow, Russia

Michael Flad Institute of Control Systems (IRS) at Karlsruhe Institute of Technology (KIT), Karlsruhe, Germany

Ekaterina Gromova Saint Petersburg State University, St. Petersburg, Russia

Elena Gubar Saint Petersburg State University, St. Petersburg, Russia

Sören Hohmann Institute of Control Systems (IRS) at Karlsruhe Institute of Technology (KIT), Karlsruhe, Germany

Igor Konnov Kazan Federal University, Kazan, Russia

Alexei Korolev National Research University Higher School of Economics, St. Petersburg, Russia

David A. Kosian Saint Petersburg State University, St. Petersburg, Russia

Nikolay A. Krasovskii Krasovskii Institute of Mathematics and Mechanics UrB RAS, Yekaterinburg, Russia

Suriya Kumacheva Saint Petersburg State University, St. Petersburg, Russia

Denis Kuzyutin Saint Petersburg State University, St. Petersburg, Russia

National Research University Higher School of Economics (HSE), St. Petersburg, Russia

Yin Li Saint Petersburg State University, St. Petersburg, Russia

Ivan Lipko Saint Petersburg State University, St. Petersburg, Russia

Anastasiya Malakhova Saint Petersburg State University, St. Petersburg, Russia

Mario Alberto Garcia Meza Universidad Juarez del Estado de Durango, Durango, Mexico

Maria Nastych National Research University Higher School of Economics, St. Petersburg, Russia

Ekaterina Orlova RANEPA, Moscow, Russia

Gennady A. Ougolnitsky I.I. Vorovich Institute of Mathematics, Mechanics and Computer Sciences of Southern Federal University, Rostov-on-Don, Russia

Yaroslavna Pankratova Saint Petersburg State University, St. Petersburg, Russia

Ovanes Petrosian Saint Petersburg State University, St. Petersburg, Russia

Leon A. Petrosyan Saint Petersburg State University, St. Petersburg, Russia

Cesar Gurrola Rios Universidad Juarez del Estado de Durango, Durango, Mexico

Dmitry B. Rokhlin I.I. Vorovich Institute of Mathematics, Mechanics and Computer Sciences of Southern Federal University and Regional Scientific and Educational Mathematical Center of Southern Federal Universit, Rostov-on-Don, Russia

Simon Rothfuß Institute of Control Systems (IRS) at Karlsruhe Institute of Technology (KIT), Karlsruhe, Germany

Igor Shevchenko Pacific Branch of the Russian Federal Research Institute of Fisheries and Oceanography, Vladivostok, Russia

Far Eastern Federal University, Vladivostok, Russia

Alexander Sidorov Sobolev Institute of Mathematics, Novosibirsk, Russia

Jannik Steinkamp Institute of Control Systems (IRS) at Karlsruhe Institute of Technology (KIT), Karlsruhe, Germany

Igor Tantlevskij Saint Petersburg State University, St. Petersburg, Russia

Svetlana Tarashnina Saint Petersburg State University, St. Petersburg, Russia

Alexander M. Tarasyev Krasovskii Institute of Mathematics and Mechanics UrB RAS, Yekaterinburg, Russia

Galina Tomilina Saint Petersburg State University, St. Petersburg, Russia

Kubra Uge Gebze Technical University, Kocaeli, Turkey

Ekaterina Zhitkova Saint Petersburg State University, St. Petersburg, Russia

Chapter 1
Competition as a Hierarchical Multistage Game

Petr Ageev and Svetlana Tarashnina

Abstract We investigate the process of competition on the market of telecommunication services between three firms: the leader, the challenger and the follower. In this work we construct a model of competition between three players in the form of a multistage hierarchical non-zero sum game and compare it with our previous model of competition between three players in the form of a multistage non-hierarchical non-zero sum game. Compared to previous model, a hierarchical component was introduced. As solution of the game we find a subgame perfect equilibrium. We illustrate and compare the results with an example for three companies working on the Saint-Petersburg telecommunications market with the same initial conditions as in the previous work.

Keywords Hierarchical game · Subgame perfect equilibrium · Competition model · Telecommunication market

1.1 Introduction

Competition between three companies on the market was investigated. All firms have different types: the leader, the challenger and the follower. The leader is a company that prevails on the market and acts in three main directions:

- expansion of the market by new customers attracting and new areas finding;
- increasing its market share;
- protecting business from attacks by defensive strategies.

The challenger firm is a company that does not lag far behind the leader of the market and tries to become the leading company by using attacking strategies. The

P. Ageev (✉) · S. Tarashnina
Saint Petersburg State University, St. Petersburg, Russia
e-mail: st012558@student.spbu.ru; s.tarashnina@spbu.ru

© The Editor(s) (if applicable) and The Author(s), under exclusive licence
to Springer Nature Switzerland AG 2020
L. A. Petrosyan et al. (eds.), *Frontiers of Dynamic Games*,
Static & Dynamic Game Theory: Foundations & Applications,
https://doi.org/10.1007/978-3-030-51941-4_1

follower firm is a company that pursues following others companies policy and does not risk achieved market positions. This firm uses strategies aimed to expand its market share, but those that do not cause active opposition to competitors.

In paper [1] the competition on the market is presented by a multistage decision-making model. At the first stage, the decision is made by the leader. At the next stage, taking into account the leader's strategy, the decision is made by the company-follower. At the same time each firms pursues its own goals during strategy selection.

First of all, we introduce a hierarchical component into the game, that was described in [2]. The hierarchy is introduced by player decision prioritizing, who is located at a higher hierarchical level relative to the opponent located at a lower level. Let the leading company be located on the first, the most important hierarchical level, the challenger company—on the second, the follower firm, respectively, on the third level. Thus, at the same hierarchical level there is only one player (firm). Mentioned hierarchy component means that the leader and the challenger choose their strategies consistently compared to the previous model, where the leader and the challenger companies act simultaneously. Due to the challenger is located at a lower hierarchical level, this company taking into account both the leader strategy and subscribers preferences, while the leader taking into account only subscribers preferences. In our model at the first stage, the leader decides which services should be offered to subscribers and chooses service price. Since the leader company is located on the first hierarchical level, we believe that when this firm make a decision, player takes into account only the preferences of subscribers and tries to satisfy of as many of them as possible. At this stage of the game, part of subscribers make their choice in favor of the leader.

At the second step of the game, the challenger company makes a decision about telecommunication services that should be offered and chooses its price. During this choice, the challenger company takes into account the preferences of subscribers and the leader strategy. At this step, part of subscribers choose the second player.

At the next step of the game, the follower firm, taking into account the choice of competitors, decides what should be offered to potential customers. At the same time, the follower company seeks to retain old subscribers and, if possible, attract a part of competitors clients. At this step, the remaining part of subscribers makes their choice. Suppose that at the start of the game, all subscribers are informed about all tariffs that operators able to offer.

We assume that each customer must choose one of the telecommunication services (tariff). If a customer decides to keep old tariff, we believe that he chooses the appropriate relevant company service. The leader and the challenger aim to maximize their profits by attracting some of the competitors' customers. The purpose of the follower is to maximize profit and to save his customers without negative reaction of competitors.

We formalize this problem of competition on the telecommunications market between three companies as a hierarchical nonzero-sum game. As a solution of this game we consider a subgame perfect equilibrium (SPE) [3–5]. Example is given and discussed in the paper. The obtained solution allows each company to develop a long-term strategy to maximize its summing payoff.

1.2   Game Formulation

We introduce the following assumptions:

1. Firms are informed about subscribers preferences.
2. As a profit we will understand the difference between the price of the service and the unit costs for it; the profit can only be positive.
3. The income from the sale of a certain service is determined by quantity of subscribers who have decided to use this service, and corresponding price.
4. Service price is the total cost of services, which should be paid by subscriber per month.
5. Telecommunication service is a certain tariff consisting of a services package. For example, a tariff consists of v minutes for all outgoing calls, b gigabytes of Internet and z outgoing SMS messages. Further, the number of outgoing SMS messages is omitted from consideration, since to date SMS messages have been replaced by so-called messengers.
6. Since quite often subscribers use the Internet for outgoing calls, and the demand for such tariffs is higher, telecommunication operators set the price for the Internet tariffs higher despite the fact that the unit costs for such services is much lower.
7. We assume that the same type services unit costs are equal for the same company.
8. Let the fixed costs for operator are equal for all of the offered tariffs.
9. The fixed costs for larger companies are higher than their competitors have, and the unit costs are lower.
10. Subscribers are informed about services which may be offered by any player.
11. The situation on the market when different firms offer similar tariffs for the same price is not possible.

We denote by F_1 the leader, F_2 is the challenger and F_3 is the follower. Let $N = \{F_1, F_2, F_3\}$ be the set of players—telecommunication companies, which provide services on the market.

Let $I = \{1, \ldots, m\}$ be the set of services (tariffs) that are offered on the telecommunications market. Each element of $i_r \in I$ is specific type of service. This service will be called the service type i_r, offered by any firm.

Denote by I_1, I_2 and I_3 subsets of I, which contain the offered services, respectively, by the leading firm, the challenger firm and the follower firm. Assume that $I_1 \cup I_2 \cup I_3 = I$ and $I_1 \cap I_2 \cap I_3 = \emptyset$.

Let the following quantities are known:

- c_i^k is the price of service i for the player F_k, where $i \in I_k$ and $k \in \{1, 2, 3\}$;
- a_i is the unit costs of service i;
- f_k is the fixed costs (i.e. costs that are not depends on the services scope) for the player service F_k, $k \in \{1, 2, 3\}$. At the same time, the fixed costs are constant.

We denote by $J = \{1, \ldots, n\}$ the set of subscribers using the services offered on the market. Each element $j \in J$ is a certain subscriber (customer). We assume, that subscriber chooses one of the firm services, based on his internal preferences, which are specified by splitting the set J into two subsets J_T and J_P. J_T includes subscribers, who mainly think about low price during selection an operator. In turn, J_T is divided into J_{T_1}, J_{T_2} and J_{T_3}. J_{T_1} consists of subscribers, who mainly think about low price and about number of minutes for outgoing calls within the tariff. J_{T_2} is subscribers, who pays attention to the low price and the volume of Internet traffic provided within the tariff. J_{T_3}—set of subscribers, who need a balanced tariff in terms of the internet traffic and the number of minutes for outgoing calls. The subset J_P contains "conservative" subscribers. They are subscribers who are not able to change operator, because it is a problem for various reasons, for example, they are corporate users. The following expressions hold:

$$J_{T_1} \cup J_{T_2} \cup J_{T_3} = J_T, \tag{1.1}$$

$$J_{T_1} \cap J_{T_2} \cap J_{T_3} = \emptyset. \tag{1.2}$$

We suppose that $J = J_T \cup J_P$, $J_T \cap J_P = \emptyset$. It means, that one subscriber able to choose only one firm. If he has several SIM-cards, we say that, in terms of our game, these are different subscribers, because spending for different SIM-cards is unique. We have

$$J^0 = J = J_1^0 \cup J_2^0 \cup J_3^0. \tag{1.3}$$

Expression (1.1) describes the distribution of the subscribers set between players at the initial stage of the game. Let the following relations hold:

$$|J_1^0 \cap J_T| \geq |J_2^0 \cap J_T| > |J_3^0 \cap J_T|,$$

$$|J_1^0 \cap J_P| > |J_2^0 \cap J_P| > |J_3^0 \cap J_P|.$$

We assume that subscribers from the $J_P \cap J_k^0$ always choose the player k, where $k \in \{1, 2, 3\}$, and the service that the operator k offers at the moment, regardless of the offered tariff.

By the strategy of player F_k, where $k \in \{1, 2, 3\}$, we define $s_k^{i_r} = (c_{i_r}^k, v_{i_r}^k, b_{i_r}^k)$, $i_r \in I_k$. We denote the set of strategies F_k by $S_k = \{s_k^{i_r} : i_r \in I_k\}$. Obviously, the player strategy is characterized by the following indicators: the price $c_{i_r}^k$, the number of minutes $v_{i_r}^k$, the number of gigabytes for the mobile Internet $b_{i_r}^k$.

Let introduce the subscriber $j \in J$ preference relationships for services offered by firms F_1 and F_2.

For subscriber $j \in J$ we say that the strategy $(c_{i_l}^1, v_{i_l}^1, b_{i_l}^1)$ of the firm F_1 is preferable to the strategy $(c_{i_p}^2, v_{i_p}^2, b_{i_p}^2)$ of the firm F_2, i.e. $(c_{i_l}^1, v_{i_l}^1, b_{i_l}^1) \succ$

$(c_{i_p}^2, v_{i_p}^2, b_{i_p}^2)$ if the following condition hold:

$$(1 - \alpha) \times \frac{c_{i_l}^1}{v_{i_l}^1} + \alpha \times \frac{c_{i_l}^1}{100 \times b_{i_l}^1} < (1 - \alpha) \times \frac{c_{i_p}^2}{v_{i_p}^2} + \alpha \times \frac{c_{i_p}^2}{100 \times b_{i_p}^2}.$$

For subscriber $j \in J \cap J_{T_1}$ $\alpha = 0$, for subscriber $j \in J \cap J_{T_2}$ $\alpha = 1$, for subscriber $j \in J \cap J_{T_3}$ $\alpha = \frac{1}{2}$. Similarly for the opposite case.

If it holds:

$$(1 - \alpha) \times \frac{c_{i_l}^1}{v_{i_l}^1} + \alpha \times \frac{c_{i_l}^1}{100 \times b_{i_l}^1} = (1 - \alpha) \times \frac{c_{i_p}^2}{v_{i_p}^2} + \alpha \times \frac{c_{i_p}^2}{100 \times b_{i_p}^2},$$

we assume that subscribers will choose an operator with a higher market status.

The process of identifying preferred strategies by subscribers is carried out in pairs for players, i.e. firstly for the leader and for the challenger, then for the leader and for the follower, as well as for the challenger and for the follower.

We introduce the switching function $V_j(s_k^{i_r})$.

$$V_j(s_k^{i_r}) = \begin{cases} 1, & \text{if } i_r \text{ is preferred service for subscriber } j; \\ 0, & \text{otherwise}, \end{cases}$$

i.e., the function characterizes the preference for the subscriber $j \in J$ of the service $i_r \in I_k$ offered by the player F_k, compared to all other types of services that are offered on the market. For regular subscribers, i.e. for $j \in J \cap J_P$:

$$V_j(s_k^i) = 1 \text{ for all } i \in I_k, \ k \in \{1, 2, 3\}.$$

We assume that the services are selected by subscriber for a month in advance. Let introduce the value $g_j(s_k^{i_r}) = (c_{i_r}^k - a_{i_r})$, which characterizes the profit of a company F_k by the subscriber j when the firm uses strategy $s_k^{i_r}$. The i_r service, which offers a larger amount of Internet traffic is designated as i_r^{gb}. The service, which offers a greater number of minutes for outgoing calls is designated as i_r^{mnt}. Balanced service is designated as i_r^{gbmnt}. Taking into account of assumption 6, we obtain the following inequality

$$g_j(s_k^{i_r^{gb}}) \geq g_j(s_k^{i_r^{gbmnt}}) \geq g_j(s_k^{i_r^{mnt}}) > 0,$$

for $k \in \{1, 2, 3\}, i_r \in I_k, j \in J$.

Denote by $G_k(s_k^{i_r})$ the total profit of firm k from customers $j \in J_P \cap J_k^0$, which choose the service i_r, i.e.

$$G_k(s_k^{i_r}) = \sum_{j \in J_P \cap J_k^0} g_j(s_k^{i_r}),$$

where $k \in \{1, 2, 3\}$ and $i_r \in I_k$.

The payoff functions in [2] were defined by the following way:

$$H_1(s_1^{i_l}, s_2^{i_p}, J^0) = -f_1 + \sum_{j \in J_T \cap J^0} g_j(s_1^{i_l}) \times V_j(s_1^{i_l}) \times (1 - V_j(s_2^{i_p})) + G_1(s_1^{i_l}),$$

where $i_l \in I_1, i_p \in I_2$.

$$H_2(s_1^{i_l}, s_2^{i_p}, J^0) = -f_2 + \sum_{j \in J_T \cap J^0} g_j(s_2^{i_p}) \times V_j(s_2^{i_p}) \times (1 - V_j(s_1^{i_l})) + G_2(s_2^{i_p}),$$

where $i_l \in I_1, i_p \in I_2$.

$$H_3(s_1^{i_l}, s_2^{i_p}, s_3^{i_s}, J^0) = -f_3+$$

$$+ \sum_{j \in J_T \cap J^0} g_j(s_3^{i_s}) \times V_j(s_3^{i_s}) \times (1 - V_j(s_1^{i_l})) \times (1 - V_j(s_2^{i_p})) + G_3(s_3^{i_s}),$$

where $i_l \in I_1, i_p \in I_2, i_s \in I_3$.

We transform these payoff functions for the case of a hierarchical non-antagonistic game. For the leader we have:

$$H_1(s_1^{i_l}, J^0) = -f_1 + \sum_{j \in J_T \cap J^0} g_j(s_1^{i_l}) \times V_j(s_1^{i_l}) + G_1(s_1^{i_l}),$$

where $i_l \in I_1$.

For the challenger we have:

$$H_2(s_2^{i_p}, J^0, J_1^1) = -f_2 + \sum_{j \in J_T \cap (J^0 \setminus J_1^1)} g_j(s_2^{i_p}) \times V_j(s_2^{i_p}) + G_2(s_1^{i_p}),$$

where $i_l \in I_1, i_p \in I_2$.

For the follower we have:

$$H_3(s_3^{i_s}, J^0, J_1^1, J_2^1) = -f_3 + \sum_{j \in J_T \cap (J^0 \setminus (J_1^1 \cup J_2^1))} g_j(s_3^{i_s}) \times V_j(s_3^{i_s}) + G_3(s_3^{i_s}),$$

where $i_l \in I_1, i_p \in I_2, i_s \in I_3$.

The payoff functions expresses the total profit of players taking into account changes in income due to the loss and acquisition of subscribers.

$$V_j(s_1^{i_l}) + V_j(s_2^{i_p}) + V_j(s_3^{i_s}) \leq 1, \ i_l \in I_1, \ i_p \in I_2, \ i_s \in I_3, \ j \in J. \qquad (1.4)$$

Inequality (1.4) shows that it is not possible that for the subscriber $j \in J$ two services are the most preferable at the same time compared to each other.

The game leader is determined by the number of subscribers available to the company at the beginning of the game. At the end of the game, in the case of equality of the subscribers for several companies, the leader is determined by the amount of total profit. Since, ceteris paribus, the follower firm can both play along with the leader firm and play along with the challenger firm, for definiteness, we assume that player F_3 plays along to player F_1.

Thus, the competition on the telecommunications market can be formalised in the form of non-zero sum game Γ:

$$\Gamma =< N, S_1, S_2, S_3, H_1, H_2, H_3 > .$$

1.3 Nash Equilibrium

We assume that for the same strategy type for the same player the value $g_j(s_k^{i_r})$ will be greater for the strategy $s_k^{i_r}$, if this tariff has greater volume of the services. That is, if i_1 and i_2 are "Internet" tariffs, than the value of $g_j(s_k^{i_1})$ will be greater if the service i_1 offers larger package of Internet traffic. This is due to the fact that when the volume of the service increases, the price for it increases, while the unit costs according to the assumptions for the same type services are the same.

We build a subgame perfect equilibrium in the same way as shown in the paper [2].

Theorem 1.1 *In a non-zero sum two-stage game* $\Gamma =< N, S_1, S_2, S_3, H_1, H_2, H_3 >$ *the strategies* s_1^*, s_2^*, s_3^* *lead to a subgame perfect equilibrium if the* *next inequalities (1.5) are fulfilled:*

$$g(s_1^*) \geq \frac{\sum\limits_{j \in J_T \cap J^0} V_j(s_1^{i_2}) + w_1}{\sum\limits_{j \in J_T \cap J^0} V_j(s_1^*) + w_1} \times g(s_1^{i_2}),$$

$$g(s_2^*) \geq \frac{\sum\limits_{j \in J_T \cap (J^0 \setminus J_1^1)} V_j(s_2^{i_2}) + w_2}{\sum\limits_{j \in J_T \cap (J^0 \setminus J_1^1)} V_j(s_2^*) + w_2} \times g(s_2^{i_2}), \tag{1.5}$$

$$g(s_3^*) \geq \frac{\sum\limits_{j \in J_T \cap (J^0 \setminus (J_1^1 \cup J_2^1))} V_j(s_3^{i_2}) + w_3}{\sum\limits_{j \in J_T \cap (J^0 \setminus (J_1^1 \cup J_2^1))} V_j(s_3^*) + w_3} \times g(s_3^{i_2}),$$

for $\forall s_1^{i_2} \in \{S_1\}, \forall s_2^{i_2} \in \{S_2\}, \forall s_3^{i_2} \in \{S_3\}$, *where* $w_k = |J_P \cap J_k^0|, k \in \{1, 2, 3\}$.

1.4 Results Comparison

In order to compare the results, we use the same strategies and the same conditions (for example, the initial distribution of subscribers between operators), presented in [2].

We assume that $I_1 = \{1, 2\}, I_2 = \{3, 4\}, I_3 = \{5, 6\}$.

- Tariff 1 contains 200 min of outgoing calls, 2 GB of Internet traffic. Fixed costs f_1^1 are equal to 70, the unit cost a_1 is equal to 60.
- Tariff 2 contains 100 min of outgoing calls, 6 GB of Internet traffic. Fixed costs f_2^1 are 70, the unit cost a_2 is 50.
- Tariff 3 contains 200 min of outgoing calls, 3 GB of Internet traffic. Fixed costs f_3^2 are equal to 60, the unit cost a_3 is equal to 70.
- Tariff 4 contains 150 min of outgoing calls, 5 GB of Internet traffic. Fixed costs f_4^2 equal to 60, the unit cost a_4 equals to 60.
- Tariff 5 contains 150 min of outgoing calls, 4 GB of Internet traffic. Fixed costs f_5^3 equal to 50, the unit cost a_5 is 70.
- Tariff 6 contains 100 min of outgoing calls, 7 GB of Internet traffic. Fixed costs f_6^3 equal to 50, the unit cost a_6 equals to 60.

Let $J = \{1, 2, 3, 4, 5, 6, 7, 8, 9, 10, 11, 12, 13, 14, 15, 16, 17\}$. Divide J_T into three sets J_{T_1}, J_{T_2} and J_{T_3}. We have $J_{T_1} = \{1, 2, 3, 4, 5\}$, $J_{T_2} = \{6, 7, 8, 9\}$, $J_{T_3} = \{\emptyset\}$.

The set J_P includes customers 10, 11, 12, 1314, 15, 16, 17.

Let $J_1^0 \cap J_P = \{10, 11, 12, 13\}$, $J_2^0 \cap J_P = \{14, 15, 16\}$, $J_3^0 \cap J_P = \{17\}$.

Assume that

$$J_1^0 \cap J_T = \{1, 4, 6, 9\}, \quad J_2^0 \cap J_T = \{2, 5, 7\}, \quad J_3^0 \cap J_T = \{3, 8\}.$$

Let us move on to the strategy sets: $S_1 = \{s_1^1, s_1^2\}$, $S_2 = \{s_2^1, s_2^2\}$, $S_3 = \{s_3^1, s_3^2\}$,

$$s_1^1 = (300, 1), \quad s_1^2 = (330, 2),$$

$$s_2^1 = (310, 3), \quad s_2^2 = (320, 4),$$

$$s_3^1 = (320, 5), \quad s_3^2 = (340, 6).$$

Thus, Table 1.1 shows the initial data and conditions.

Then, according to the tariff preference rules, we get the following results. Table 1.2 presents the results of the subscribers distribution between players, depends on used strategies.

Figure 1.1 shows the constructed subgame perfect equilibrium and players payoffs.

Then, the game situation, when in the node 1 player F_1 chooses the strategy s_1^1, in nodes 2 and 3 player F_2 chooses the strategy s_2^2, and in nodes 4, 5, 6, 7 player F_3 chooses the strategy s_3^2, is the subgame perfect equilibrium, which is written as $[(s_1^1); (s_2^2); (s_2^2); (s_3^2); (s_3^2); (s_3^2); (s_3^2)]$.

Table 1.1 Baseline data

Firm	Strategy	Tariff	Fixed costs	Unit costs	Minutes	Gigabytes	Price
Leader	s_1^1	1	70	60	200	2	300
	s_1^2	2	70	50	100	6	330
Challenger	s_2^1	3	60	70	200	3	310
	s_2^2	4	60	60	150	5	320
Follower	s_3^1	5	50	70	150	4	320
	s_3^2	6	50	60	200	7	340

Table 1.2 Distribution results

Strategy profile	Customers of firm F_1	Customers of firm F_2	Customers of firm F_3	Payoffs (H_1, H_2, H_3)
s_1^1, s_2^1, s_3^1	{1, 2, 3, 4, 5, 10, 11, 12, 13}	{14, 15, 16}	{6, 7, 8, 9, 17}	(2090, 660, 1200)
s_1^2, s_2^1, s_3^1	{6, 7, 8, 9, 10, 11, 12, 13}	{1, 2, 3, 4, 5, 14, 15, 16}	{17}	(2170, 1860, 270)
s_1^1, s_2^2, s_3^1	{1, 2, 3, 4, 5, 10, 11, 12, 13}	{6, 7, 8, 9, 14, 15, 16}	{17}	(2090, 1760, 200)
s_1^2, s_2^2, s_3^1	{6, 7, 8, 9, 10, 11, 12, 13}	{1, 2, 3, 4, 5, 14, 15, 16}	{17}	(2170, 2020, 270)
s_1^1, s_2^1, s_3^2	{1, 2, 3, 4, 5, 10, 11, 12, 13}	{14, 15, 16}	{6, 7, 8, 9, 17}	(2090, 660, 1350)
s_1^2, s_2^1, s_3^2	{10, 11, 12, 13}	{1, 2, 3, 4, 5, 14, 15, 16}	{6, 7, 8, 9, 17}	(1050, 1860, 1350)
s_1^1, s_2^2, s_3^2	{1, 2, 3, 4, 5, 10, 11, 12, 13}	{14, 15, 16}	{6, 7, 8, 9, 17}	(2090, 720, 1350)
s_1^2, s_2^2, s_3^2	{10, 11, 12, 13}	{1, 2, 3, 4, 5, 14, 15, 16}	{6, 7, 8, 9, 17}	(1050, 2020, 1350)

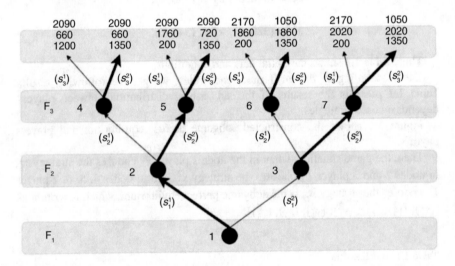

Fig. 1.1 Subgame perfect equilibrium

1.5 Conclusion

Thus, we obtain that the strategies included in the subgame perfect equilibrium coincide with those that were found in [2]. One of the reasons for this result is the fact that in our previous work the J_T set was divided into J_{T_1} and J_{T_2}. In our new game formulation, the set J_{T_3} was added, but according to the initial conditions, it is initially empty. However, from the result interpretation point of view, it is fundamentally different compared to the previous one, since in this case we see that the challenger company generally loses achieved market position and becomes a follower on the market, while the follower firm becomes a challenger.

References

1. Stackelberg, H.V.: The theory of the market economy, 328 p. Oxford University, England (1952)
2. Ageev, P.V., Pankratova, Y.B., Tarashnina, S.I.: On competition in the telecommunications market. Contribut. Game Theory Manag. **11**, 7–21 (2018)
3. Hellwig, M., Leininger, W.: On the existence of subgame-perfect equilibrium in infinite-action games of perfect information. J. Econom. Theory **43**(1), 55–75 (1987)
4. Yeung, D.W.K., Petrosyan, L.A., Zenkevich, N.A.: Dynamic games and applications in management, 415 p. GSOM Press, St Petersburg (2009)
5. Nessah, R., Tian, G.: On the existence of strong Nash equilibria. J. Math. Anal. Appl. **414**(2), 871–885 (2014)

Chapter 2
Information Exchange in Price Setting Mixed Duopoly

Sadettin Haluk Citci and Kubra Uge

Abstract Consider a dynamic Bertrand competition in mixed oligopoly, where a private firm competes with a social welfare maximizing public firm. Firms produce substitute products, face stochastic demand and each firm receive noisy signals on common stochastic demand. In this mixed oligopoly, we examine incentives of public and private firms to share their private signals through an independent trade agency and we characterize equilibrium outcomes. We established two main effects of information sharing: information sharing increases production efficiency by enabling firms to predict stochastic demand shocks better. However, more precise signals increase power of private firm to capture consumer surplus and lowers social welfare. In Perfect Bayesian equilibrium of the mixed oligopoly game, private firm shares all signals it receives with the public firm, whereas public firm shares no information with the private firm. The market outcome is never optimal: it satisfy neither of informational efficiency, production efficiency and allocative efficiency.

Keywords Mixed oligopoly · Information sharing · Information acquisition · State owned enterprises · Stochastic demand

2.1 Introduction

Do firms in oligopolistic markets share information sufficiently or are oligopolistic markets informationally efficient? This question has become more important in the last decade as firms have started to use big data more intensely and competitive intelligence activities have started to be more central in firm activities. Framingham [3] reports that big data and business analytics revenues were 189 billion dollar in 2019. An example better crystallizing the argument is competitive intelligence

S. H. Citci (✉) · K. Uge
Gebze Technical University, Kocaeli, Turkey
e-mail: hcitci@gtu.edu.tr

© The Editor(s) (if applicable) and The Author(s), under exclusive licence
to Springer Nature Switzerland AG 2020
L. A. Petrosyan et al. (eds.), *Frontiers of Dynamic Games*,
Static & Dynamic Game Theory: Foundations & Applications,
https://doi.org/10.1007/978-3-030-51941-4_2

system adoption of the European unit of Cisco Systems [1]. Cisco initially started to use this system to acquire information on firm demand, industry demand and competitors' activities. Meanwhile, Cisco voluntarily disclosed some of information is gathered.

Acquiring and sharing information on demand and cost conditions is extensively analyzed for pure oligopolies in the literature. Early contributions to the literature were made by Ponssard [10], Novshek and Sonnenschein [9], Clarke [2], Fried [4], Vives [11] and Gal-Or [5]. In the following years, many other studies have been added to the literature such as Haraguchi and Matsumura [6], Myatt and Wallace [8]. The literature showed that information sharing behaviors of firms extensively depend on the type of competition (Cournot or Bertrand), types of products and nature of uncertainty (whether it is cost of demand uncertainty). When uncertainty is on common demand parameters, firms tend to share all information they have with each other if they compete in quantities and goods are complements or if they compete in prices and goods are substitutes [11]. The results are reversed if uncertainty is on cost parameters.

This study extends prior literature by examining firms' incentives to acquire and share information on stochastic demand in mixed oligopolies. Mixed oligopolies, characterized as the competition between private firms with a public firm whose objective is not solely profit maximization, exist in many oligopolistic markets. 10% of world 2000 largest publicly listed firms are identified as state-owned enterprises [7]. State owned enterprises represent 62% of Russia's stock market capitalization. Therefore, a sizeable part of economic activities are maintained in mixed oligopolies rather than in pure oligopolies. This study focuses on mixed oligopolies to extend the information sharing literature in this direction.

Specifically, we consider a mixed duopoly where a private firm competes a la Bertrand with a social welfare maximizing public firm. Firms produce substitute products under common demand uncertainty and each firm receives an observation sample on uncertain demand parameter. In this model economy, firms play two-stage game and decide the extent of information to reveal with the other firm in the first stage and decide pricing in the second stage. In equilibrium, private firm always shares all information it has, whereas social welfare maximizing public firm shares no information. These actions are dominant strategies for both firms. Moreover, we established that the market outcome is never optimal: it does not satisfy informational efficiency, production efficiency and allocative efficiency.

2.2 Model

We study a dynamic model economy where a social welfare maximizing public firm (firm 1) competes with a private firm (firm 2) in prices. In the economy, there are two differentiated, substitute goods, produced by each firm. To derive closed form solutions, utility function of consumers is assumed to be quadratic, strictly concave and symmetric in the quantity of the goods.

$$U(q_1, q_2) = \alpha \times (q_1 + q_2) - \frac{\beta q_1^2 + 2\gamma q_1 q_2 + \beta q_2^2}{2}. \tag{2.1}$$

The specified utility function yields linear demand functions:

$$p_1 = \alpha - \beta q_1 - \gamma q_2,$$
$$p_2 = \alpha - \gamma q_1 - \beta q_2, \tag{2.2}$$

where $\alpha > 0$, $\beta > \gamma > 0$ and the assumption $\gamma > 0$, guarantees that the goods are substitutes. Accompanying consumer surplus can be defined as follows:

$$CS = U(q_1, q_2) - \sum_{i=1}^{2} p_i q_i,$$
$$CS = \alpha \times (q_1 + q_2) - \frac{\beta q_1^2 + 2\gamma q_1 q_2 + \beta q_2^2}{2} - p_1 q_1 - p_2 q_2. \tag{2.3}$$

To simplify the model, without loss of generality, marginal costs of both firms are assumed to be zero. As a result, the profit function of the private firm is equal to:

$$\pi_2 = p_2 q_2,$$
$$\pi_2 = (\alpha - \gamma q_1 - \beta q_2) \times q_2, \tag{2.4}$$

and total producer surplus is equal to:

$$PS = p_1 q_1 + p_2 q_2. \tag{2.5}$$

Both firms are assumed to be risk neutral. The private firm aims to maximize solely its profit function. However, the public firm takes into account both its own profit function, private firm's profit function (producer surplus) and consumer surplus. Public firm aims to maximize social welfare, defined by the following equation

$$SW = \frac{2\alpha^2(\beta - \gamma) - \beta(p_i^2 + p_j^2) + 2\gamma p_i p_j}{2(\beta^2 - \gamma^2)}. \tag{2.6}$$

The crucial part of the model is that we model uncertainty in demand and allow firms to choose information sharing about this uncertainty. Specifically, following to Vives (1984), we assume that demand intercept, α is a random variable and is normally distributed with mean $\bar{\alpha}$ and variance $V(\alpha)$. Each firm starts to game with n_i independent observation sample $(t_{i1}, t_{i2}, t_{i3}, \ldots, t_{in})$, where $t_{ik} = \alpha + u_{ik}$ and u_{ik}'s are independent and identically distributed random variables. Their mean is zero, variance σ_u^2 and independent with α.

There is an independent trade agency that collects the observation samples. Firm 1 (public firm) receives n_1 observation sample and allows the trade agency to reveal $\lambda_1 n_1$ observation where $0 \leq \lambda_1 \leq 1$. Also, Firm 2 (private firm) receives n_2 observation sample and allows the trade agency to reveal $\lambda_2 n_2$ observation where $0 \leq \lambda_2 \leq 1$. There are $\lambda_1 n_1 + \lambda_2 n_2$ observation sample in the common pool after each firm shares $\lambda_i n_i$ observation it has.

As a result of this information sharing process, each firm observes a private noisy signal for the random variable α. The equation of signals is given as

$$s_i = \alpha + \frac{1}{(n_i + \lambda_j n_j)} \left(\sum_{k-1}^{n_i} u_{ik} + \sum_{k=1}^{\lambda_j n_j} u_{jk} \right), \ i = 1, 2. \tag{2.7}$$

We have bivariate, normally distributed error terms with zero means for s_i, where $v_i = \sigma_u^2 / (n_i + \lambda_j n_j)$ and $\sigma_{12} = ((\lambda_1 n_1 + \lambda_2 n_2) / (n_1 + \lambda_2 n_2)(n_2 + \lambda_1 n_1)) \sigma_u^2$, implies that $v_i \geq \sigma_{12} \geq 0$, $i = 1, 2$.

With these assumptions, we define following equations: $E(\alpha | s_i) = (1 - t_i)\bar{\alpha} + t_i s_i$ and $E(s_j | s_i) = (1 - d_i)\bar{\alpha} + d_i s_i$, with $t_i = V(\alpha)/(V(\alpha) + v_i)$ and $d_i = (V(\alpha) + \sigma_{12})/(V(\alpha) + v_i)$, i=1,2, $i \neq j$, where $1 \geq d_i \geq t_i \geq 0$ since $v_i \geq \sigma_{12} \geq 0$.

The equations imply that signals give more precise information about the demand intercept as the variance decreases. The conditional expectation formula is as the following:

$$E(\alpha | s_i) = (1 - t_i)\bar{\alpha} + t_i s_i. \tag{2.8}$$

If the precision of the signals increase, t_i increases because when t_i increases $E(\alpha | s_i)$ gets closer to s_i than $\bar{\alpha}$. Also, t_i increases as v_i increases because $t_i = V(\alpha)/(V(\alpha) + v_i)$. While the signal goes from being perfectly precise to being completely imprecise, v_i goes from 0 to ∞ and t_i goes from 1 to 0. Last, all of these are common knowledge.

Public and private firms play two-stage game. Timing of the game is as follows: in the first stage, both firms receive private noisy signal about the uncertain demand parameter. Each firm decides the amount of observation to share with its competitor. Then, the independent trade agency collects these observations and distributes these observations. In the second stage, given their received signal s_i, based on their collected and received information about uncertain demand, each firm decides price to charge. The game ends at the end of the second stage.

2.3   Analysis

In this section, we determine Perfect Bayesian Equilibrium of the model. We start to solve model using backward induction. Equilibrium price strategies in the second stage are derived by establishing convergence points of "I think that he thinks that I think. . ." type model.

Public firm maximizes following expected social welfare function with respect to p_1.

$$E(SW|s_1) = E\left(\frac{2\alpha^2(\beta - \gamma) - \beta(p_i^2 + p_j^2) + 2\gamma p_i p_j)}{2(\beta^2 - \gamma^2)}|s_1\right). \quad (2.9)$$

While private firm maximizes following expected profit function with respect to p_2.

$$E(\pi_2|s_2) = E\left(\frac{p_2(\alpha(\beta - \gamma) - \beta p_2 + \gamma p_1)}{(\beta^2 - \gamma^2)}|s_2\right). \quad (2.10)$$

So, the best response functions for each firm are as follows:

$$p_1(s_1) = \tfrac{\gamma}{\beta} E(p_2(s_2)|s_1)$$

$$and$$

$$(p_2|s_2) = E\left(\tfrac{\alpha(\beta-\gamma)+\gamma p_1}{2\beta}|s_2\right). \quad (2.11)$$

This yields following equations:

$$p_1(s_1) = \frac{\gamma}{\beta} E_1\left(\frac{\beta - \gamma}{2\beta} E_2(\alpha|s_2) + \frac{\gamma}{2\beta} E_2(p_1(s_1)|s_2)\middle| s_1\right). \quad (2.12)$$

After some messy calculations, we obtain following best response function in terms of exogenous variables:

$$p_1(s_1) = \frac{\gamma(\beta - \gamma)}{2\beta^2}\left(\frac{2\beta^2}{2\beta^2 - \gamma^2}\right)\bar{\alpha} + \frac{\gamma(\beta - \gamma)}{2\beta^2}\left(\frac{2\beta^2}{2\beta^2 - \gamma^2 d_1 d_2}\right)d_1 t_2(s_1 - \bar{\alpha}) \quad (2.13)$$

and we can re-write expected social welfare function as the following:

$$E(SW|s_1) = \frac{1}{2(\beta^2 - \gamma^2)}\Big[2(\beta - \gamma) E(\alpha^2) - \beta E(p_1(s_1)) - \beta E(p_2(s_2)|s_1^2)$$

$$+ 2\gamma E(p_1(s_1) p_2(s_2) | s_1)\Big]. \quad (2.14)$$

Again, after proper substitution and calculations, we derive expected social function in exogenous terms:

$$(SW|s_1) = \frac{1}{2(\beta^2 - \gamma^2)}\Big[2(\beta - \gamma)(V(\alpha) + \bar{\alpha}^2)$$

$$- \beta\left(X_1^2 + X_2^2 t_2^2(V(\alpha) + v_1) + Z_1^2 + Z_2^2 t_2^2 d_1^2(V(\alpha) + v_1)\right)$$

$$+ 2\gamma\left(X_1 Z_1 + X_2 Z_2 t_2^2 d_1(V(\alpha) + v_1)\right)\Big], \quad (2.15)$$

where $X_1 = \frac{\gamma(\beta-\gamma)}{2\beta^2} \frac{2\beta^2}{2\beta^2-\gamma^2}\bar{\alpha}$, $x_2 = \frac{\gamma(\beta-\gamma)}{2\beta^2} \frac{2\beta^2}{2\beta^2-\gamma^2 d_1 d_2} d_1$, and $Z_1 = \frac{\beta-\gamma}{2\beta}\bar{\alpha} \frac{2\beta^2}{2\beta^2-\gamma^2}$,
$Z_2 = \frac{\beta-\gamma}{2\beta} \frac{2\beta^2}{2\beta^2-\gamma^2 d_1 d_2}$.

Then, we make a similar analysis to derive best response function and expected profit of the private firm in exogenous terms. We already derived the following function for the best response function of the private firm:

$$p_2(s_2) = E\left(\frac{\alpha(\beta-\gamma)+\gamma p_1}{2\beta}\bigg|s_2\right). \tag{2.16}$$

This yields,

$$p_2(s_2) = \frac{\beta-\gamma}{2\beta} E_2(\alpha) + \frac{\gamma}{2\beta} E_2\left(\frac{\gamma}{\beta} E_1(p_2)\right). \tag{2.17}$$

After proper substitution and calculations, we derive best response function of the private firm in exogenous terms.

$$p_2(s_2) = \frac{\beta-\gamma}{2\beta}\bar{\alpha}\frac{2\beta^2}{2\beta^2-\gamma^2} + \frac{\beta-\gamma}{2\beta}t_2(s_2-\bar{\alpha}^2)\frac{2\beta^2}{2\beta^2-\gamma^2 d_1 d_2}. \tag{2.18}$$

Or we can re-write this function in linear form as follows:

$$p_2(s_2) = Z_1 + t_2 Z_2(s_2 - \bar{\alpha}). \tag{2.19}$$

Then, the expected profit of the private will be equal to:

$$E(\pi_2|s_2) = \frac{1}{\beta^2-\gamma^2}\left(Z_1^2 + Z_2^2 t_2^2(V(\alpha)+v_2)\right). \tag{2.20}$$

So far, we derived best response functions of both firms and corresponding expected values of their objective functions by solving each firm's equilibrium price decisions in the second period. The derived conditional expected values of both firms' objective functions, $E(SW|s_1)$ and $E(\pi_2|s_2)$, show how these functions are related to the signals s_1 and s_2. Now, following the backward induction, we turn to the first period and derive their information sharing decisions in the first period of the game. The following lemma summarizes our first result.

Lemma 2.1 *An increase in the precision of private firm's information, or in the precision of public firm's information, or correlation of signals unconditionally raises expected profit of private firm.*

Proof The following equations show the derivative of expected profit of private firm with respect to the precision of private firm's information (v_2), the precision of public firm's information (v_1), and correlation of signals (σ_{12}).

$$\frac{\partial E(\pi_2|s_2)}{\partial v_2} = -\frac{1}{4}\left(\frac{\zeta + \eta(\beta-\gamma) V(\alpha)^2 (V(\alpha)+v_1)^2 \beta^2)}{(\zeta-\eta)^3 (\beta+\gamma)}\right), \tag{2.21}$$

where $\zeta = \beta^2 \left(V(\alpha)^2 + (v_1 + v_2)V(\alpha) + v_1 v_2 \right)$ and $\eta = \frac{1}{2}\gamma^2 \left(V(\alpha) + \sigma_{12} \right)^2$

The derivative, $\frac{\partial E(\pi_2 | s_2)}{\partial v_2}$, is always negative for all values of the parameters. As a result, an increase in the precision of private firm's information, the decrease of the variance (v_2), raises the expected profit of the private firm.

$$\frac{\partial E(\pi_2 | s_2)}{\partial v_1} = -\frac{1}{4}\left(\frac{\beta^2 \gamma^2 V(\alpha)^2 (\beta - \gamma)(V(\alpha) + v_2)(V(\alpha) + v_1)(V(\alpha) + \sigma_{12})}{(\zeta - \eta)^3 (\beta + \gamma)} \right).$$

(2.22)

Then, as long as $\gamma \neq 0$, $\frac{\partial E(\pi_2 | s_2)}{\partial v_1} < 0$. As we assumed that goods are substitutes ($\gamma > 0$), then the derivative is negative.

Last,

$$\frac{\partial E(\pi_2 | s_2)}{\partial \sigma_{12}} = \frac{1}{2}\left(\frac{\beta^2 \gamma^2 V(\alpha)^2 (\beta - \gamma)(V(\alpha) + v_2)(V(\alpha) + v_1)^2 (V(\alpha) + \sigma_{12})^2}{(\zeta - \eta)^3 (\beta + \gamma)} \right).$$

(2.23)

The assumption $\beta > \gamma > 0$ guarantees that the derivative is positive, $\frac{\partial E(\pi_2 | s_2)}{\partial \sigma_{12}} > 0$. As a result, the increase of the correlation of the signals raises expected profit of the private firm. ∎

The first lemma summarizes that the increase of all kind of information in the pool has a positive effect on the expected profit of the firm. The reason of this result is simple. There are two effects of the increase of the information: the first one is the output adjustment effect. More information leads firms to adjust to shocks and increases efficiency. Evidently, through this effect, more information in the market has tendency to increase the expected profit of the private firm. The second effect is that: when the firm is price setter, more information leads greater scope to extract consumer surplus. As, the private firm solely aims to increase its profit, through this effect, more information again has tendency to increase the expected profit. Hence, as more information has positive impact on profit through both channels, the net effect is also positive on the expected profit.

The next lemma summarizes the effect of more information in the market on social welfare.

Lemma 2.2 *An increase in the precision of private firm's information, or in the precision of public firm's information, or correlation of signals unconditionally lowers expected social welfare.*

Proof The derivative of the expected social welfare with respect to relevant parameters are as follows:

$$\frac{\partial E(SW | s_1)}{\partial v_1} = \frac{1}{8}\left(\frac{(\zeta + \eta)(\beta - \gamma)^2 V(\alpha)^2 (V(\alpha) + \sigma_{12})^2 \beta}{(\zeta - \eta)^3} \right),$$

(2.24)

where $\beta^2(V(\alpha)^2 + (v_1 + v_2)V(\alpha) + v_1v_2) = \zeta$ and $\frac{1}{2}\gamma^2(V(\alpha) + \sigma_{12})^2 = \eta$

Then the derivative with respect to v_1 is positive, $\frac{\partial E(SW|s_1)}{\partial v_1} > 0$.

$$\frac{\partial E(SW|s_1)}{\partial v_2} = \frac{1}{4}\left(\frac{(V(\alpha) + v_1)^2 (\beta - \gamma)^2 V(\alpha)^2 (V(\alpha) + \sigma_{12})^2 \beta^3}{(\zeta - \eta)^3}\right).$$

(2.25)

Similarly, the derivative with respect to v_2 is positive, $\frac{\partial E(SW|s_1)}{\partial v_2} > 0$. That implies an increase of the precision of information, a decrease in v_1 or v_2, lowers expected social welfare.

$$\frac{\partial E(SW|s_1)}{\partial \sigma_{12}} = -\frac{1}{4}\left(\frac{(\zeta + \eta)(V(\alpha) + \sigma_{12})(V(\alpha) + v_1)(\beta - \gamma)^2 V(\alpha)^2 \beta}{(\zeta - \eta)^3}\right).$$

(2.26)

Last, the derivative with respect to σ_{12} is negative, $\frac{\partial E(SW|s_1)}{\partial \sigma_{12}} < 0$, which completes the proof. ∎

The reason behind the results summarized in Lemma II stems from the trade-off between increasing production efficiency versus increasing power of capturing consumer surplus. As the Lemma I shows more information in the market leads greater scope to extract consumer surplus of the private firm, this leads overall fall in the social welfare.

Lemma 2.3 *An increase in λ_j lowers v_i and v_i is independent of λ_i, $i = 1, 2$, $j \neq i$.*

Proof Equation is as follows:

$$v_i = \frac{\sigma_u^2}{n_i + \lambda_j n_j},$$

(2.27)

v_i is inversely related to λ_j, while independent of λ_i. ∎

Lemma 2.4 *If $\lambda_j < 1, i = 1, 2, j \neq i$, then σ_{12} increases with λ_i, while if $\lambda_j = 1$, then σ_{12} is independent of λ_i*

Proof Equation for σ_{12} is as follows:

$$\sigma_{12} = \frac{(\lambda_1 n_1 + \lambda_2 n_2)}{(n_1 + \lambda_2 n_2)(n_2 + \lambda_1 n_1)}\sigma_u^2.$$

(2.28)

Now for $i = 1$

$$\frac{\partial \sigma_{12}}{\partial \lambda_1} = -\frac{(\lambda_2 - 1)n_1 n_2}{(n_1 + \lambda_2 n_2)(n_2 + \lambda_1 n_1)^2}\sigma_u^2.$$

(2.29)

If $\lambda_2 = 1$, the derivative is 0, if $\lambda_2 < 1$, then the derivative is positive. Similarly for $i = 2$

$$\frac{\partial \sigma_{12}}{\partial \lambda_2} = -\frac{(\lambda_1 - 1)n_1 n_2}{(n_1 + \lambda_2 n_2)(n_2 + \lambda_1 n_1)^2} \sigma_u^2. \tag{2.30}$$

If $\lambda_1 = 1$, the derivative is 0, if $\lambda_1 < 1$, then the derivative is positive. ∎

The previous two lemmas, Lemma III and Lemma IV show effects of increasing the number of observations in the pool on variance, v_i and correlation of signals, σ_{12}. After these lemmas, the next corollary summarizes net effect of information sharing on expected profit of the private firm and expected social welfare.

Corollary 1 *More information sharing in the pool increases expected profit of the private firm while decreases expected social welfare.*

Proof Lemma I together with Lemma III and Lemma IV imply that increase in λ_i, $i = 1, 2$, has increasing effect on expected profit of the private firm. Similarly, Lemma II together with Lemma III and Lemma IV imply that increase in λ_i, $i = 1, 2$, has increasing effect on expected social welfare. ∎

Table 2.1 summarizes the net effects of more information sharing and channels in which more information affects expected profit and social welfare.

The next proposition summarizes main result of the paper.

Proposition 1 *Suppose goods are substitutes. Then the two-stage Bertrand game has a unique Perfect Bayes Equilibrium in dominant strategies. There is partial information pooling: the public firm does not share any information, while private firm completely shares all information it has.*

Proof As corollary shows, more information in the pool increases expected profit of the firm in all cases, independently of the best response of the public firm. As a result, dominant strategy of the private firm is to share all information it has. Conversely, more information in the pool decreases overall expected social welfare in all cases. Thus, no information sharing is the dominant strategy for social welfare maximizing public firm. ∎

An important remark is that although one of the firms is social welfare maximizing public firm, the market outcome is never first-best optimal. As there is partial information sharing, the market outcome does not maximize production efficiency and does not enhance informational efficiency.

Table 2.1 Effects of information sharing on expected profit and social welfare

$\lambda_2 \uparrow$	$v_1 \downarrow$	$E\pi_2 \uparrow$	$ESW \downarrow$
	$\sigma_{12} \uparrow$	$E\pi_2 \uparrow$	$ESW \downarrow$
$\lambda_1 \uparrow$	$v_2 \downarrow$	$E\pi_2 \uparrow$	$ESW \downarrow$
	$\sigma_{12} \uparrow$	$E\pi_2 \uparrow$	$ESW \downarrow$

A relevant question here might be whether public firm may choose not to use the information that the private firm provides. As less information in the information pool has always social welfare increasing effect, one may argue that public firm may choose not to receive information provided by the private firm. If ex ante the public firm could guarantee that it would not use the information shared by the private firm, it would choose not to use the information shared by the private firm. However, once the private firm puts observations it has to the information pool and shares them, the public firm ex post cannot guarantee that it will not use the information that the private firm shares. Because as more information has always positive effect on enhancing production efficiency, once received the information shared, the best the public firm can do is to use this information. Hence, the public firm uses all information shared in equilibrium.

Last, we compare equilibrium outcomes established in mixed oligopoly with the ones in pure oligopolies. In a very similar two stage duopoly setting, Vives [11] analyzes informational equilibrium outcomes for a pure duopoly. The information setting and production functions of the firms are same in that paper with the ones assumed in this study. Vives [11] establishes that if goods are substitutes and firms compete a la Bertrand, in Perfect Bayesian equilibrium of the game, both private firms share all information that they have with each other. This implies that information sharing behavior of private firm does not change according to whether it competes in a pure oligopoly market or in a mixed oligopoly market. However, behavior of the public firm is totally different as we showed in this study. In sum, although there is full information sharing in a pure oligopoly with Bertrand competition, informational equilibrium is characterized with partial information sharing in mixed oligopoly. Thus, comparing with equilibrium outcomes in pure oligopoly, mixed oligopoly yields less information efficiency and less production efficiency but still enhances social welfare.

2.4 Conclusion

We have considered informational outcomes in a mixed oligopoly, where a private firm competes with a social welfare maximizing public firm. We analyzed firms' incentives to share and diffuse information when firms produce substitute products and compete in prices under stochastic common demand shocks. It seemed there are two main effects: information sharing increases production efficiency by enabling firms to predict stochastic demand shocks better. The other effect is to increase power of capturing consumer surplus. For private firm, both effects work in the same direction and gives incentive to private firm to share all private signals it received with the public firm. As a result, in equilibrium, private firm always share all information it has. For the public firm, which aims to maximize social welfare, production efficiency motive gives it tendency to share information. However, as more information also increases private firm's power of capturing consumer surplus, this second effect leads public firm to not to share the information it has. In equilibrium, the second effect dominates the first one and as a result, in order to reduce private firm's power of capturing consumer surplus, the public firm shares no information with the private firm. Hence, the equilibrium is characterized with partial information sharing.

There are several lines to extend the analysis of this study. We have considered only Bertrand competition with substitute goods. The prior studies on the information sharing in pure oligopolies show that equilibrium outcomes heavily depend on type of goods and type of competition. Thus, extending analyzes for Cournot competition and for complement goods will be beneficial. Another line of research may focus on the effects of privatization with taking informational outcomes into account. This line of research may establish important policy implications.

Acknowledgments Haluk would like to acknowledge financial support from the Scientific and Technological Research Council of Turkey (TUBITAK International Cost Grant 217K428) and COST action European Network for Game Theory (GAMENET). The authors thank Hasan Karaboga, co-editors Leon Petrosyan, Vladimir Mazalov and Nikolay Zenkevich and seminar participants at International Meeting on Game Theory (ISDG12-GTM2019) Saint-Petersburg.

References

1. Bagnoli, M., Watts, S.G.: Competitive intelligence and disclosure. RAND J. Econom. **46**(4), 709–729 (2015)
2. Clarke, R.N.: Collusion and the incentives for information sharing. Bell J. Econom. **14**(2), 383–394 (1983)
3. Framingham, M.: IDC forecasts revenues for big data and business analytics solutions will reach $189.1 Billion this year with double-digit annual growth through 2022. IDC, Framingham (2019). https://www.idc.com/getdoc.jsp?containerId=prUS44998419. Cited 29 Jan 2020
4. Fried, D.: Incentives for information production and disclosure in a duopolistic environment. Q. J. Econ. **99**(2), 367–381 (1984)
5. Gal-Or, E.: Information sharing in oligopoly. Econometrica **53**, 329–343 (1985)
6. Haraguchi J., Matsumura T.: Cournot-Bertrand comparison in a mixed oligopoly. J. Econom. **117**(2), 117–136 (2016)
7. Kowalski, P., Büge, M., Sztajerowskai, M., Egeland, M.: State-Owned Enterprises: Trade Effects and Policy Implications. OECD Trade Policy Papers, No. 147, OECD Publishing, Paris. https://doi.org/10.1787/5k4869ckqk7l-en
8. Myatt D.P., Wallace C.: Cournot competition and the social value of information. J. Econom. Theory **158**, 466–506 (2015)
9. Novshek, W., Sonnenschein, H.: Fulfilled expectations cournot duopoly with information acquisition and release. Bell J. Econom. **13**, 214–218 (1982)
10. Ponssard, J.P.: The strategic role of information on the demand functions in an oligopolistic market. Manag. Sci. **25**(3), 243–250 (1979)
11. Vives, X.: Duopoly information equilibrium: Cournot and Bertand. J. Econom. Theory **34**, 71–94 (1984)

Chapter 3
Reflexive and Epistemic Properties of the Tullock Rent-Seeking Game

Denis Fedyanin

Abstract This study sets out to investigate the impact of information control in the Tullock rent-seeking game. The game itself is constructed by using a normal form and making suggestions on the agents' believes and knowledge. We found domains of parameters where monotonicity of the impact holds too. Together, these results provide valuable insights into the effects of reflexive analysis on the properties of information control.

Keywords Parametrized equilibrium · Epistemic models · Informational structure · Social interaction · Opinion · Tullock · Rent-seeking · Control

3.1 Introduction

Players discuss parameters of the competition for a prize and compete by making costly investments. Players have initial beliefs, types. Communication between them could be fruitful or not.

We analyze how we should change beliefs or parameters of the competition (game) to get higher investment (action) or utility for the given player. Our investigation splits into several parts. We should be able to model the beliefs, should be able to predict the results of the negotiations of players based on their primary types, expectations and available communication, and should be able to find the equilibria [4, 5].

The critical part is types and uncertainties. Epistemic games [10], epistemic logic [2, 14, 15], k-level epistemic models [7], reflexive models, fuzzy logic [8], and numerous other directions have been developed during years and have many results. In our case, we use a particular type of structures. The structures are simple,

D. Fedyanin (✉)
V.A.Trapeznikov Institute of Control Sciences, Moscow, Russia
e-mail: dfedyanin@inbox.ru

L. A. Petrosyan et al. (eds.), *Frontiers of Dynamic Games*,
Static & Dynamic Game Theory: Foundations & Applications,
https://doi.org/10.1007/978-3-030-51941-4_3

symmetric, and fruitful. We avoid fuzziness by using a specific variety of standard uncertainties.

Communication and belief interaction is also an object of investigations in opinion dynamics models, and they could be considered sometimes like judgement aggregation [11]. There are many problems and successes with the aggregates of preferences like the social choice domain [1]. Social cognitive maps are also close to our method. The uniqueness of our paper is in a sophisticated approach. Other models of negotiations like auctions could be used here too. We have chosen the simplest from the unusual for us models.

The game-theoretical part is based on the known Nash equilibriums of the purest form of our game. It is an exciting game by itself, and we are sure that it still hides many secrets to be discovered. Our uniqueness is that we reshape basic games with three parameters into a game with more settings. Though they have symmetric properties, the standard methods of solving these games could not be directly applied without preliminary investigations.

3.2   The Model

First of all, our research is based on a fundamental concept on the Nash equilibrium and rent-seeking game where higher investments increase the probability of winning the prize [12, 13]. There are applications: competition for monopoly rents, investments in $R\&D$, competition for a promotion/bonus, political contests. A formal model is the following. Set of players $N = \{1, \ldots, n\}$, strategies $x_i > 0, \forall i \in N$, utility function

$$f_i = \frac{x_i^{\alpha}}{\sum_{j \in N} x_j^{\alpha}} M - x_i, \forall i \in N.$$

Restrictions

$$M \geq 1, 0 < \alpha < 1.$$

The parameter M is the amount of the resource that is given and should be split among agents; the parameter α is the measure for the probability of winning for the agents with small actions. If α is very large, then the probability of winning for the agents with small actions is smaller.

The very vital part is that we allow payers do not have the same belief about the parameters of the game. It makes the difference with a classic Tullock game. We add the classic game as one of the cases (Game 1) to compare the results and show how our method of using control could be applied to this case.

We have two types of players and two types of communication (available or not), and we analyzed all possible combinations. It leads us to the four unique blends of types and communications for analysis. We have suggested epistemic models for all of them and calculated equilibriums for the first three of them.

We suggest the following algorithm to classify which model suits specific cases in the real world.

1. Is there a difference in beliefs and real value of the parameters α, M, n. If there is no, then the Game 1 should be used, and Nash equilibrium will be the solution for parametrized equilibrium.
2. If there is a meaningful social interaction among agents? If there is no, then we should use Game 2 to find a solution since the game splits into the separate Games of type 1. The difference with Game 1 is that parameters α, M, n in these games will differ, and the combined output actions would not necessarily form a Nash equilibrium in Game 1. Thus the values of utilities would not obligatorily coincide with the values of utility functions in Game 1. Trust is crucial since it could be communication, but all lie. It means that there is no meaningful communication since agents could choose a strategy just to discard incoming messages.
3. Are all agents stubborn, or they all want to come to a consensus? It is not a complete list of alternatives, but our next models correspond only to these two extreme cases. If the agents want to come to a consensus, we suggest using a model 3 that is a De Groot model—the linear model of opinion modification. It leads us to a Game 1 -like a model when we have a single belief about the parameter, but since it is the result of some negotiations, we should take into account the social influences of agents. In case all agents are stubborn, then we should use Game 4. It is the most complicated model. We do not investigate this game in this paper but mention it since it is a part of a general method.

We have considered the decision rules to choose which game should be used. These decisions are schematically shown in Fig. 3.1.

We will use a more generic version of the Nash equilibrium that is informational. It is very similar to the standard Nash equilibrium but takes into account informational structure in agents' representations of the game. The formal definition can be found in [9].

Reflective game Γ_I is a game described by the following tuple:

$$\Gamma_I = \{N, (X_i)_{i \in N}, f_i(\cdot)_{i \in N}, I\},\tag{3.1}$$

Fig. 3.1 The decision tree representations of the classification of the games by the types of belief dynamics

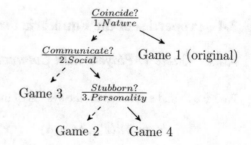

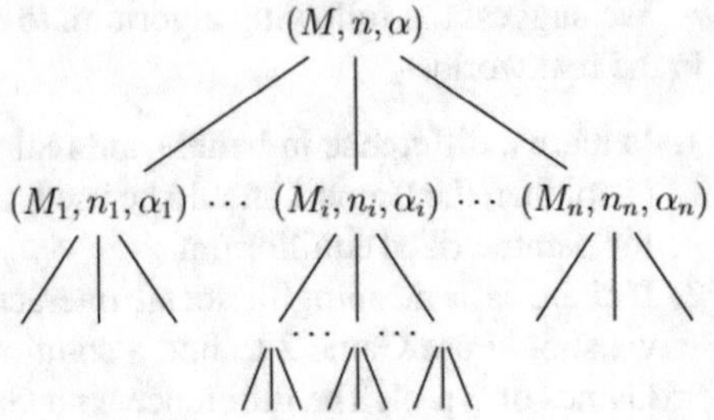

Fig. 3.2 An example of a
simple informational
structure that is similar to
Harsanyie types approach [6]

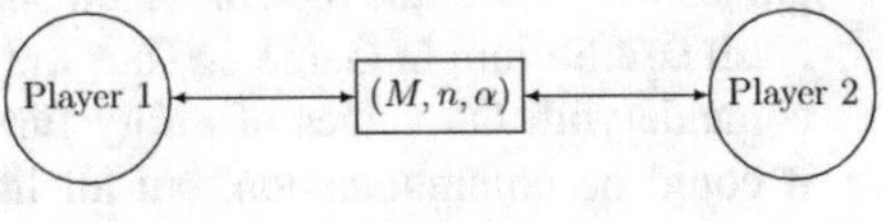

$$(M, n, \alpha)$$

$$(M_1, n_1, \alpha_1) \quad \cdots \quad (M_i, n_i, \alpha_i) \quad \cdots \quad (M_n, n_n, \alpha_n)$$

Fig. 3.3 The informational
structure for the Game 1

Player 1 ⟷ (M, n, α) ⟷ Player 2

where N is the set of real agents, X_i is the set of valid actions of the i th agent, $f_i(\cdot)$ is its target function, $i \in N$, I—informational structure or belief representations (Fig. 3.2).

3.3 Epistemic Models of Beliefs

We use Chkhartishvili-Novikov notations for belief representations [9]. In brief, it is the graph of possible worlds (rectangles) and images of agents (circles). Some images coincide with real agents, and others are just phantom agents. Term 'phantom agent' is a feature of Chkhartishvili-Novikov notations for belief representations. A possible world is connected to an image of an agent iff the world contains the image of the agent. An image of an agent is connected to a potential world iff the agent considers this world possible. We will use this notation in this paper. For example, two agents and a single possible world are shown in the Fig. 3.1. It is the simplest case—the informational structure for a game in the normal form where Nash equilibrium directly coincides with equilibrium in the reflexive game (Fig. 3.3).

3.4 Properties of the Equilibria for the Games 1–3

3.4.1 Game 1: Players with Common Knowledge

We have to find a solution for the system of the best responses (BR) of the players.

$$x_1^* = BR(x_{-1}^*, M, n, \alpha); \ldots; x_n^* = BR(x_{-n}^*, M, n, \alpha).$$

This solution gives us equilibrium.

Actions of agents are

$$x_i^* = \frac{n-1}{n^2}\alpha M; \forall i \in N.$$

Furthermore, monotonicity could be found by an analysis of the following derivatives.

$$\frac{\partial}{\partial M}x_i^* = \frac{n-1}{n^2}\alpha > 0; \frac{\partial}{\partial \alpha}x_i^* = \frac{n-1}{n^2}M > 0; \frac{\partial}{\partial n}x_i^* = \frac{2-n}{n^3}\alpha M < 0.$$

Utility functions are

$$f_i(x^*) = \frac{n-(n-1)\alpha}{n^2}M.$$

Futhermore, monotonicity could be found by an analysis of the following derivatives.

$$\frac{\partial}{\partial M}f_i(x^*) = \frac{n-(n-1)\alpha}{n^2} > 0; \frac{\partial}{\partial \alpha}f_i(x^*) = -\frac{n-1}{n^2}M < 0.$$

$$\frac{\partial}{\partial n}f_i(x^*) = \frac{(1-\alpha)n - 2(n-(n-1)\alpha)}{n^3}M < 0.$$

3.4.2 Game 2: Players Without Communication

A brief example of this model is the following. Let there are Ann and Bob. Ann watches the TV channel, and there is a claim that there is a storm nearby. She could think that it is such important news that everyone should know it. Bob does not know anything about the storm and feels that nobody thinks that there is a storm. Both of them are wrong in detail but make actions as they are right. We can model this situation to see Fig. 3.4 for Chkhartishvili-Novikov representations of beliefs of such agents.

Fig. 3.4 The informational structure in Chkhartishvili-Novikov form for players without communication

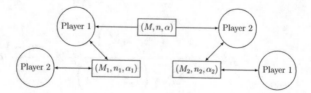

We have to find a solution for the system of the best responses (BR) of the players.

$$x_1^* = BR(x_{-1}^*, M, n, \alpha); \ldots; x_n^* = BR(x_{-n}^*, M, n, \alpha);$$

$$x_1^{*1} = BR(x_{-1}^{*1}, M_1, n_1, \alpha_1); \ldots; x_n^{*1} = BR(x_{-n}^{*1}, M_1, n_1, \alpha_1);$$

$$x_1^{*j} = BR(x_{-1}^{*j}, M_j, n_j, \alpha_j); \ldots; x_n^{*j} = BR(x_{-n}^{*j}, M_j, n_j, \alpha_j);$$

$$x_1^{*n} = BR(x_{-1}^{*n}, M_n, n_n, \alpha_n); \ldots; x_n^{*n} = BR(x_{-n}^{*n}, M_n, n_n, \alpha_n);$$

$$x_1^* = x_1^{*1}; \ldots; x_n^* = x_n^{*n}.$$

This solution gives us equilibrium.

Actions of agents are

$$x_i^* = \frac{n_i - 1}{n_i^2} \alpha_i M_i.$$

Moreover, monotonicity could be found by an analysis of the following derivatives.

$$\frac{\partial}{\partial M_i} x_i^* = \frac{n_i - 1}{n_i^2} \alpha_i > 0; \quad \frac{\partial}{\partial M_j} x_i^* = 0; \quad \frac{\partial}{\partial M} x_i^* = 0.$$

$$\frac{\partial}{\partial \alpha_i} x_i^* = \frac{n_i - 1}{n_i^2} M_i > 0; \quad \frac{\partial}{\partial \alpha_j} x_i^* = 0; \quad \frac{\partial}{\partial \alpha} x_i^* = 0.$$

$$\frac{\partial}{\partial n_i} x_i^* = \frac{2 - n_i}{n_i^3} \alpha_i M_i < 0; \quad \frac{\partial}{\partial n_j} x_i^* = 0; \quad \frac{\partial}{\partial n} x_i^* = 0.$$

Utility functions are

$$f_i(x^*) = \frac{x_i^\alpha}{\sum_{j \in N} x_j^\alpha} M - x_i;$$

$$f_i(x^*) = \frac{\left(\frac{n_i - 1}{n_i^2} \alpha_i M_i \right)^\alpha}{\sum_{j \in N} \left(\frac{n_j - 1}{n_j^2} \alpha_j M_j \right)^\alpha} M - \frac{n_i - 1}{n_i^2} \alpha_i M_i, \forall i \in N.$$

Futhermore, monotonicity could be found by an analysis of the following derivatives.

Real-world properties are the following.

$$\frac{\partial}{\partial M} f_i(x^*) = \frac{\left(\frac{n_i-1}{n_i^2}\alpha_i M_i\right)^\alpha}{\sum_{j\in N}\left(\frac{n_j-1}{n_j^2}\alpha_j M_j\right)^\alpha} > 0;$$

$$\frac{\partial}{\partial \alpha} f_i(x^*) = \frac{\left(\frac{n_i-1}{n_i^2}\alpha_i M_i\right)^\alpha \sum_{j\in N}\left(\frac{n_j-1}{n_j^2}\alpha_j M_j\right)^\alpha \left(\ln\alpha - \sum_{j\in N}\ln\alpha_j\right)}{\left(\sum_{j\in N}\left(\frac{n_j-1}{n_j^2}\alpha_j M_j\right)^\alpha\right)^2} M > 0; , \forall i \in N.$$

Belief properties are the following.

$$\frac{\partial}{\partial M_i} f_i(x^*) =$$

$$\frac{\alpha\left(\frac{n_i-1}{n_i^2}\alpha_i\right)^\alpha M_i^{\alpha-1}\left(\sum_{j\in N}\left(\frac{n_j-1}{n_j^2}\alpha_j M_j\right)^\alpha\right) - \alpha\left(\frac{n_i-1}{n_i^2}\alpha_i\right)^\alpha M_i^{2\alpha-1}}{\left(\sum_{j\in N}\left(\frac{n_j-1}{n_j^2}\alpha_j M_j\right)^\alpha\right)^2} M$$

$$+\frac{n_i-1}{n_i^2}\alpha_i;$$

$$\frac{\partial}{\partial M_j} f_i(x^*) = \left(\frac{n_j-1}{n_j^2}\alpha_j\right)^\alpha M_j^{\alpha-1}\frac{\alpha\left(\frac{n_i-1}{n_i^2}\alpha_i M_i\right)^\alpha}{\left(\sum_{j\in N}\left(\frac{n_j-1}{n_j^2}\alpha_j M_j\right)^\alpha\right)^2} M > 0;$$

$$\frac{\partial}{\partial \alpha_i} f_i(x^*) =$$

$$\frac{\alpha\left(\frac{n_i-1}{n_i^2}M_i\right)^\alpha \alpha_i^{\alpha-1}\left(\sum_{j\in N}\left(\frac{n_j-1}{n_j^2}\alpha_j M_j\right)^\alpha\right) - \alpha\left(\frac{n_i-1}{n_i^2}M_i\right)^\alpha \alpha_i^{2\alpha-1}}{\left(\sum_{j\in N}\left(\frac{n_j-1}{n_j^2}M_j\alpha_j\right)^\alpha\right)^2} M$$

$$+\frac{n_i-1}{n_i^2}M_i;$$

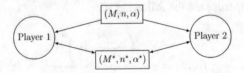

Fig. 3.5 The informational structure in Chkhartishvili-Novikov form for players with communication and consensus

$$\frac{\partial}{\partial \alpha_j} f_i(x^*) = \left(\frac{n_j - 1}{n_j^2} M_j\right)^\alpha \alpha_j^\alpha \frac{2\alpha \left(\frac{n_i-1}{n_i^2}\alpha_i M_i\right)^\alpha}{\left(\sum_{j\in N}\left(\frac{n_j-1}{n_j^2}\alpha_j M_j\right)^\alpha\right)^2} M > 0.$$

Properties of beliefs about a number of agents are the following (Fig. 3.5).

$$\frac{\partial}{\partial n_j} f_i(x^*) = -\frac{\left(\frac{n_i-1}{n_i^2}\alpha_i M_i\right)^\alpha (\alpha_j M_j)^\alpha \alpha \left(\frac{n_j-1}{n_j^2}\right)^{\alpha-1}(2-n_j)}{n_j^3 \left(\sum_{j\in N}\left(\frac{n_j-1}{n_j^2}\alpha_j M_j\right)^\alpha\right)^2} M < 0;$$

$$\frac{\partial}{\partial n_i} f_i(x^*) = \frac{(\alpha_i M_i)^\alpha \alpha \left(\frac{n_i-1}{n_i^2}\right)^{\alpha-1}\frac{(2-n_i)}{n_i^3}\left(\sum_{j\in N}\left(\frac{n_j-1}{n_j^2}\alpha_j M_j\right)^\alpha - \left(\frac{n_i-1}{n_i^2}\alpha_i M_i\right)^\alpha\right)}{\left(\sum_{j\in N}\left(\frac{n_j-1}{n_j^2}\alpha_j M_j\right)^\alpha\right)^2} M -$$

$$-(\alpha_i M_i)^\alpha \alpha \left(\frac{n_i-1}{n_i^2}\right)^{\alpha-1}\frac{(2-n_i)}{n_i^3}.$$

3.4.3 Game 3: Players with Communication and Consensus

There could be communication between agents, and they can communicate according to the de Groot model [3]. There is no difference if the existence of such communication is common knowledge among all agents, or it is not.

$$M^* = \sum_{i\in N} w_i^M M_i; \quad \alpha^* = \sum_{i\in N} w_i^\alpha \alpha_i; \quad n^* = \sum_{i\in N} w_i^n n_i,$$

where w_i^M, w_i^α, w_i^n is the influence of the agent i on a social network consensus opinion about M, α, n

We have to find a solution for the system of best responses (BR) of players.

$$x_1^* = BR(x_{-1}^*, \sum_i w_i^M M_i, \sum_i w_i^n n, \sum_i w_i^\alpha \alpha);$$

$$\cdots$$

$$x_n^* = BR(x_{-n}^*, \sum_i w_i^M M_i, \sum_i w_i^n n, \sum_i w_i^\alpha \alpha).$$

This solution gives us an equilibrium.
  Actions of agents are

$$x_i^* = \frac{n^* - 1}{(n^*)^2} \alpha^* M^*.$$

Moreover, monotonicity could be found by an analysis of the following derivatives.
  Real-world properties are the following.

$$\frac{\partial}{\partial M} x_j^* = 0; \quad \frac{\partial}{\partial \alpha} x_j^* = 0; \quad \frac{\partial}{\partial n} x_j^* = 0.$$

Belief properties are the following

$$\frac{\partial}{\partial M_i} x_j^* = \frac{n^* - 1}{(n^*)^2} \alpha^* w_i^M > 0; \quad \frac{\partial}{\partial \alpha_i} x_j^* = \frac{n^* - 1}{(n^*)^2} w_i^\alpha M^* > 0; \quad \frac{\partial}{\partial n_i} x_j^* = \frac{2 - n^*}{(n^*)^3} w_i^n \alpha^* M^* < 0.$$

Dependencies of social influences in network are

$$\frac{\partial}{\partial w_i^M} x_j^* = \frac{n^* - 1}{(n^*)^2} \alpha^* M_i > 0; \quad \frac{\partial}{\partial w_i^\alpha} x_j^* = \frac{n^* - 1}{(n^*)^2} \alpha_i M^* > 0; \quad \frac{\partial}{\partial w_i^n} x_j^* = \frac{2 - n^*}{(n^*)^3} n_i \alpha^* M^* < 0.$$

Utility functions are

$$f_i(x^*) = \frac{M}{n} - \frac{(n^* - 1)\alpha^* M^*}{(n^*)^2}.$$

Furthermore, monotonicity could be found by an analysis of the following derivatives.
  Real-world properties are the following.

$$\frac{\partial}{\partial M} f_j(x^*) = \frac{2}{n} > 0; \quad \frac{\partial}{\partial \alpha} f_i(x^*) = 0; \quad \frac{\partial}{\partial n} f_i(x^*) = -\frac{M}{n^2} < 0.$$

Fig. 3.6 The informational
structure in
Chkhartishvili-Novikov form
for stubborn players with
communication without
consensus

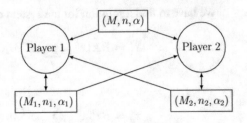

Belief properties are the following

$$\frac{\partial}{\partial M_i} f_j(x^*) = w_j^M \frac{n^* - (n^* - 1)\alpha^*}{(n^*)^2} > 0; \quad \frac{\partial}{\partial \alpha_j} f_i(x^*) = -\frac{M^*(n^* - 1)w_j^\alpha}{(n^*)^2} < 0;$$

$$\frac{\partial}{\partial n_j} f_i(x^*) = \frac{(1 - \alpha^*)w_j^n n^* - 2w_j^n (n^* - (n^* - 1)\alpha^*)}{(n^*)^3} M^* < 0.$$

Dependencies of network social influences are (Fig. 3.6)

$$\frac{\partial}{\partial w_i^M} f_j(x^*) = M_j \frac{(n^* - (n^* - 1)\alpha^*)}{(n^*)^2} > 0; \quad \frac{\partial}{\partial w_j^\alpha} f_i(x^*) = -\frac{M^*(n^* - 1)\alpha_j}{(n^*)^2} < 0;$$

$$\frac{\partial}{\partial w_j^n} f_i(x^*) = \frac{(1 - \alpha^*)n_j n^* - 2n_j (n^* - (n^* - 1)\alpha^*)}{(n^*)^3} M^* < 0.$$

3.4.4 Game 4: Stubborn Players with Communication Without Consensus

If there is communication with no trust at all, then all agents 'become stubborn', and other opinions do not change their views. There is no difference if the existence of such communication is common knowledge among all agents, or it is not.

We have to find a solution for the system of th best responses (BR) of the players.

$$x_1^* = BR(x_{-1}^*, M_1, n_1, \alpha_1); \dots; x_n^* = BR(x_{-n}^*, M_n, n_n, \alpha_n).$$

This solution gives us an equilibrium.

Table 3.1 The monotonicity of the actions and the utilities for the games (columns) for the controls of beliefs and the real world (rows)

Control	Game 1 Strategy x_i	Game 1 Utility f_i	Game 2 Strategy x_i	Game 2 Utility f_i	Game 3 Strategy x_i	Game 3 Utility f_i
M	↗	↗	0	↗	0	↗
M_i	NA	NA	↗	↗	↗	?
M_j	NA	NA	↗	↗	0	↗
n	↘	↘	0	↘	0	NA
n_i	NA	NA	↘	?	↘	?
n_j	NA	NA	↘	?	0	↘
α	↗	↘	0	0	0	↗
α_i	NA	NA	↗	↘	↘	?
α_j	NA	NA	↗	↘	0	↗
w_i^M	NA	NA	↗	↗	NA	NA
w_j^M	NA	NA	↗	↗	NA	NA
w_i^n	NA	NA	↘	?	NA	NA
w_j^n	NA	NA	↘	?	NA	NA
w_i^α	NA	NA	↗	↘	NA	NA
w_j^α	NA	NA	↗	↘	NA	NA

3.5 Conclusion

In this paper, we considered a Tullock rent-seeking game with parameters: benefits for large players α, budget M, number of players n, and suggested that this are uncertain parameters for agents. We applied our previously developed method and found useful preliminary information for monotonicity analysis and further control analysis (Table 3.1).

Acknowledgments The article was partially supported by RSCF project 16-19-10609. The support was used to prepare Sect. 3.3.

References

1. Arrow, K.J.: Social Choice and Individual Values, vol. 12. Yale University, New Haven (2012)
2. Aumann, R.J.: Interactive epistemology I: knowledge. Int. J. Game Theory **28**(3), 263–300 (1999)
3. DeGroot, M.H.: Reaching a consensus. J. Am. Stat. Assoc. **69**, 118–121 (1974)
4. Fedyanin, D.N.: An example of Reflexive Analysis of a game in normal form. In: Petrosyan, L., Mazalov, V., Zenkevich, N. (eds.) Frontiers of Dynamic Games. Static & Dynamic Game Theory: Foundations and Applications, pp. 1–11 (2019)
5. Fedyanin, D.: The complex mechanism of belief control for cost reduction under stability restriction. Cournout competition example. In: Proceedings of the 19th IFAC Conference on Technology, Culture and International Stability (2020)

6. Harsanyi, J.C.: Games with incomplete information played by bayesian players, I-III. Manag. Sci. **14**(3), 159–183 (Part I), **14**(5), 320–334 (Part II), **14**(7), 486–502 (Part III) (1967–1968)
7. Kneeland, T.: Testing behavioral game theory: higher-order rationality and consistent beliefs. In: Mimeo (2013)
8. Massad, E. et al.: Fuzzy Logic in Action: Applications in Epidemiology and Beyond, vol. 232. Springer, Berlin (2009)
9. Novikov D., Chkhartishvili, A.: Reflexion Control: Mathematical models. Series: Communications in Cybernetics, Systems Science and Engineering (Book 5), p. 298. CRC, New York (2014)
10. Perea, A.: Epistemic game theory: reasoning and choice. Cambridge University, Cambridge (2012)
11. Rahwan, I., Fernando T.: Collective argument evaluation as judgement aggregation. In: Proceedings of the 9th International Conference on Autonomous Agents and Multiagent Systems (AAMAS 2010), pp. 417–424 (2010)
12. Tullock, G.: The welfare cost of tariffs, monopolies, and theft. Western Econom. J. **5**, 224–232 (1967)
13. Tullock, G., Buchanan, J.M., Tollison, R.D.: Toward a Theory of the Rent-seeking Society. In: Efficient rent seeking (1980)
14. Van Benthem, J.: Games in dynamic-epistemic logic. Bull. Econ. Res. **53**(4), 219–248 (2001)
15. Van Benthem, J. et al.: Logic in Action. Universiteit van Amsterdam/Institute for Logic, Amsterdam (2001)

Chapter 4
Solution of Differential Games with Network Structure in Marketing

Ekaterina Gromova and Anastasiya Malakhova

Abstract A marketing network model of goodwill accumulation with spillover effect is analysed in a differential game theory framework. Cooperative form of the game is considered under α-characteristic function. An approach is illustrated on a numerical example with particular values of the model parameters fixed.

Keywords Network differential games · Characteristic function · Differential game theory

4.1 Introduction

Modern mathematical game theory sets out to model, analyse and resolve various issues associated with conflict-controlled processes. Of particular interest are dynamic processes, the conflict processes developing over time, which could be well described in differential games terms [6].

Another essential branch of mathematical game theory covers network models. The models taking place under an assumption of some network structure among players. Differential games on networks were widely studied in [11]. Moreover, such game formulation found its place in economic and marketing issues [5].

In recent literature [8, 9], dynamic processes in marketing, which evolves over time, are often described in the framework of differential game theory. But there are only a few papers in which marketing is considered with the network structure of participants, especially in the continuous-time formulation [1].

In this paper differential game with network structure applied for a marketing model of goodwill accumulation is considered [7]. Additionally, the model includes the spillover effect [2] that accounts for the influence of other players' decisions on the total payoff of the players.

E. Gromova (✉) · A. Malakhova
Saint Petersburg State University, St. Petersburg, Russia

L. A. Petrosyan et al. (eds.), *Frontiers of Dynamic Games*,
Static & Dynamic Game Theory: Foundations & Applications,
https://doi.org/10.1007/978-3-030-51941-4_4

The paper is organised as follows. Section 4.2 is dedicated to the game formulation with all the additional necessary assumptions given. The cooperative setup is proposed in Sect. 4.3. In the following section α-characteristic function of the game is calculated in the form of *maxmin* problem. In Sect. 4.4 the proposed approach is illustrated by a numerical example.

4.2 Game Formulation

Consider a differential game of three ($N = 3$) players $\Gamma(t_0, x_0)$. The game starts from the initial time instant t_0 and initial state x_0 and supposed to proceed on the infinite interval. The game is assumed to have a network structure represented by the non-oriented graph illustrated by the Fig. 4.1.

Assume that the dynamic of the common state variable for each player takes form of the following differential equation (4.1)

$$\dot{x}_i = \alpha_i u_i(t) - \delta_i x_i(t), \quad i = \{1, 2, 3\}, \quad x_i(t_0) = x_0^i. \tag{4.1}$$

The state variables $x_i(t)$ refer to the amount of stock of the player (advertising, technology, resource, capital). The control variables $u_i(t)$ are the open-loop strategies of the player i and represent the investment/extraction effort of the player (firm). In addition, both the controls and the state variables are required to be (almost everywhere) differentiable.

The player's payoff consists of components depending on his network connections with other players. If the player i is connected with the player j then the payoff component takes form (4.2)

$$h_{ij}(t) = e^{-\rho t}(a_i x_i(t) + c_j x_i(t) u_j(t) - \frac{1}{2} u_i^2(t)). \tag{4.2}$$

Thus, the payoff of the player is the sum of connection components

$$J_i\left(x_0^i, x_0^{K(i)}; u^i, u^{K(i)}\right) = \int_{t_0}^{\infty} \sum_{j \in K(i)} h_{ij}(\tau) d\tau, \tag{4.3}$$

Fig. 4.1 The game's network structure

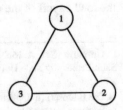

here $K(i) = \{(i, j), j \in N, (i, j) \in L\}$—the set of all connections of the player i, $L = \{i, j\}$—edges of the graph.

The payoff functional of the considered model (4.2) has a linear-quadratic form. This fact implements a number of valuable properties. In particular, Hamilton-Jacobi-Bellman approach and Maximum Principle yield the same decision if they are restricted to linear-feedback forms, see Dockner et al. (2000) [3].

Assume the following restriction which are standard for economical applications.

- Non-negative constrains on the controls' value due to its nature as an effort level, thus

$$u_i \in U_i \subseteq \mathbb{R}_+, \quad i = \{1, 2, 3\}$$

- Open-loop strategies is taken from the closed compact set

$$u_i(t) \in \hat{U}_i \subset Comp\mathbb{R}, \quad i = \{1, 2, 3\}$$

- Non-negative constrains on the common state's value due to it's nature as a stock level, thus

$$x_i \in X_i \subseteq \mathbb{R}_+ \quad i = \{1, 2, 3\}.$$

This game has a linear-quadratic structure which takes place commonly among advertising and marketing models (see Deal et al. (1979) [2] and He et al. (2007) [5]) and includes a spillover effect represented by the term $c_j x_i(t) u_j(t)$. This effect represents a specific economic behaviour in the form of positive or negative impact on the value of the economic agent i by state and investments product of the firm j. Such phenomenon is widespread for advertising and goodwill models, where the value of advertising for one firm positively depends on advertising efforts of the other firm provided they have similar products.

4.3   Differential Game in the Form of Characteristic Function

To define the cooperative game the characteristic function (ch.f.) $V(S, x_0, t_0)$ should be constructed for every coalition $S \subset N$ in the game $\Gamma(x_0, t_0)$. In the modern literature under the characteristic function in cooperative games is implied a mapping from the set of all possible coalitions to real set:

$$V(\cdot) : 2^N \to R,$$

$$V(\emptyset) = 0.$$

Note that the value of the characteristic function for the grand coalition N is equal to $V(N, x_0, t_0)$. There are several main approaches to the construction of the characteristic function which show the power of the coalition S (see, for example, [4, 14]). The most commonly used classes of characteristic functions can be indicated in the order that they appeared in literature as α-, β-, γ-, δ-, ζ-characteristic function.

The value $V(S, x_0, t_0)$ can be interpreted as a power of the coalition S. The essential property is the property of superadditivity:

$$V(S_1 \cup S_2, x_0, t_0) \geq V(S_1, x_0, t_0) + \tag{4.4}$$

$$+V(S_2, x_0, t_0), \forall S_1, S_2 \subseteq N, \ S_1 \cap S_2 = \emptyset.$$

However, the use of superadditive characteristic function in solving various problems in the field of cooperative game theory in static and dynamic setting, provides a number of advantages such as:

1. provides the individual rationality property for cooperative solutions,
2. encourages players to sustain large coalitions and eventually unite into a Grand coalition N,
3. delivers clear meaning to the Shapley value (a component of the division for each player is equal to its average contribution to the payoff of the Grand coalition under a certain mechanism of its formation),
4. necessary when you build a strongly dynamically stable optimality principles.

Thus, in many aspects more useful to have superadditive characteristic function.

It is rather easy to construct the characteristic function $V(S, x_0, t_0)$ in the form of α—ch.f. [12]. The characteristic function of coalition S is constructed through the classical approach of Neumann, Morgenstern, formulated in 1944 in [10]. According to this approach, under $V^\alpha(S, x_0, t_0)$ is understood the maximum guaranteed payoff of coalition S, and the value $V^\alpha(S)$ can be calculated on the basis of the auxiliary zero-sum game $\Gamma_{S, N \setminus S}(t_0, x_0)$ between the coalition S and anti-coalition $N \setminus S$.

$$V^\alpha(S, x_0, t_0) = \begin{cases} 0, & S = \{\emptyset\}, \\ val\,\Gamma_{S, N \setminus S}(x_0, t_0,), & S \subset N, \\ \max_{u_1, u_2, \ldots u_n} \sum_{i=1}^{n} J_i(x_0, t_0, u(t)), & S = N. \end{cases} \tag{4.5}$$

In this paper, without loss of generality, characteristic function calculation could be divided into three main steps: ch.f. for coalition which consists of the only one individual player, two players coalition and grand coalition.

4.3.1   One Player Coalition Characteristic Function

Calculate the value of the characteristic function for a coalition consisting only of the player $\{1\}$. Characteristic function would be calculated as a function of time moment θ, where θ is the initial moment. We assume that $\theta = 0$. The following maximisation problem is settled

$$V(\{1\}, x_0, t_0) = \max_{u_1} \min_{u_2, u_3} \int_{\theta}^{\infty} e^{-pt}(2a_1 x_1(t) + c_2 x_1(t)u_2(t) + c_3 x_1(t)u_3(t) - 2\frac{1}{2}u_1^2(t))dt.$$

(4.6)

Minimisation by $u_2(t)$ and $u_3(t)$ will result in zero controls. Ultimately, we need to solve the following maximisation problem

$$\begin{cases} \int_{\theta}^{\infty} e^{-pt}(2a_1 x_1(t) - u_1^2(t))dt \to \max_{u_1}, \\ \dot{x}_1(t) = \alpha_1 u_1(t) - \delta_1 x_1(t), \\ x_1(t_0) = x_0^1. \end{cases}$$

(4.7)

Using Maximum Principal [13] the following form of optimal control depending on adjoint variable is obtained

$$u_1^*(t) = 0.5\alpha_1 \psi_1(t)e^{pt}.$$

Corresponding differential equation for adjoint variable is

$$\dot{\psi}_1(t) = \delta_1 \psi_1(t) - 2a_1 e^{-pt}.$$

Under transversality conditions

$$\lim_{t \to \infty} \psi_1(t) = 0.$$

To simplify denote the variable

$$\lambda_1(t) = e^{pt}\psi_1(t).$$

Thus, by solving the system below a final form for optimal control could be obtained

$$
\begin{cases}
\dot{\lambda}_1(t) = (p + \delta_1)\lambda_1(t) - 2a_1, \\
\dot{x}_1(t) = \alpha_1 u_1(t) - \delta_1 x_1(t), \\
x_1(t_0) = x_0, \\
\lim_{t \to \infty} e^{-pt}\lambda(t) = 0.
\end{cases}
\tag{4.8}
$$

As the result

$$
u_1^*(t) = \frac{\alpha_1 a_1}{p + \delta_1} = Const.
\tag{4.9}
$$

The optimal trajectory could be derived under the assumption that the game was started from the point $(\theta, x^*(\theta))$.

$$
x^*(t) = \frac{\alpha_1^2 a_1}{\delta_1(p + \delta_1)} + x^*(\theta)e^{\delta_1(\theta - t)} - \frac{\alpha_1^2 a_1}{\delta_1(p + \delta_1)}e^{\delta_1(\theta - t)}.
$$

In particular, if $\theta = t_0 = 0$ and $x(t_0) = x_0^1$

$$
x^*(t) = \frac{\alpha_1^2 a_1}{\delta_1(p + \delta_1)}(1 - e^{-\delta_1 t}) + x_0^1 e^{-\delta_1 t}.
$$

Characteristic function for coalition consisting of the only player one is

$$
V(\{1\}, x_0, t_0) = \int_\theta^\infty e^{-pt}(2a_1 x_1^*(t) - u_1^{*2}(t))dt =
$$

$$
= \int_\theta^\infty e^{-pt}(2a_1 \frac{\alpha_1^2 a_1}{\delta_1(p + \delta_1)} + x^*(\theta)e^{\delta_1(\theta - t)} - \frac{\alpha_1^2 a_1}{\delta_1(p + \delta_1)}e^{\delta_1(\theta - t)} - (\frac{\alpha_1 a_1}{p + \delta_1})^2)dt =
$$

$$
= (\frac{2\alpha_1^2 a_1^2}{\delta_1(p + \delta_1)} - \frac{\alpha_1^2 a_1^2}{(p + \delta_1)^2})(-\frac{1}{p})e^{-pt} +
$$

$$
+ (x^*(\theta) - \frac{\alpha_1^2 a_1}{\delta_1(p + \delta_1)})e^{\delta_1\theta}(-\frac{1}{\delta_1 + p})e^{-(\delta_1 + p)t}\Big|_\theta^\infty =
$$

$$
= -\frac{\alpha_1^2 a_1^2(2p + \delta_1)}{p\delta_1(p + \delta_1)^2}e^{-p\theta} + (x^*(\theta) - \frac{\alpha_1^2 a_1}{\delta_1(p + \delta_1)})e^{\delta_1\theta}(-\frac{1}{\delta_1 + p})e^{-(\delta_1 + p)\theta}.
$$

4.3.2   Two Player Coalition Characteristic Function

Without loss of generality, consider the coalition of $\{1, 2\}$ as an example of two players coalition. In other cases calculation would be the same accurate to indexes.

$$V(\{1, 2\}, x_0, t_0) = \max_{u_1, u_2} \min_{u_3} \int_\theta^\infty e^{-pt}(h_{12}(t) + h_{13}(t) + h_{21}(t) + h_{23}(t))dt =$$

$$= \max_{u_1, u_2} \min_{u_3} \int_\theta^\infty e^{-pt}(2a_1x_1(t) - u_1^2(t) + c_1x_2(t)u_1(t) + 2a_2x_2(t) - u_2^2(t)+$$

$$+c_2x_1(t)u_2(t) + (x_1(t) + x_2(t))c_3u_3(t))dt.$$

Minimisation by $u_3(t)$ will result in zero controls. Ultimately, we need to solve the following maximisation problem

$$\begin{cases} \int_\theta^\infty e^{-pt}(2a_1x_1(t) - u_1^2(t) + c_1x_2(t)u_1(t) + 2a_2x_2(t) - u_2^2(t)+ \\ +c_2x_1(t)u_2(t))dt \to \max_{u_1, u_2}, \\ \dot{x}_1(t) = \alpha_1u_1(t) - \delta_1x_1(t), \\ \dot{x}_2(t) = \alpha_2u_2(t) - \delta_2x_2(t), \\ x_1(t_0) = x_0^1, \\ x_2(t_0) = x_0^2. \end{cases} \qquad (4.10)$$

Using Maximum principle for the first player

$$u_1^*(t) = 0.5\psi_1(t)\alpha_1e^{pt} + 0.5c_1x_2(t).$$

Due to the index symmetry among players both optimal control forms could be derived

$$\begin{cases} u_1^*(t) = 0.5\psi_1(t)\alpha_1e^{pt} + 0.5c_1x_2(t), \\ u_2^*(t) = 0.5\psi_2(t)\alpha_2e^{pt} + 0.5c_2x_1(t). \end{cases} \qquad (4.11)$$

Denote again for the first player and for the second player correspondingly $\lambda_i(t) = e^{pt}\psi_i(t)$, $i = \{1, 2\}$.

$$\begin{cases} \dot{\lambda}_1(t) = (p + \delta_1)\lambda_1(t) - (2a_1 + c_2 u_2(t)), \\ \dot{\lambda}_2(t) = (p + \delta_2)\lambda_2(t) - (2a_2 + c_1 u_1(t)). \end{cases} \quad (4.12)$$

Corresponding differential equations for adjoint variables in aggregate with dynamic equations lead to the system

$$\begin{cases} \dot{\lambda}_1(t) = (p + \delta_1)\lambda_1(t) - 0.5c_2\alpha_2\lambda_2(t) - 0.5c_2^2 x_1(t) - 2a_1, \\ \dot{\lambda}_2(t) = (p + \delta_2)\lambda_2(t) - 0.5c_1\alpha_1\lambda_1(t) - 0.5c_1^2 x_2(t) - 2a_2, \\ \dot{x}_1(t) = 0.5\alpha_1^2\lambda_1(t) - \delta_1 x_1(t) + 0.5\alpha_1 c_1 x_2(t), \\ \dot{x}_2(t) = 0.5\alpha_2^2\lambda_2(t) - \delta_2 x_2(t) + 0.5\alpha_2 c_2 x_1(t). \end{cases} \quad (4.13)$$

In matrix form

$$\begin{pmatrix} \dot{\lambda}_1(t) \\ \dot{\lambda}_2(t) \\ \dot{x}_1(t) \\ \dot{x}_2(t) \end{pmatrix} = \begin{pmatrix} \delta_1 + p & -0.5\alpha_2 c_2 & -0.5c_2^2 & 0 \\ -0.5\alpha_1 c_1 & \delta_2 + p & 0 & -0.5c_1^2 \\ 0.5\alpha_1^2 & 0 & -\delta_1 & 0.5\alpha_1 c_1 \\ 0 & 0.5\alpha_2^2 & 0.5\alpha_2 c_2 & -\delta_2 \end{pmatrix} \times \begin{pmatrix} \lambda_1(t) \\ \lambda_2(t) \\ x_1(t) \\ x_2(t) \end{pmatrix} + \begin{pmatrix} -2a_1 \\ -2a_2 \\ 0 \\ 0 \end{pmatrix}.$$

To solve the system the corresponding homogeneous system.

Denote A as

$$A = \begin{pmatrix} \delta_1 + p & -0.5\alpha_2 c_2 & -0.5c_2^2 e^{pt} & 0 \\ -0.5\alpha_1 c_1 & \delta_2 + p & 0 & -0.5c_1^2 e^{pt} \\ 0.5\alpha_1^2 & 0 & -\delta_1 & 0.5\alpha_1 c_1 e^{pt} \\ 0 & 0.5\alpha_2^2 & 0.5\alpha_2 c_2 e^{pt} & -\delta_2 \end{pmatrix}.$$

To obtain the decision, eigen values and eigen vectors of the matrix A should be derived and analysed to understand if the decision is in real or complex surface. Nevertheless, the above described operation are rather computationally complex to obtain the decision in an analytical form. However, an approach could be illustrated on the simplified system in case of constant values of some of the system parameters which are denote in the way not being in contradiction with the economical meaning of the model. This result is shown in the section below.

4.3.3 Grand Coalition Characteristic Function

Calculate the characteristic function for a coalition consisting of three players (Grand coalition).

$$V(\{1, 2, 3\}, x_0, t_0) = \max_{u_1, u_2, u_3} \int_\theta^\infty e^{-pt}(h_{12}(t) + h_{13}(t) + h_{21}(t) + h_{23}(t) + h_{31}(t) +$$

$$+h_{32}(t))dt = \max_{u_1, u_2, u_3} \int_\theta^\infty e^{-pt}(2a_1x_1(t) + 2a_2x_2(t) + 2a_3x_3(t) + (x_2(t) + x_3(t))c_1u_1(t) +$$

$$+(x_1(t) + x_3(t))c_2u_2(t) + (x_1(t) + x_2(t))c_3u_3(t) - u_1^2(t) - u_2^2(t) - u_1^3(t))dt.$$

The following maximisation problem is needed to be solved

$$\begin{cases} \int_\theta^\infty e^{-pt}(2a_1x_1(t) + 2a_2x_2(t) + 2a_3x_3(t) + (x_2(t) + x_3(t))c_1u_1(t) + (x_1(t) + \\ +x_3(t))c_2u_2(t) + (x_1(t) + x_2(t))c_3u_3(t) - u_1^2(t) - u_2^2(t) - u_1^3(t))dt \to \max_{u_1, u_2, u_3}, \\ \dot{x}_1(t) = \alpha_1 u_1(t) - \delta_1 x_1(t), \\ \dot{x}_2(t) = \alpha_2 u_2(t) - \delta_2 x_2(t), \\ \dot{x}_3(t) = \alpha_3 u_3(t) - \delta_3 x_3(t), \\ x_1(t_0) = x_0^1, \\ x_2(t_0) = x_0^2, \\ x_3(t_0) = x_0^3. \end{cases}$$

$$(4.14)$$

Using Maximum Principle derive the optimal controls depending on adjoint variables

$$\begin{cases} u_1^*(t) = 0.5\psi_1(t)\alpha_1 e^{pt} + 0.5c_1(x_2(t) + x_3(t)), \\ u_2^*(t) = 0.5\psi_2(t)\alpha_2 e^{pt} + 0.5c_2(x_1(t) + x_3(t)), \\ u_3^*(t) = 0.5\psi_3(t)\alpha_3 e^{pt} + 0.5c_3(x_1(t) + x_2(t)). \end{cases} \qquad (4.15)$$

Denote again for every player $\lambda_i(t) = e^{pt}\psi_i(t), \quad i \in \{1, 2, 3\}$.

Corresponding differential equations for adjoint variables in aggregate with dynamic equations lead to the system

$$
\begin{cases}
\dot{\lambda}_1(t) = (p + \delta_1)\lambda_1(t) - 0.5c_2\alpha_2\lambda_2(t) - 0.5c_3\alpha_3\lambda_3(t) - \\
\quad -0.5(c_2^2 + c_3^2)x_1(t) - 0.5c_3^2x_2(t) - 0.5c_2^2x_3(t) - 2a_1, \\
\dot{\lambda}_2(t) = (p + \delta_2)\lambda_2(t) - 0.5c_1\alpha_1\lambda_1(t) - 0.5c_3\alpha_3\lambda_3(t) - \\
\quad -0.5c_1^2x_3(t) - 0.5c_3^2x_1(t) - 0.5(c_1^2 + c_3^2)x_2(t) - 2a_2, \\
\dot{\lambda}_3(t) = (p + \delta_3)\lambda_3(t) - 0.5c_1\alpha_1\lambda_1(t) - 0.5c_2\alpha_2\lambda_2(t) - \\
\quad -0.5c_1^2x_2(t) - 0.5c_2^2x_1(t) - 0.5(c_1^2 + c_2^2)x_3(t) - 2a_3, \\
\dot{x}_1(t) = 0.5\alpha_1^2\lambda_1(t) - \delta_1x_1(t) + 0.5\alpha_1c_1x_2(t) + 0.5\alpha_1c_1x_3(t), \\
\dot{x}_2(t) = 0.5\alpha_2^2\lambda_2(t) - \delta_2x_2(t) + 0.5\alpha_2c_2x_1(t) + 0.5\alpha_2c_2x_3(t), \\
\dot{x}_3(t) = 0.5\alpha_3^2\lambda_3(t) - \delta_3x_3(t) + 0.5\alpha_3c_3x_1(t) + 0.5\alpha_3c_3x_2(t).
\end{cases}
\tag{4.16}
$$

Denote the system matrix $\hat{A}$

$$
\hat{A} =
\begin{pmatrix}
\delta_1 + p & -0.5\alpha_2c_2 & -0.5\alpha_3c_3 & -0.5(c_2^2 + c_3^2) & -0.5c_3^2 & -0.5c_2^2 \\
-0.5\alpha_1c_1 & \delta_2 + p & -0.5\alpha_3c_3 & -0.5c_3^2 & -0.5(c_1^2 + c_3^2) & -0.5c_1^2 \\
-0.5\alpha_1c_1 & -0.5\alpha_2c_2 & \delta_3 + p & -0.5c_2^2 & -0.5c_1^2 & -0.5(c_1^2 + c_2^2) \\
0.5\alpha_1^2 & 0 & 0 & -\delta_1 & 0.5\alpha_1c_1 & 0.5\alpha_1c_1 \\
0 & 0.5\alpha_2^2 & 0 & 0.5\alpha_2c_2 & -\delta_2 & 0.5\alpha_2c_2 \\
0 & 0 & 0.5\alpha_3^2 & 0.5\alpha_3c_3 & 0.5\alpha_3c_3 & -\delta_3
\end{pmatrix}.
$$

As in the case of two player there is a computational complexity on the way to analytical solution which depends on all the system parameters. However, the approach is the same as for the case of two players.

4.4 Numerical Example

To show the existence of the feasible solution of such a system as (4.13), denote the parameters of the model in the following way, so the analytical form of the decision could take reasonable view

$$
\alpha_1 = \alpha_2 = \alpha = 1,
$$

$$
c_1 = c_2 = c = 1.
$$

Therefore, A matrix takes form

$$A = \begin{pmatrix} \delta_1 + p & -0.5 & -0.5 & 0 \\ -0.5 & \delta_2 + p & 0 & -0.5 \\ 0.5 & 0 & -\delta_1 & 0.5 \\ 0 & 0.5 & 0.5 & -\delta_2 \end{pmatrix}$$

Eigen values for this matrix could be simplified to the following form

$$\begin{pmatrix} 0.5\left(p - \sqrt{\left(\sqrt{(\delta_1 - \delta_2)^2 + 1} - (\delta_1 + \delta_2 + p)\right)^2 - 1}\right) \\ 0.5\left(p + \sqrt{\left(\sqrt{(\delta_1 - \delta_2)^2 + 1} - (\delta_1 + \delta_2 + p)\right)^2 - 1}\right) \\ 0.5\left(p - \sqrt{\left(\sqrt{(\delta_1 - \delta_2)^2 + 1} + (\delta_1 + \delta_2 + p)\right)^2 - 1}\right) \\ 0.5\left(p + \sqrt{\left(\sqrt{(\delta_1 - \delta_2)^2 + 1} + (\delta_1 + \delta_2 + p)\right)^2 - 1}\right) \end{pmatrix}$$

All the eigen values of A are different. However, there is an issue if they are on the real surface of they are complex ones.

If the following conditions are held then the eigen values of A are real numbers and the decision of the system exists on the real surface.

$$\begin{cases} \left(\sqrt{(\delta_1 - \delta_2)^2 + 1} - (\delta_1 + \delta_2 + p)\right)^2 \geq 1, \\ \left(\sqrt{(\delta_1 - \delta_2)^2 + 1} + (\delta_1 + \delta_2 + p)\right)^2 \geq 1. \end{cases} \tag{4.17}$$

4.5 Conclusion

We proposed an analysis of the marketing network model in the form of differential game. An approach is presented for the calculation of the α-characteristic function of the game and illustration is given for the numerical example with particular values of the number of model parameters.

Acknowledgments The work has been supported by Russian Scientific Foundation (grant N 17-11-01079)

References

1. Bondarev, A., Gromov, D.: On the regularity of optimal solutions in a differential game with regime-switching dynamics. In: Submitted to Dynamic Economic Problems with Regime Switches (2019)
2. Deal, K.R., Sethi, S.P., Thompson, G.L.: A bilinear-quadratic game in advertising. In: Liu, P.T., Sutinen, J.G. (eds.), Control Theory in Mathematical Economics, pp. 91–109. Marcel Dekker, New York (1979)
3. Dockner, E., Jorgensen, S., Long, N., Sorger, G.: Differential Games in Economics and Management Sciences. Cambridge University, Cambridge (2000)
4. Gromova, E., Petrosyan, L.: On a approach to the construction of characteristic function for cooperative differential games. Mat. Teor. Igr Pril. **7**(4), 19–39 (2015)
5. He, X., Prasad, A., Sethi, S.P., Gutierrez, G.J.: A survey of Stackelberg differential game models in supply and marketing channels. J. Syst. Sci. Syst. Eng. **16**(4), 385–413 (2007)
6. Isaacs, R.: Differential Games. Wiley, New York (1965)
7. Jorgensen, S., Gromova, E.: Sustaining cooperation in a differential game of advertising goodwill accumulation. Eur. J. Oper. Res. **254**(1), 294–303 (2016)
8. Jorgensen, S., Zaccour, G.: Differential Games in Marketing. Springer, New York (2004)
9. Jorgensen, S., Zaccour, G.: A survey of game-theoretic models of cooperative advertising. Eur. J. Oper. Res. **237**(1), 1–14 (2014)
10. Neumann, J.V., Morgenstern, O.: Theory of Games and Economic Behavior. Princeton University, Princeton (1944)
11. Petrosyan, L.: Cooperative differential games on networks. Trudy Inst. Mat. i Mekh. UrO RAN **16**(5), 143–150 (2010)
12. Petrosyan, L., Zaccour, G.: Time-consistent Shapley value allocation of pollution cost reduction. J. Econ. Dyn. Control **27**(3), 381–398 (2003)
13. Pontryagin, L., Boltyanskii, V., Gamkrelidze, R., Mishchenko, E.: The mathematical theory of optimal processes. Interscience, New York (1962)
14. Reddy, P., Zaccour, G.: A friendly computable characteristic function. Math. Soc. Sci. **82**(C), 18–25 (2016)

Chapter 5
Penalty Method for Games of Constraints

Igor Konnov

Abstract We define a game problem for evaluation of composite system performance under possible external interference and in the presence of protection resources. The guaranteed joint system side decision is suggested to be found by an inexact penalty method. This enables one to essentially simplify the solution process in comparison with finding worst case strategies in the custom zero-sum game.

Keywords Game of constraints · Evaluation of system performance · Zero-sum games · Inexact penalty method

5.1 Introduction

We first describe a game problem for evaluation of composite system performance in the case where capacities of its subsystems may be changed by some influence from interference/protection sides. The problem somewhat extends those suggested in [1, 2]. We suppose that the system contains m subsystems and can carry out various works (or produce various goods) by implementing n working technologies, so that each technology is accomplished with some fixed collection of the subsystems at proper levels of their capacities. Let $Y \subset \mathbb{R}^n$ denote the set of all the technology levels profiles. Next, suppose that there are t points within the system where the restriction activity is possible and s points within the system where the protection activity is possible. These restriction and protection activities have certain impact on subsystems' capacity. More precisely, let $V \subset \mathbb{R}^t$ define the set of all the possible restriction interference volume profiles and $U \subset \mathbb{R}^s$ define the set of all

I. Konnov (✉)
Department of System Analysis and Information Technologies, Kazan Federal University,
Kazan, Russia
e-mail: konn-igor@ya.ru

© The Editor(s) (if applicable) and The Author(s), under exclusive licence
to Springer Nature Switzerland AG 2020
L. A. Petrosyan et al. (eds.), *Frontiers of Dynamic Games*,
Static & Dynamic Game Theory: Foundations & Applications,
https://doi.org/10.1007/978-3-030-51941-4_5

possible protection activity volume profiles. For a given pair $(u, v) \in U \times V$ and a given technology levels profile $y \in Y$ the capacity excess for the i-th subsystem is determined as value of the function $\beta_i(y, u, v)$, $i = 1, \ldots, m$. Let the value of a function $f(y, u, v)$ determine the estimate of the system utility at the technology levels profile y, protection activity volume u, and restriction activity volume v. This estimate reflects utility of the corresponding works, proper protection costs, and possible interference costs. For a given pair $(u, v) \in U \times V$ we can define the feasible set

$$D(u, v) = \{y \in Y \mid \beta_i(y, u, v) \leq 0, \ i = 1, \ldots, m\}$$

and the following problem of performance evaluation for the whole system:

$$\max_{y \in D(u,v)} \rightarrow f(y, u, v). \tag{5.1}$$

We denote by $H(u, v)$ the optimal value of the goal function in (5.1) and define the antagonistic "attack-defense" game with the utility function H and strategy sets U and V.

We observe that most works devoted to investigations of "attack-defense" type games involve the assumption that there is a simple formula for calculation of the value of the utility function for any pair of strategies; see e.g. [3–6]. This assumption seems rather restrictive for complex systems as above, where each calculation of the value of the utility function requires a solution of rather complex optimization problem (5.1). Hence, the streamlined way of finding a solution of the game above becomes very difficult for implementation.

In [1], it was suggested to find first the guaranteed system performance with the help of the non-smooth penalty function approach. A simplified formulation of the optimal guaranteed system performance problem was suggested in [2]. It enables one to apply custom penalty methods to the problem of the optimal guaranteed system performance and to find solutions of the above game. In this work we propose an inexact penalty method for the more general problem of the optimal guaranteed system performance, which enables us to simplify the calculation of its solutions essentially.

5.2 Guaranteed System Performance

We will utilize the following general assumptions.

(A1) Y, U and V are nonempty compact sets in $\mathbb{R}^n$, $\mathbb{R}^s$ and $\mathbb{R}^t$, respectively.
(A2) $f : Y \times U \times V \rightarrow \mathbb{R}$ and $\beta_i : Y \times U \times V \rightarrow \mathbb{R}$, $i = 1, \ldots, m$, are continuous functions.

The usual worst case protection strategy in the game can be found from the problem

$$\max_{u \in U} \ \to \ \min_{v \in V} H(u, v). \tag{5.2}$$

Clearly, problem (5.2) is very difficult for direct solution in the general case. We recall that the indicator function for a set Z is defined by

$$\delta(z|Z) = \begin{cases} 0, & \text{if } z \in Z, \\ +\infty, & \text{if } z \notin Z. \end{cases}$$

Then we can rewrite problem (5.2) equivalently as follows:

$$\max_{u \in U} \ \to \ \min_{v \in V} \max_{y \in Y} \eta(y, u, v), \tag{5.3}$$

where

$$\eta(y, u, v) = f(y, u, v) - \delta((y, u, v)|\tilde{D}),$$
$$\tilde{D} = \{(y, u, v) \mid \beta_i(y, u, v) \leq 0, \ i = 1, \ldots, m\}.$$

Following [1], we treat the pair (y, u) as a joint protection strategy for the system side and replace (5.3) with the guaranteed system performance problem:

$$\max_{(y,u) \in Y \times U} \ \to \ \min_{v \in V} \eta(y, u, v). \tag{5.4}$$

For the sake of brevity, we will set $x = (y, u) \in \mathbb{R}^n \times \mathbb{R}^s$, $X = Y \times U$, hence we will write $f(x, v) \equiv f(y, u, v)$, $\eta(x, v) \equiv \eta(y, u, v)$, $\beta_i(x, v) \equiv \beta_i(y, u, v)$, etc. Therefore, we propose now to solve the zero-sum game with the utility function $\eta(x, v)$ and strategy sets X and V.

We now show that problem (5.4) can be simplified as follows:

$$\max_{x \in X} \ \to \ \sigma(x), \tag{5.5}$$

where

$$\sigma(x) = \mu(x) - \delta(x|D), \quad D = \{x \mid h_i(x) \leq 0, \ i = 1, \ldots, m\},$$
$$\mu(x) = \min_{v \in V} f(x, v), \quad h_i(x) = \max_{v \in V} \beta_i(x, v), \quad i = 1, \ldots, m.$$

Proposition 5.1 *Suppose that assumptions (A1) and (A2) are fulfilled. Then problems (5.4) and (5.5) are equivalent.*

Proof It suffices to show that, for each $x \in X$ we have

$$\sigma(x) = \min_{v \in V} \eta(x, v). \tag{5.6}$$

In case $x \notin D$ we have $\delta(x|D) = +\infty$, hence $\sigma(x) = -\infty$. Next, there exists an index l and a point $v' \in V$ such that $\beta_l(x, v') > 0$, hence $\delta((x, v')|\tilde{D}) = +\infty$ and $\eta(x, v') = -\infty$. Then (5.6) holds.

In case $x \in D$ we have $\delta(x|D) = 0$ and $\sigma(x) = \mu(x)$. Also, we now have $\delta((x, v)|\tilde{D}) = 0$ for each $v \in V$, hence $\eta(x, v) = f(x, v)$ for each $v \in V$. It follows that

$$\min_{v \in V} \eta(x, v) = \min_{v \in V} f(x, v).$$

Then (5.6) also holds true.   ∎

The above property extends that in [2, Proposition 2.1]. We observe that problem (5.5) can be written in the standard optimization form:

$$\max_{x \in B} \to \mu(x), \tag{5.7}$$

where

$$B = \{x \in X \,|\, h_i(x) \leq 0, \ i = 1, \ldots, m\}.$$

We say that the system is *reliable* if $B \neq \varnothing$. We denote by μ^* the optimal value in (5.5) or (5.7). This means that $\mu^* = -\infty$ if $B = \varnothing$, hence $\mu^* > -\infty$ for reliable systems. Therefore, we can find the basic protection strategy from this optimization problem. We observe that a problem similar to (5.7) was used in [7, Chapter IV] for evaluation of the guaranteed system survivability with applications in energy sector. Besides, similar formulations were used for obtaining the so-called robust solutions of optimization problems; see e.g. [8, 9].

5.3 Descent Method for Auxiliary Problems

We suggest to solve the above problem with the help of an inexact penalty method. In order to substantiate the method we have to specialize the basic assumptions.

(B1) Y, U and V are nonempty, convex and compact sets in $\mathbb{R}^n$, $\mathbb{R}^s$ and $\mathbb{R}^t$, respectively.

(B2) $f : Y \times U \times V \to \mathbb{R}$ is a continuous function, which is concave in the variables y and u, $\beta_i : Y \times U \times V \to \mathbb{R}$, $i = 1, \ldots, m$, are continuous functions, which are convex in the variables y and u.

Then conditions (A1) and (A2) are fulfilled, (5.2) and (5.7) are convex optimization problems. For a point $h \in \mathbb{R}^m$ we denote by $[h]_+$ its projection onto the non-negative orthant

$$\mathbb{R}^m_+ = \left\{ v \in \mathbb{R}^m \mid v_i \geq 0 \quad i = 1, \ldots, m \right\}.$$

For the sake of simplicity, we take the most popular quadratic penalty function for the set B defined by

$$P(x) = \|[h(x)]_+\|^2, \ h(x) = (h_1(x), \ldots, h_m(x))^\top. \tag{5.8}$$

Then the original problem (5.5) (or (5.7)) is replaced by a sequence of auxiliary penalized problems of the form

$$\max_{x \in X} \to \Psi_\tau(x), \tag{5.9}$$

where

$$\Psi_\tau(x) = \mu(x) - 0.5\tau P(x), \ \tau > 0. \tag{5.10}$$

We intend to find only approximate solutions of problem (5.8)–(5.10) with an iterative method without line-search, which is based on those suggested in [10, 11]. It follows from Proposition 5 in [11] that problem (5.8)–(5.10) is now equivalent to the mixed variational inequality (MVI for short): Find a point $x(\tau) \in X$ such that

$$\mu(x(\tau)) - \mu(\tilde{x}) + \langle [\tau h(x(\tau))]_+, h(\tilde{x}) - h(x(\tau)) \rangle \geq 0 \quad \forall \tilde{x} \in X. \tag{5.11}$$

Let us fix $\tau > 0$ and define the gap function

$$\varphi_\tau(x) = \max_{\tilde{x} \in X} \Phi_\tau(x, \tilde{x}) = \Phi_\tau(x, \tilde{x}(x)),$$

where

$$\Phi_\tau(x, \tilde{x}) = \mu(\tilde{x}) - \mu(x) + \langle [\tau h(x)]_+, h(x) - h(\tilde{x}) \rangle - 0.5 \|x - \tilde{x}\|^2$$

at any point x. Under the assumptions made the point $\tilde{x}(x)$ is defined uniquely. This point is also a solution of the optimization problem

$$\max_{\tilde{x} \in X} \to \left\{ \mu(\tilde{x}) - \langle [\tau h(x)]_+, h(\tilde{x}) \rangle - 0.5 \|\tilde{x} - x\|^2 \right\}. \tag{5.12}$$

Hence, we can define the single-valued mapping $x \mapsto \tilde{x}(x)$. We give now its basic properties as proper adjustment of those in [11, Lemma 3] with respect to problem (5.8)–(5.10).

Lemma 5.1 *Let conditions (B1) and (B2) be satisfied. Then the following statements are true.*

(a) *The mapping $x \mapsto \tilde{x}(x)$ is continuous on the set X.*
(b) *At any point $x \in X$ the inequality holds*

$$\Psi'_\tau(x; \tilde{x}(x) - x) \geq \|\tilde{x}(x) - x\|^2.$$

(c) *The set of fixed points of the mapping $x \mapsto \tilde{x}(x)$ on X coincides with the set of solutions of problem (5.8)–(5.10).*

Now we can apply an adaptive composite step method without line-search based on the general scheme (SBM) in [10] for problem (5.8)–(5.10).

Method (ACS)

Step 0: Choose a point $z^0 \in X$, a number $\theta \in (0, 1)$ and a sequence $\{\alpha_l\} \to 0$, $\alpha_l \in (0, 1)$. Set $i = 0, l = 0$, choose a number $\lambda_0 \in (0, \alpha_0]$.
Step 1: Take the point $\tilde{z}^i = \tilde{x}(z^i)$. If $\tilde{z}^i = z^i$, stop. Otherwise set $d^i = \tilde{z}^i - z^i$ and $z^{i+1} = z^i + \lambda_i d^i$. If

$$\Psi_\tau(z^{i+1}) \geq \Psi_\tau(z^i) + \theta \lambda_i \|d^i\|^2,$$

take $\lambda_{i+1} \in [\lambda_i, \alpha_l]$, set $i = i + 1$ and go to Step 1.
Step 2: Set $\lambda'_{i+1} = \min\{\lambda_i, \alpha_{l+1}\}, l = l + 1$, take $\lambda_{i+1} \in (0, \lambda'_{i+1}]$, set $i = i + 1$ and go to Step 1.

Unlike the usual descent methods, the direction finding procedure of (ACS) is based on MVI (5.10), rather than the initial problem (5.9), although its descent condition involves the auxiliary function Ψ_τ. Besides, (ACS) does not involve any line-search.

Due to Lemma 5.1, termination of (ACS) yields a solution of problem (5.8)–(5.10). Hence, we will consider only the case where the sequence $\{z^i\}$ is infinite. Convergence properties of (ACS) are obtained directly from Lemma 5.1 and Theorem 3.1 in [10].

Proposition 5.2 *Let assumptions (B1) and (B2) be fulfilled. Then all the limit points of the sequence $\{z^i\}$ are solutions of problem (5.8)–(5.10), besides, we have*

$$\lim_{i \to \infty} \Psi_\tau(z^i) = \Psi^*_\tau,$$

where $\Psi^*_\tau = \max\{\Psi_\tau(x) \mid x \in X\}$.

We intend to find an approximate solution of problem (5.8)–(5.10) in a finite number of iterations and need additional properties of the function φ_τ. First we note that by definition $\varphi_\tau(x) \geq 0$ for any $x \in X$. Next, due to Lemma 5.1, the function φ_τ is continuous on the set X. It follows from Lemma 4 in [11] that the following relations are equivalent:

$$\varphi_\tau(x) = 0 \quad \text{and} \quad x = \tilde{x}(x).$$

We can now conclude from Lemma 5.1(c) that the optimization problem

$$\min_{x \in X} \ \rightarrow \ \varphi_\tau(x)$$

is equivalent to problem (5.8)–(5.10) and to MVI (5.11). Therefore, the value $\varphi_\tau(x)$ gives an error estimate for these problems at $x \in X$ and Proposition 5.2 now implies

$$\lim_{i \to \infty} \varphi_\tau(z^i) = 0.$$

This property enables us to attain any approximation with respect to the gap function.

Corollary 5.1 *Let all the conditions of Proposition 5.2 be satisfied. Then, for any number $\varepsilon > 0$ there exists an iteration number $i = i(\varepsilon)$ of Method (ACS) such that $\varphi_\tau(z^i) \leq \varepsilon$.*

We now give additional error estimates for the gap function from [11, Lemma 4 and Proposition 6].

Proposition 5.3 *Let assumptions (B1) and (B2) be fulfilled. Then, for any point $\in X$, the following inequalities hold:*

$$\varphi_\tau(x) \geq 0.5 \|x - \tilde{x}(x)\|^2 \tag{5.13}$$

and

$$\Psi_\tau(x) - \Psi_\tau(\tilde{x}) + \langle \tilde{x}(x) - x, \tilde{x} - x \rangle \geq -\varphi_\tau(x) + 0.5 \|x - \tilde{x}(x)\|^2 \quad \forall \tilde{x} \in X. \tag{5.14}$$

5.4 Inexact Penalty Method

We now define the optimization problem:

$$\max_{x \in B'} \ \rightarrow \ \mu(x), \tag{5.15}$$

where

$$B' = \left\{ x \in X \mid P(x) = P^* \right\}, \ P^* = \min_{x \in X} P(x),$$

and set

$$\mu' = \max_{x \in B'} \mu(x).$$

From the definitions it follows that $\mu' > -\infty$. Besides, $\mu' = \mu^*$ if $B \neq \varnothing$ since $P^* = 0$. Unlike (5.7), formulation (5.15), (5.8) can be used even if the system is non-reliable. It enables one to reveal "bottle necks" in the system structure whose protection should be strengthened.

We now describe the two-level implementable penalty method, which uses approximate solutions of problems (5.8)–(5.10) based on the gap function φ_τ.

Method (PCS) Choose a point $x^0 \in X$ and sequences of positive numbers $\{\varepsilon_k\}$, $\{\tau_k\}$.

At the k-th stage, $k = 1, 2, \ldots$, we have a point $x^{k-1} \in X$ and numbers ε_k, τ_k. Applying Method (ACS) to problem (5.8)–(5.10) with the starting point $z^0 = x^{k-1}$ and $\tau = \tau_k$, we obtain the point $\tilde{z} = z^i$ such that

$$\varphi_{\tau_k}(\tilde{z}) \leq \varepsilon_k, \tag{5.16}$$

and set $x^k = \tilde{z}$.

Theorem 5.1 *Let assumptions (B1) and (B2) be fulfilled and let the parameters $\{\varepsilon_k\}$ and $\{\tau_k\}$ satisfy the conditions:*

$$\{\varepsilon_k\} \searrow 0, \quad \{\tau_k\} \nearrow +\infty. \tag{5.17}$$

Then the following assertions are true.

(a) The number of iterations at each stage of Method (PCS) is finite.
(b) The sequence $\{z^l\}$ generated by Method (PCS) has limit points and all these points are solutions to problem (5.15), besides,

$$\lim_{k \to \infty} \mu(x^k) = \mu'. \tag{5.18}$$

Proof Assertion (a) follows from Corollary 5.1. Since the set X is compact, the sequence $\{x^k\}$ is bounded, hence it has limit points. For brevity, denote by $\bar{x}^k = \tilde{x}(x^k)$ the solution of problem (5.12) at $\tau = \tau_k$, $x = x^k$. Now from (5.14) it follows that

$$\Psi_{\tau_k}(x^k) - \Psi_{\tau_k}(\tilde{x}) + \langle \bar{x}^k - x^k, \tilde{x} - x^k \rangle \geq -\varphi_{\tau_k}(x^k) \quad \forall \tilde{x} \in X.$$

Taking into account (5.13) and (5.16) we obtain

$$\Psi_{\tau_k}(x^k) - \Psi_{\tau_k}(\tilde{x}) + \sqrt{2\varepsilon_k}\|\tilde{x} - x^k\| \geq -\varepsilon_k \quad \forall \tilde{x} \in X. \tag{5.19}$$

Setting $\tilde{x} = x^* \in B'$ in this inequality and using (5.17) give

$$0 \leq P(x^k) \leq P(x^*) + 2[\mu(x^k) - \mu(x^*) + \varepsilon_k + \sqrt{2\varepsilon_k}\|\tilde{x} - x^k\|]/\tau_k \to P^*$$

as $k \to \infty$. If $\bar{x}$ is an arbitrary limit point of $\{x^k\}$, then $\bar{x} \in B'$. Next, relation (5.19) also gives

$$\mu(x^k) \geq \mu(x^*) - \varepsilon_k - \sqrt{2\varepsilon_k}\|x^* - x^k\| + 0.5\tau_k[P(x^k) - P^*]$$
$$\geq \mu(x^*) - \varepsilon_k - \sqrt{2\varepsilon_k}\|x^* - x^k\|,$$

hence $\mu(\bar{x}) = \mu'$. It follows that (5.18) holds. ∎

We notice that the case $P^* = 0$ corresponds to the usual penalty method. Then relation (5.18) can be strengthened. In fact, setting $\tilde{x}$ in (5.19) to be a solution $x(\tau_k)$ of (5.8)–(5.10) and taking a point $x^* \in B'$ will yield

$$\mu(x^k) \geq \Psi_{\tau_k}(x^k) \geq \Psi_{\tau_k}(x(\tau_k)) - \varepsilon_k - \sqrt{2\varepsilon_k}\|x(\tau_k) - x^k\|$$
$$= \Psi_{\tau_k}^* - \varepsilon_k - \sqrt{2\varepsilon_k}\|x(\tau_k) - x^k\| \geq \Psi_{\tau_k}(x^*) - \varepsilon_k - \sqrt{2\varepsilon_k}\|x(\tau_k) - x^k\|$$
$$= \mu^* - \varepsilon_k - \sqrt{2\varepsilon_k}\|x(\tau_k) - x^k\|,$$

hence

$$\lim_{k\to\infty} \mu(x^k) = \lim_{k\to\infty} \Psi_{\tau_k}(x^k) = \lim_{k\to\infty} \Psi_{\tau_k}^* = \mu^*.$$

5.5 Some Application Issues

After finding a joint protection strategy $x^* = (y^*, u^*)$ for the system side from the guaranteed system performance problem (5.5) one can calculate solution strategies of the interference side. These strategies appear to be mixed; see [2]. Next, since the interference side may have incomplete and inexact information about the system, its real strategy may differ from the optimal one. Hence, the protection (system) side may take the strategy (y^*, u^*) as a basis, but change it properly after the strategy deviations of the interference side.

Let us now consider the custom approach based on the antagonistic "attack-defense" game with the utility function H and strategy sets U and V. Then each calculation of the value of the utility function will require a solution of optimization problem (5.1). Moreover, the utility function $H(u, v)$ does not possess the concavity-convexity property even under additional assumptions, hence solution of this game is very difficult especially for high-dimensional problems.

In order to solve problem (5.8)–(5.10) we can in principle apply some other suitable convex optimization method instead of (ACS); see e.g. [12, 13]. Following [11], we can also apply a similar method with line-search. We observe that both (ACS) and that method solve the optimization problem (5.12) at each iteration for finding the point $\tilde{x}(x)$ at x. If we add the convexity assumption for the functions

f and β_i, $i = 1, \ldots, m$, in the variable v, then the calculations of values of the functions μ and h_i, $i = 1, \ldots, m$, also reduce to convex optimization problems.

In order to illustrate some preferences of the proposed method we describe applications to special cases of problem (5.1).

Example 5.1 We take the special case where the functions f and β_i are affine in y. More precisely, we are given the parametric linear programming problem of the form:

$$\min \to \sum_{j=1}^{n} c_j y_j$$

subject to

$$\sum_{j=1}^{n} a_{ij} y_j \leq \tilde{b}_i(u, v), \quad i = 1, \ldots, m;$$

$$y_j \geq 0, \quad j = 1, \ldots, n;$$

$$u \in U, \quad v \in V;$$

or briefly,

$$\max \to \{\langle c, y \rangle \mid Ay \leq \tilde{b}(u, v), \ y \geq \mathbf{0}\}, \ u \in U, \ v \in V, \tag{5.20}$$

where A is an $m \times n$ matrix, $c, y \in \mathbb{R}^n$, $\tilde{b}(u, v) \in \mathbb{R}^m$. This case corresponds to the linear working technology. Hence, c_j is the system utility per unit level of the j-th working technology, a_{ij} is the loading level of the i-th subsystem per unit level of the j-th working technology, whereas $\tilde{b}_i(u, v)$ is the maximal capacity of the i-th subsystem at the activity profile (u, v). Here the system utility does not depend on both restriction and protection activity. We also suppose that U and V are nonempty convex and compact sets in $\mathbb{R}^s$ and $\mathbb{R}^t$, respectively, and that $\tilde{b}_i$, $i = 1, \ldots, m$, are continuous functions, which are convex in v and concave in u.

First we take the custom antagonistic game approach. Then due to (5.20) we define its utility function

$$H(u, v) = \max\{\langle c, y \rangle \mid Ay \leq \tilde{b}(u, v), \ y \geq \mathbf{0}\}$$

and strategy sets U and V. We can show that $H(u, v)$ is concave in u under the above assumptions. In fact, fix any $v \in V$, take arbitrary points $u', u'' \in U$ and denote by y' and y'' the corresponding solutions of the inner problem (5.20), i.e. $\langle c, y' \rangle = H(u', v)$ and $\langle c, y'' \rangle = H(u'', v)$. Choose an arbitrary number $\lambda \in (0, 1)$ and set $u(\lambda) = \lambda u' + (1 - \lambda)u''$ and $y(\lambda) = \lambda y' + (1 - \lambda)y''$. Then $y(\lambda) \geq \mathbf{0}$ and due to the concavity of $\tilde{b}(\cdot, v)$ we have

$$\tilde{b}(u(\lambda), v) \geq \lambda \tilde{b}(u', v) + (1 - \lambda)\tilde{b}(u'', v) \geq Ay(\lambda),$$

i.e. $y(\lambda)$ is a feasible point in (5.20) at $(u(\lambda), v)$. It follows that

$$H(u(\lambda), v) \geq \langle c, y(\lambda) \rangle = \lambda H(u', v) + (1 - \lambda)H(u'', v),$$

hence, $H(\cdot, v)$ in concave. However, $H(\cdot, v)$ is not convex in general. Even if $\tilde{b}(u, \cdot)$ is affine we can only prove as above that $H(u, \cdot)$ is concave. Similarly, if $\tilde{b}$ is jointly concave we can prove that so is H. Therefore, the utility function $H(u, v)$ is not concave-convex, hence solution of the antagonistic game is very difficult.

Next we take the proposed penalty approach. Set $b_i(u) = \min_{v \in V} \tilde{b}_i(u, v)$ for $i = 1, \dots, m$. Then the guaranteed system performance problem (5.7) is written as follows:

$$\max \rightarrow \{\langle c, y \rangle \mid Ay \leq b(u), \ y \geq \mathbf{0}, \ u \in U\},$$

it is clearly a convex optimization problem. It is replaced by a sequence of auxiliary penalized problems of the form

$$\max_{y \in \mathbb{R}^n_+, u \in U} \rightarrow \{\langle c, y \rangle - 0.5\tau \|[Ay - b(u)]_+\|^2\};$$

cf. (5.8)–(5.10). Each penalized problem is solved approximately with Method (ACS). The main part of this method consists of finding the value of the single-valued mapping $x \mapsto \tilde{x}(x)$ at $x = (y, u)$, where x stands for the current iterate. In other words, we have to solve the auxiliary optimization problem (5.12), which is now written as follows:

$$\max_{\tilde{y} \in \mathbb{R}^n_+, \tilde{u} \in U} \rightarrow \{\langle c, \tilde{y} \rangle - \langle \tau[Ay - b(u)]_+, A\tilde{y} - b(\tilde{u}) \rangle - 0.5(\|\tilde{y} - y\|^2 + \|\tilde{u} - u\|^2)\}.$$

Clearly, it can be solved separately in each vector variable, i.e. it reduces to the independent optimization problems:

$$\max_{\tilde{y} \in \mathbb{R}^n_+} \rightarrow \{\langle c, \tilde{y} \rangle - \langle \tau[Ay - b(u)]_+, A\tilde{y} \rangle - 0.5\|\tilde{y} - y\|^2\} \tag{5.21}$$

and

$$\max_{\tilde{u} \in U} \rightarrow \{\langle \tau[Ay - b(u)]_+, b(\tilde{u}) \rangle - 0.5\|\tilde{u} - u\|^2\}. \tag{5.22}$$

Since the pair (y, u) is fixed, problem (5.21) decomposes into n simple independent one-dimensional quadratic optimization problems, their solutions are found by an explicit formula. Problem (5.22) has a unique solution, its calculation can be simplified after further specialization of the set U and functions b_i, $i = 1, \ldots, m$. The other parts of Method (PCS) can be implemented easily.

Example 5.2 We now take the example of an optimal flow distribution problem in computer and telecommunication data transmission networks, which was described in [1] and is based on the network flow distribution model from [14].

This model describes a network that contains m transmission links (arcs) and accomplishes some submitted data transmission requirements from n selected pairs of origin-destination vertices within a fixed time period. Denote by y_j and d_j the current and maximal value of data transmission for pair demand j, respectively, and by c_i the capacity of link i, which depends on network protection-interference profiles $(u, v) \in U \times V$, i.e. $c_i = c_i(u, v)$. Each pair demand is associated with a unique data transmission path, hence each link i is associated uniquely with the set $N(i)$ of pairs of origin-destination vertices, whose transmission paths contain this link. For each pair demand j we denote by $\mu_j(y_j)$ the utility value at the data transmission volume y_j. Then we can write the parametric utility maximization problem as follows:

$$\max \to \mu(y) = \sum_{j=1}^{n} \mu_j(y_j)$$

subject to

$$\sum_{j \in N(i)} y_j \leq c_i(u, v), \ i = 1, \ldots, m;$$

$$0 \leq y_j \leq d_j, \ j = 1, \ldots, n;$$

$$u \in U, \ v \in V.$$

If the functions $\mu_j(y_j)$ are concave, this is a parametric convex optimization problem. The sets U, V of protection-interference activity profiles are usually determined as polyhedra, for instance, we take

$$U = \left\{ u \in \mathbb{R}^s \, \middle| \, 0 \leq u_i \leq \alpha_i, \ i = 1, \ldots, s, \ \sum_{i=1}^{s} u_i \leq C' \right\},$$

$$V = \left\{ v \in \mathbb{R}^t \, \middle| \, 0 \leq v_j \leq \beta_j, \ j = 1, \ldots, t, \ \sum_{j=1}^{t} v_j \leq C'' \right\}.$$

In addition we suppose that the functions $c_i(u, v)$ are concave-convex. Then as above we can show that the custom antagonistic game approach leads to the utility function $H(u, v)$, which is concave in u under the above assumptions, but is not convex in v in general. Therefore, solution of this antagonistic game will be very difficult.

In [1], the guaranteed system performance problem was replaced by a sequence of general convex non-smooth optimization problems. Now we describe application of the proposed method. Set $d = (d_1, \ldots, d_1)^\top$ and $b_i(u) = \min\limits_{v \in V} c_i(u, v)$ for $i = 1, \ldots, m$. Then the guaranteed system performance problem (5.7) is written as follows:

$$\max \to \left\{ \mu(y) \;\middle|\; \sum_{j \in N(i)} y_j \leq b_i(u), \; i = 1, \ldots, m, \; 0 \leq x \leq d \right\}.$$

It is clearly a convex optimization problem. We find its solution with taking a sequence of auxiliary penalized problems of the form

$$\max_{y \in [0,d], u \in U} \to \{\mu(y) - 0.5\tau P(y, u)\}, \tag{5.23}$$

where

$$P(y, u) = \sum_{i=1}^{m} \left[\sum_{j \in N(i)} y_j - b_i(u) \right]_+^2 ;$$

cf. (5.8)–(5.10). Each penalized problem (5.23) is solved approximately with Method (ACS). The main part of this method consists of finding the point $\tilde{x}(x)$ at $x = (y, u)$, where x stands for the current iterate. It is a solution of problem (5.12), which is now written as follows:

$$\max_{\tilde{y} \in [0,d], \tilde{u} \in U} \to \left\{ \begin{array}{l} \mu(\tilde{y}) - \tau \sum\limits_{i=1}^{m} \left[\sum\limits_{j \in N(i)} y_j - b_i(u) \right]_+ \left(\sum\limits_{j \in N(i)} \tilde{y}_j - b_i(\tilde{u}) \right) \\ -0.5(\|\tilde{y} - y\|^2 + \|\tilde{u} - u\|^2) \end{array} \right\}.$$

Again, this problem reduces to the independent convex optimization problems:

$$\max_{\tilde{y} \in [0,d]} \to \left\{ \mu(\tilde{y}) - \tau \sum_{i=1}^{m} \left[\sum_{j \in N(i)} y_j - b_i(u) \right]_+ \sum_{j \in N(i)} \tilde{y}_j - 0.5\|\tilde{y} - y\|^2 \right\}$$

and

$$\min_{\tilde{u} \in U} \to \left\{ 0.5\|\tilde{u} - u\|^2 - \tau \sum_{i=1}^{m} \left[\sum_{j \in N(i)} y_j - b_i(u) \right]_+ b_i(\tilde{u}) \right\}.$$

Both the problems admit further decomposition with easy solution. Hence, this method seems more efficient than that in [1].

Therefore, the proposed method admits a decomposition technique which simplifies the implementation essentially.

5.6   Conclusions

We considered a general problem of performance evaluation for some composite system involving subsystems. This system can produce several kinds of commodities having different load volumes, so that each kind of commodity (or each kind of work) may be accomplished with some collections of the subsystems. The subsystems capacity may be changed under influence of some activity both for restriction of their performance and for protection from this restriction. In such a way, we obtained an "attack-defense" type antagonistic game where calculation of the value of the utility function requires a solution of the optimization problem.

We proposed to modify the formulation of the above game problem in order to evaluate guaranteed system performance in the general case. The problem suggested to be solved by an inexact penalty optimization method. This enables one to essentially simplify the solution process in comparison with finding worst case strategies in the custom zero-sum game. An example of applications that shows the efficiency of the proposed approach was also described.

Acknowledgments The results of this work were obtained within the state assignment of the Ministry of Science and Education of Russia, project No. 1.460.2016 /1.4. In this work, the author was also supported by Russian Foundation for Basic Research, project No. 19-01-00431.

References

1. Konnov, I.V.: Game of constraints for evaluation of guaranteed composite system performance. Russ. Math. (Iz. VUZ) **63**(4), 79–83 (2019)
2. Konnov, I.V.: Games of constraints for complex systems. Lobachevskii J. Math. **40**(4), 660–666 (2019)
3. Dresher, M.: Games of Strategy: Theory and Applications. Englewood Cliffs, Prentice-Hall (1961)
4. Germeyer, Y.B.: Introduction to Operations Research Theory. Nauka, Moscow (1971, in Russian)
5. Vorob'yev, N.N.: The Principles of Game Theory. In: Non-Cooperative Games. Nauka, Moscow (1984, in Russian)
6. Davydov, E.G.: Operations Research. Vysshaya Shkola, Moscow (1990, in Russian)
7. Antsiferov, E.G., Ashchepkov, L.T., Bulatov, V.P.: Optimization Methods and Their Applications. Part I: Mathematical Programming. Nauka, Novosibirsk (1990, in Russian)
8. Ben-Tal, A., Nemirovski, A.: Robust convex optimization. Math. Oper. Res. **23**(4), 769–805 (1998)

9. García, J., Peña, A.: Robust optimization: concepts and applications. In: Del Ser Lorente, J. (ed.) Nature-inspired Methods for Stochastic, Robust and Dynamic Optimization, Chapter II. Intechopen, London (2018)

10. Konnov, I.V.: A simple adaptive step-size choice for iterative optimization methods. Adv. Model Optim. **20**(2), 353–369 (2018)

11. Konnov, I.V.: An approximate penalty method with descent for convex optimization problems. Russ. Math. (Iz. VUZ) **63**(7), 41–55 (2019)

12. Polyak, B.T.: Introduction to Optimization. Nauka, Moscow (1983); English Translation in Optimization Software, New York (1987)

13. Konnov, I.V.: Nonlinear Optimization and Variational Inequalities. Kazan University, Kazan (2013, in Russian)

14. Kelly F.P., Maulloo A., Tan D.: Rate control for communication networks: shadow prices, proportional fairness and stability. J. Oper. Res. Soc. **49**(3), 237–252 (1998)

Chapter 6
Adjustment Dynamics in Network Games with Stochastic Parameters

Alexei Korolev

Abstract In this paper we introduce stochastic parameters into the network game model with production and knowledge externalities. This model was proposed by V. Matveenko and A. Korolev as a generalization of the two-period Romer model. Agents differ in their productivities which have deterministic and stochastic (Wiener) components. We study the dynamics of a single agent and the dynamics of a dyad where two agents are aggregated. We derive explicit expressions for the dynamics of a single agent and dyad dynamics in the form of Brownian random processes, and qualitatively analyze the solutions of stochastic equations and systems of stochastic equations.

Keywords Network games · Differential games · Brownian motion · Stochastic differential equations · Ito's lemma · Heterogeneous agents · Productivity

6.1 Introduction

Recent decades have seen an increase in research on social networks, economics of networks and games on networks (e.g. [2–8, 10]). Numerous theoretical results in these areas are widely used in the analysis of real-life networks such as the Internet, social interactions, foreign relations, etc. However, the existing literature pays much less attention to production networks. In [11] a model with production and knowledge externalities with two time periods is considered which generalizes the Romer model [14], where essentially a special case of the complete network is being examined. Agents are located in the nodes of a network of arbitrary form and derive utility from consumption in both periods. In the first period an agent receives an endowment which can be allocated between investment in knowledge and consumption. The consumption in the second period is determined

A. Korolev (✉)
National Research University Higher School of Economics, St. Petersburg, Russia

L. A. Petrosyan et al. (eds.), *Frontiers of Dynamic Games*,
Static & Dynamic Game Theory: Foundations & Applications,
https://doi.org/10.1007/978-3-030-51941-4_6

by production, which depends on her own investment and investments made by her closest neighbors in the network.

In [11] the concept of Nash equilibrium with externalities is introduced. As in the usual Nash equilibrium, agents maximize their gain (utility) and none of the agents find it beneficial to deviate if others do not change their behavior. However, this model assumes that the agent cannot change its behavior arbitrarily, as the concept of Nash equilibrium implies, but to a certain extent is bounded by the equilibrium situation in the game. Namely, in [11] it is assumed that the agent makes a decision taking into account a certain environment formed by her and her neighbors in the network, and although she affects the environment, it is taken as exogenously given when making the decision.

However, [11] considers only networks with homogeneous agents. In [12] a generalization of the model [11] is studied where the productivities of the agents may be different. The dynamics of networks in discrete time is introduced and the concept of dynamic stability of equilibria is defined. The authors characterize equilibrium behavior of the agents and provide conditions under which in equilibrium agent is passive (does not invest), active (invests part of the income), hyperactive (invests all income). Conditions for the existence of internal equilibrium (i.e., equilibrium with active agents) are established for several networks, and a theorem about comparison of agents' utilities is proved.

Also, [12] considers dynamics in discrete time which occur when networks are combined. In [13] the dynamics of networks with production and knowledge externalities is studied, and the concept of dynamic stability of equilibria is defined in continuous time framework. However, in all of the above papers, the network parameters were deterministic.

The contribution of this paper is in describing transition dynamics in the stochastic case where agent productivity has both deterministic and Brownian components. We study the behavior of a single agent and of a dyad. It turns out that the threshold values of parameters under which agent's equilibrium behavior changes in the stochastic case are shifted compared to the deterministic case.

The rest of the paper is organized as follows. Section 6.2 describes the main model and reviews some of the previous results. We define Nash equilibrium in the network with production and knowledge externalities, define dynamic stability of equilibrium and characterize the equilibrium behavior of agents. Section 6.3 considers the dynamics of a single agent in deterministic and stochastic cases, i.e., when agent has constant productivity and when her productivity consists of two terms (deterministic and a stochastic processes). We derive explicit expression for the dynamics of a single agent in the form of Brownian random processes (see Proposition 6.2) and provide a qualitative analysis of the solution to the stochastic equation (see Corollary 6.2). Section 6.4 compares the dynamics in the dyad for deterministic and stochastic cases. We derive explicit expressions for the dyad dynamics in the form of Brownian random processes (see Theorem 6.3) and provide a qualitative analysis of the solutions of stochastic equations systems (see Corollary 6.3). Section 6.5 concludes and discusses possible topics for further research.

6.2 Deterministic Model and Review of Previous Results

We begin by describing our main model and reviewing some of its properties, since our new model differs from the original model in the stochastic nature of the parameters. Our main model, formulated in [11], is deterministic and is the generalization of the Romer model [14] for networks.

Consider a network (undirected graph) with n nodes, $i = 1, 2, \ldots, n$; each node represents an agent. In period 1, each agent i has initial endowment of good, ε, and can use it for consumption in period 1, c_1^i, and for investment into knowledge, k_i. Knowledge is used in the production of good for consumption c_2^i in the period 2:

$$c_1^i + k_i = \varepsilon, \ i = 1, 2, \ldots, n.$$

Preferences of agent i are described by quadratic utility function:

$$U_i\left(c_1^i, c_2^i\right) = c_1^i\left(\varepsilon - ac_1^i\right) + b_i c_2^i,$$

where a is a satiation coefficient, $b_i > 0$ characterizes the value of comfort and health in period 2 compared to consumption in period 1. It is assumed that $c_1^i \in [0, \varepsilon]$, the utility increases in c_1^i, and is concave in c_1^i (i.e., the marginal utility of consumption decreases). These assumptions hold in particular when $0 < a < 1/2$.

By environment of agent i we mean the sum of investments by the agent herself and her neighbors:

$$K_i = k_i + \tilde{K}_i, \quad \tilde{K}_i = \sum_{j \in N(i)} k_j,$$

where $N(i)$ is the set of neighboring nodes of node i, $\tilde{K}_i$ we will call the *pure externality*. Production in node i is described by production function:

$$F(k_i, K_i) = B_i k_i K_i, \ B_i > 0$$

which depends on the state of knowledge in node i, k_i, and on *environment*, K_i, while B_i is a technological coefficient.

We will denote the product $b_i B_i$ by A_i and assume that $a < A_i$. Since increase in any of parameters b_i, B_i leads to increase of the second period consumption, we will call A_i "productivity". We will assume that $A_i \neq 2a, i = 1, 2, \ldots, n$. If $A_i > 2a$, we will say that agent i is *productive*, and if $A_i < 2a$, we will say that agent i is *unproductive*.

There are three possible types of agent's behavior: agent i is called *passive* if she makes no investment, $k_i = 0$ (i.e. consumes the whole endowment in period 1); *active* if $0 < k_i < \varepsilon$; *hyperactive* if she makes maximally possible investment ε (i.e. consumes nothing in period 1).

Consider the following game. Players are agents $i = 1, 2, \ldots, n$. Possible actions (strategies) of player i are the values of investment k_i from the interval $[0, \varepsilon]$. *Nash equilibrium with externalities* (hereinafter referred to as *equilibrium*) is a profile of knowledge levels (investments) $(k_1^*, k_2^*, \ldots, k_n^*)$, such that each k_i^* is a solution of the following problem $P(K_i)$ where player i maximizes her utility given environment K_i:

$$U_i\left(c_1^i, c_2^i\right) \xrightarrow[c_1^i, c_2^i, k_i]{} \max$$

$$\begin{cases} c_1^i \leq \varepsilon - k_i, \\ c_2^i \leq F(k_i, K_i), \\ c_1^i \geq 0, \ c_2^i \geq 0, \ k_i \geq 0, \end{cases}$$

and the environment K_i is defined by the profile $(k_1^*, k_2^*, \ldots, k_n^*)$:

$$K_i = k_i^* + \sum_{j \in N(i)} k_j^*$$

The first two constraints in problem $P(K_i)$ are evidently binding in the solution. Substituting the equalities into the objective function, we obtain a *payoff function*, or *indirect utility function*:

$$V_i(k_i, K_i) = U_i\left(\varepsilon - k_i, F_i(k_i, K_i)\right) = (\varepsilon - k_i)\left(\varepsilon - a(\varepsilon - K_i)\right) + A_i k_i K_i =$$

$$= \varepsilon^2 (1 - a) - k_i \varepsilon (1 - 2a) - ak_i^2 + A_i k_i K_i. \tag{6.1}$$

If all players' solutions are internal $(0 < k_i < \varepsilon)$, i.e. all players are active, the equilibrium is called *inner*. Otherwise it is called *corner* equilibrium. A corner equilibrium in which the level of knowledge at each node is 0 or ε, i.e. all players are passive or hyperactive, we will call *purely corner* equilibrium.

Clearly, the inner equilibrium (if it exists for given values of parameters) is defined by the system

$$D_1 V_i(k_i, K_i) = 0, \ i = 1, 2, \ldots, n, \tag{6.2}$$

or according to (6.1) it is

$$D_1 V_i(k_i, K_i) = \varepsilon(2a - 1) - 2ak_i + A_i K_i = 0, \ i = 1, 2, \ldots, n. \tag{6.3}$$

Let us introduce the following notations: $\tilde{A}$—diagonal matrix, which has numbers $A_1, A_2, \ldots, A_n$ on the main diagonal, I—unit $n \times n$ matrix, M—network adjacency matrix, i.e. $M_{ij} = M_{ji} = 1$, if there is the edge connecting nodes i and

j, and $M_{ij} = M_{ji} = 0$ otherwise. It is assumed that $M_{ii} = 0$ for all $i = 1, 2, \ldots, n$. The system of Eqs. (6.3) takes the form:

$$\left(\tilde{A} - 2aI\right) k + \tilde{A} M k = \bar{\varepsilon},\tag{6.4}$$

where $k = (k_1, k_2, \ldots, k_n)^T$, $\bar{\varepsilon} = \left(\varepsilon\left(1 - 2a\right), \varepsilon\left(1 - 2a\right), \ldots, \varepsilon\left(1 - 2a\right)\right)^T$.

Theorem 6.1 ([12], Theorem 1.1) *The system of Eqs. (6.4) for a complete network has a unique solution.*

Thus, the system of Eqs. (6.3) for a complete network always has a unique solution k^s, whose components we will call the stationary values of the investment. In the inner equilibrium $k_i^* = k_i^s$, $i = 1, 2, \ldots, n$.

The following proposition plays a central role in the analysis of equilibria in deterministic version of model.

Proposition 6.1 ([12], Lemmas 2.1, 2.2 and Corollary 2.1) *In equilibrium, the agent i is passive if and only if*

$$K_i \leq \frac{\varepsilon\left(1 - 2a\right)}{A_i};$$

the agent i is active if and only if

$$\frac{\varepsilon\left(1 - 2a\right)}{A_i} < K_i < \frac{\varepsilon}{A_i},$$

or, equivalently,

$$k_i = \frac{\varepsilon(2a - 1) + A_i \tilde{K}_i}{2a - A_i};$$

the agent i is hyperactive if and only if

$$K_i \geq \frac{\varepsilon}{A_i}.$$

In [12] the adjustment dynamics in discrete time is introduced, which begins after a small deviation of the agents' strategies from the equilibrium, or when the networks which were in equilibrium, are combined together. In [13] adjustment dynamics is studied in continuous time.

Definition 6.1 ([13], Definition 5) In the adjustment process, each agent maximizes her utility by choosing a level of investment; when she makes her decision, she treats her environment as exogenously given. Therefore, if $k_i(t_0) = 0$, where t_0 is an arbitrary moment, and $D_1 V_i(k_i, K_i)|_{k_i=0} \leq 0$, then $k_i(t) = 0$ for any $t > t_0$, and if $k_i(t_0) = \varepsilon$ and $D_1 V_i(k_i, K_i)|_{k_i=\varepsilon} \geq 0$, then $k_i(t) = \varepsilon$ for any $t > t_0$; in all other cases, $k_i(t)$ solves the differential equation:

$$\dot{k}_i = \frac{A_i}{2a}\tilde{K}_i + \frac{A_i - 2a}{2a}k_i - \frac{\varepsilon(1 - 2a)}{2a}.$$

Definition 6.2 ([13], Definition 6) The equilibrium is called *dynamically stable* if, after a small deviation of one of the agents from the initial equilibrium, the adjustment dynamics returns the network back to the initial equilibrium. Otherwise, the equilibrium is called *dynamically unstable*.

The natural generalization of the deterministic model described above is to introduce stochastic parameters. The assumption that agents' productivities may have stochastic components seems quite realistic, while the endowments of agents are constants for this concept.

In new version of the model, the productivity of each agent has not only deterministic A, but also Brownian (Wiener) component αW_t ($W_0 = 0$). Thus, the total productivity of agent i is now equal to $A_i + \alpha W_t^i$. All other assumptions remain the same as in our main deterministic model.

In this paper we lay the groundwork for the stochastic model by considering the behavior of a single agent and dyad agents in the stochastic case.

6.3 The Stochastic Extension of Model for Single Agent

Consider first the dynamics of a single agent. We assume that the deterministic component of her productivity A does not depend on time, and her initial investment at time $t = 0$ is $k(0) = k_0$. Then the dynamics of a single agent is described by the following equation:

$$\dot{k} = \left(\frac{A + \alpha W_t}{2a} - 1\right)k - \frac{\varepsilon(1 - 2a)}{2a},$$

or in differential form,

$$dk = \left(\frac{A}{2a} - 1\right)k\,dt + \frac{\alpha}{2a}k\,dW_t - \frac{\varepsilon(1 - 2a)}{2a}dt. \tag{6.5}$$

Proposition 6.2 *The dynamics of investments in knowledge of a single agent is*

$$k(t) = e^{\lambda t + \mu W_t - \frac{\mu^2}{2}t}k_0 - \frac{\varepsilon(1 - 2a)}{2a}\int_0^t e^{\lambda(t-\tau) - \frac{\mu^2}{2}(t-\tau) + \mu(W_t - W_\tau)}d\tau.$$

Proof Introducing the notation

$$\lambda = \frac{A}{2a} - 1, \quad \mu = \frac{\alpha}{2a}, \quad \Psi_t = -\lambda t - \mu W_t + \frac{1}{2}\mu^2 t \tag{6.6}$$

and multiplying equation (6.5) by e^{Ψ_t}, we get

$$e^{\Psi_t} dk = \lambda k e^{\Psi_t} dt + \mu k e^{\Psi_t} dW_t - \frac{\varepsilon(1-2a)}{2a} e^{\Psi_t} dt. \tag{6.7}$$

Note that by Ito's lemma

$$de^{\Psi_t} = -\lambda e^{\Psi_t} dt - \mu e^{\Psi_t} dW_t + \frac{1}{2}\mu^2 e^{\Psi_t} dt + \frac{1}{2}\mu^2 e^{\Psi_t} dt = e^{\Psi_t}(-\lambda dt - \mu dW_t + \mu^2 dt), \tag{6.8}$$

and therefore given (6.7) and (6.8):

$$d\left(k e^{\Psi_t}\right) = e^{\Psi_t} dk + k de^{\Psi_t} + dk de^{\Psi_t} =$$

$$= \lambda k e^{\Psi_t} dt + \mu k e^{\Psi_t} dW_t - \frac{\varepsilon(1-2a)}{2a} e^{\Psi_t} dt + k e^{\Psi_t}\left(-\lambda dt - \mu dW_t + \mu^2 dt\right) - \mu^2 k e^{\Psi_t} dt =$$

$$= -\frac{\varepsilon(1-2a)}{2a} e^{\Psi_t} dt.$$

So

$$k(t) = e^{-\Psi_t}\left(k_0 - \frac{\varepsilon(1-2a)}{2a}\int_0^t e^{\Psi_\tau} d\tau\right) =$$

$$= e^{\lambda t + \mu W_t - \frac{\mu^2}{2}t} k_0 - \frac{\varepsilon(1-2a)}{2a}\int_0^t e^{\lambda(t-\tau)-\frac{\mu^2}{2}(t-\tau)+\mu(W_t-W_\tau)} d\tau. \tag{6.9}$$

∎

Proposition 6.3 *The mathematical expectation of geometric Brownian motion without drift, i.e. $e^{\mu W_t}$, where $W_0 = 0$, is equal to $e^{\frac{\mu^2}{2}t}$.*

Proof By Ito's lemma

$$de^{\mu W_t} = \mu e^{\mu W_t} dW_t + \frac{1}{2}\mu^2 e^{\mu W_t} dt,$$

or in finite form

$$e^{\mu W_t} = e^{\mu W_0} + \mu \int_0^t e^{\mu W_s} dW_s + \frac{1}{2}\mu^2 \int_0^t e^{\mu W_s} ds. \tag{6.10}$$

Taking the mathematical expectation from both parts of (6.10) and taking into account the property of Brownian processes $E\left[\int_0^t e^{\mu W_s} dW_s\right] = 0$, we obtain

$$E\left[e^{\mu W_t}\right] = E\left[e^{\mu W_0}\right] + \frac{1}{2}\mu^2 \int_0^t E\left[e^{\mu W_s}\right] ds,$$

or

$$\frac{d}{dt}E\left[e^{\mu W_t}\right] = \frac{1}{2}\mu^2 E\left[e^{\mu W_t}\right], \quad E\left[e^{\mu W_0}\right] = 1.$$

Solving this differential equation, we find

$$E\left[e^{\mu W_t}\right] = e^{\frac{\mu^2}{2}t}.$$

∎

Remark 6.1 The mathematical expectation of the stochastic process $k(t)$ according to (6.9) is equal to

$$E\left[k(t)\right] = e^{\lambda t} k_0 + \frac{\varepsilon(1-2a)}{2a} \int_0^t e^{\lambda(t-\tau)} d(t-\tau) = \left(k_0 - k^s\right) e^{\frac{A-2a}{2a}t} + k^s,$$

where

$$k^s = \frac{\varepsilon(1-2a)}{A-2a}$$

is the stationary value of the investment. Thus, the dynamics of the mathematical expectation of the value of investments in the knowledge of a single agent coincides with the dynamics of the value of her investments in knowledge in the deterministic case, when $\alpha = 0$, i.e.

$$k(t) = \left(k_0 - k^s\right) e^{\frac{A-2a}{2a}t} + k^s. \tag{6.11}$$

However, the boundaries of various scenarios of the agent's behavior (and the behavior itself) in the stochastic case are obviously shifted compared to the deterministic case.

Remark 6.2 In the deterministic case, among the two possible corner equilibria, the equilibrium $k = 0$ is always possible and stable because, according to (6.3),

$$D_1 V(0, 0) = \varepsilon(2a - 1) < 0,$$

and it follows from Proposition 6.1 that the equilibrium $k = \varepsilon$ is possible if $A \geq 1$ and is stable if $A > 1$ since according to (6.3),

$$D_1 V(\varepsilon, \varepsilon) = \varepsilon(2a - 1) - 2a\varepsilon + A\varepsilon = A\varepsilon - \varepsilon.$$

Inner equilibrium

$$k^* = k^s = \frac{\varepsilon(1 - 2a)}{A - 2a},$$

in accordance with the Proposition 6.1 is possible if and only if $0 < k^s < \varepsilon$ i.e. when $A > 2a$, $A > 1$. The inner equilibrium is unstable, since the root of the characteristic equation $\lambda = (A - 2a)/2a > 0$.

Corollary 6.1 *In the deterministic case, the following cases take place.*

(1) $A < 2a < 1$. *Then* $k^s < 0$, $\lambda < 0$, *and for any initial value of* $k_0 \in [0, \varepsilon]$, *we have*

$$\lim_{t \to \infty} k(t) = 0.$$

(2) $2a < A < 1$. *Then* $k^s > \varepsilon$, $\lambda > 0$, *and for any initial value of* $k_0 \in [0, \varepsilon]$, *we have*

$$\lim_{t \to \infty} k(t) = 0.$$

(3) $A = 1$. *Then* $k^s = \varepsilon$, $\lambda > 0$, *and for any initial value of* $k_0 \in [0, \varepsilon)$, *we have*

$$\lim_{t \to \infty} k(t) = 0,$$

and for the initial value $k_0 = \varepsilon$, *the agent remains in unstable equilibrium* $k = \varepsilon$, *but after a small deviation she begins to decrease her investment and converges to a stable equilibrium* $k = 0$.

(4) $A > 1 > 2a$. *Then* $k^s \in (0, \varepsilon)$, $\lambda > 0$, *and three cases are possible. If* $k_0 \in [0, k^s)$, *then*

$$\lim_{t \to \infty} k(t) = 0.$$

If $k_0 \in (k^s, \varepsilon]$, then

$$\lim_{t \to \infty} k(t) = \varepsilon.$$

If $k_0 = k^s$ then the agent remains in this unstable equilibrium $k = k^s$, but after a small deviation she will continue to move in the same direction and converge to one of two stable equilibria, respectively, $k = 0$ or $k = \varepsilon$.

In the stochastic case, we will use the law of the iterated logarithm.

Theorem 6.2 ([9], The Law of the Iterated Logarithm) *For a one-dimensional Brownian motion W_t, the following equality holds:*

$$\limsup_{t \to \infty} \frac{W_t}{\sqrt{2t \ln \ln t}} = 1 \quad a.s.$$

We rewrite expression (6.9) taking into account (6.6) as follows:

$$k(t) = e^{\frac{A-2a}{2a}t + \frac{\alpha}{2a}W_t - \frac{\alpha^2}{8a^2}t} \left(k_0 - \frac{\varepsilon(1-2a)}{2a} \int_0^t e^{-\frac{A-2a}{2a}\tau - \frac{\alpha}{2a}W_\tau + \frac{\alpha^2}{8a^2}\tau} d\tau \right).$$

$$(6.12)$$

To perform a qualitative analysis of the behavior of the solution, it is important to know in the case $A > 2a + \frac{\alpha^2}{4a}$ whether the value of last integral term in the right hand side of the Eq. (6.12) reaches the value k_0 when $t \to \infty$.

Remark 6.3 Random density of

$$\int_0^\infty e^{-\frac{A-2a}{2a}\tau - \frac{\alpha}{2a}W_\tau + \frac{\alpha^2}{8a^2}\tau} d\tau$$

if $A > 2a + \frac{\alpha^2}{4a}$ according to [1] is

$$f(k) = \frac{\left(\frac{\alpha^2}{8a^2}\right)^{-\frac{4a(A-2a)}{\alpha^2}+1}}{\Gamma\left(\frac{4a(A-2a)}{\alpha^2} - 1\right)} \cdot \frac{\exp\left(-\frac{8a^2}{\alpha^2 k}\right)}{k^{\frac{4a(A-2a)}{\alpha^2}}},$$

where Γ is the gamma function. Then the probability that

$$\frac{\varepsilon(1-2a)}{2a} \cdot \int_0^\infty e^{-\frac{A-2a}{2a}\tau - \frac{\alpha}{2a}W_\tau + \frac{\alpha^2}{8a^2}\tau} d\tau$$

does not reach k_0 is equal to

$$\hat{P} = \frac{\left(\frac{\alpha^2}{8a^2}\right)^{-\frac{4a(A-2a)}{\alpha^2}+1}}{\Gamma\left(\frac{4a(A-2a)}{\alpha^2}-1\right)} \cdot \int_0^{\frac{2ak_0}{e(1-2a)}} \frac{\exp\left(-\frac{8a^2}{\alpha^2 k}\right)}{k^{\frac{4a(A-2a)}{\alpha^2}}}dk. \tag{6.13}$$

Corollary 6.2 *In the stochastic case, the following cases take place.*

(1) $A < 2a + \frac{\alpha^2}{4a}$. *Then according to the law of the iterated logarithm we have*

$$\lim_{t\to\infty} k(t) = 0 \quad a.s.$$

(2) $A = 2a + \frac{\alpha^2}{4a}$. *Then the process $k(t)$ will fluctuate between 0 and ε.*
(3) $A > 2a + \frac{\alpha^2}{4a}$. *Then with probability $\hat{P}$ (see (6.13)) we get*

$$\lim_{t\to\infty} k(t) = \varepsilon$$

and with probability $1 - \hat{P}$ respectively

$$\lim_{t\to\infty} k(t) = 0.$$

6.4   The Stochastic Extension of Model for Dyad

Definition 6.3 A dyad is a network consisting of two nodes connected by an arc.

Suppose that two agents with different productivity and with initial values of investment in knowledge k_1^0 and k_2^0 are aggregated into a dyad. The productivity of each agent has both a constant and a random (Brownian) component. We assume that changes in the productivity of dyad agents are caused by the same random influences in this network, and the sizes of random components are proportional to the constant components of productivities. In other words, the productivity of the first agent is $A_1 + \alpha_1 W_t$, the productivity of the second agent is $A_2 + \alpha_2 W_t$, while

$$\frac{A_1}{A_2} = \frac{\alpha_1}{\alpha_2}. \tag{6.14}$$

The dynamics in such a dyad is described by a system of stochastic equations

$$\begin{cases} \dot{k}_1 = \left(\frac{A_1}{2a} - 1\right)k_1 + \frac{\alpha_1}{2a}W_t k_1 + \frac{A_1}{2a}k_2 + \frac{\alpha_1}{2a}W_t k_2 - \frac{\varepsilon(1-2a)}{2a}, \\ \dot{k}_2 = \frac{A_2}{2a}k_1 + \frac{\alpha_2}{2a}W_t k_1 + \left(\frac{A_2}{2a} - 1\right)k_2 + \frac{\alpha_2}{2a}W_t k_2 - \frac{\varepsilon(1-2a)}{2a}, \end{cases}$$

or, in differential form,

$$\begin{cases} dk_1 = \left(\frac{A_1}{2a} - 1\right)k_1dt + \frac{\alpha_1}{2a}k_1dW_t + \frac{A_1}{2a}k_2dt + \frac{\alpha_1}{2a}k_2dW_t - \frac{\varepsilon(1-2a)}{2a}dt, \\ dk_2 = \frac{A_2}{2a}k_1dt + \frac{\alpha_2}{2a}k_1dW_t + \left(\frac{A_2}{2a} - 1\right)k_2dt + \frac{\alpha_2}{2a}k_2dW_t - \frac{\varepsilon(1-2a)}{2a}dt. \end{cases}$$
(6.15)

The matrix notation of system (6.15) has the form

$$dk = Akdt + \alpha kdW + \bar{E}dt,$$
(6.16)

where

$$dk = \begin{pmatrix} dk_1 \\ dk_2 \end{pmatrix}, \quad A = \begin{pmatrix} \frac{A_1}{2a} - 1 & \frac{A_1}{2a} \\ \frac{A_2}{2a} & \frac{A_2}{2a} - 1 \end{pmatrix}, \quad k = \begin{pmatrix} k_1 \\ k_2 \end{pmatrix},$$

$$\alpha = \begin{pmatrix} \frac{\alpha_1}{2a} & \frac{\alpha_1}{2a} \\ \frac{\alpha_2}{2a} & \frac{\alpha_2}{2a} \end{pmatrix}, \quad \bar{E} = \begin{pmatrix} -\frac{\varepsilon(1-2a)}{2a} \\ -\frac{\varepsilon(1-2a)}{2a} \end{pmatrix}.$$

We first consider the deterministic case when $\alpha_1 = \alpha_2 = 0$.

Proposition 6.4 *In the deterministic case, the dynamics of investments in knowledge of dyad agents is given by*

$$\begin{cases} k_1 = \left(\frac{A_2k_1^0 - A_1k_2^0}{4\bar{A}} - \frac{e(1-2a)(A_1-A_2)}{8a\bar{A}}\right)e^{-t} + A_1\frac{\bar{k}^0 - \bar{D}}{\bar{A}}e^{\left(\frac{\bar{A}}{a}-1\right)t} + D_1, \\ k_2 = -\left(\frac{A_2k_1^0 - A_1k_2^0}{4\bar{A}} - \frac{e(1-2a)(A_1-A_2)}{8a\bar{A}}\right)e^{-t} + A_2\frac{\bar{k}^0 - \bar{D}}{\bar{A}}e^{\left(\frac{\bar{A}}{a}-1\right)t} + D_2, \end{cases}$$
(6.17)

where D_1 and D_2 are determined by the expression (6.18)

$$D_1 = \frac{\varepsilon(1-2a)}{2a} \cdot \frac{A_1 - A_2 + 2a}{A_1 + A_2 - 2a}, \quad D_2 = \frac{\varepsilon(1-2a)}{2a} \cdot \frac{A_2 - A_1 + 2a}{A_1 + A_2 - 2a}.$$
(6.18)

The proof of this statement is in Appendix.

Theorem 6.3 *In the stochastic case, the dynamics of investments in knowledge of dyad agents is given by*

$$k_1(t) = \frac{\varepsilon(1-2a)(A_1 - A_2)}{2a(A_1 + A_2)} + \left[\frac{A_2k_1^0 - A_1k_2^0}{A_1 + A_2} + \frac{\varepsilon(1-2a)(A_2 - A_1)}{2a(A_1 + A_2)}\right]e^{-t} +$$

$$+ \left[\frac{A_1(k_1^0 + k_2^0)}{A_1 + A_2} - \frac{A_1\varepsilon(1-2a)}{a(A_1 + A_2)}\int_0^t e^{\left(-\frac{\bar{A}}{a} + \frac{(\alpha_1+\alpha_2)^2}{8a^2} + 1\right)\tau - \frac{\alpha_1+\alpha_2}{2a}W_\tau}d\tau\right] \times$$

$$\times e^{\left(\frac{\bar{A}}{a} - \frac{(\alpha_1 + \alpha_2)^2}{8a^2} - 1\right)t + \frac{\alpha_1 + \alpha_2}{2a} W_t}, \tag{6.19}$$

$$k_2(t) = \frac{\varepsilon(1-2a)(A_2 - A_1)}{2a(A_1 + A_2)} + \left[\frac{A_1 k_2^0 - A_2 k_1^0}{A_1 + A_2} + \frac{\varepsilon(1-2a)(A_1 - A_2)}{2a(A_1 + A_2)}\right] e^{-t} +$$

$$+ \left[\frac{A_2(k_1^0 + k_2^0)}{A_1 + A_2} - \frac{A_2 \varepsilon(1-2a)}{a(A_1 + A_2)} \int_0^t e^{\left(-\frac{\bar{A}}{a} + \frac{(\alpha_1 + \alpha_2)^2}{8a^2} + 1\right)\tau - \frac{\alpha_1 + \alpha_2}{2a} W_\tau} d\tau\right] \times$$

$$\times e^{\left(\frac{\bar{A}}{a} - \frac{(\alpha_1 + \alpha_2)^2}{8a^2} - 1\right)t + \frac{\alpha_1 + \alpha_2}{2a} W_t}. \tag{6.20}$$

The proof of this theorem is in Appendix.

Remark 6.4 Deterministic case solution

$$k(t) = e^{At} k_0 + e^{At} \left(\int_0^t e^{-A\tau} d\tau\right) \bar{E} \tag{6.21}$$

is obtained from (6.25) by putting $\alpha = 0$. However, in the deterministic case, the solution obtained by the Euler method is better suited to analyze the behavior of the function $k(t)$.

Remark 6.5 Note that we could obtain the same expression (6.17) which is convenient for a qualitative analysis by substituting the expressions for S, S^{-1}, e^{tJ} in (6.21), i.e. calculating

$$\begin{pmatrix} k_1(t) \\ k_2(t) \end{pmatrix} = \frac{1}{A_1 + A_2} \begin{pmatrix} 1 & A_1 \\ -1 & A_2 \end{pmatrix} \begin{pmatrix} e^{-t} & 0 \\ 0 & e^{\left(\frac{\bar{A}}{a} - 1\right)t} \end{pmatrix} \begin{pmatrix} A_2 & -A_1 \\ 1 & 1 \end{pmatrix} +$$

$$+ \frac{1}{A_1 + A_2} \begin{pmatrix} 1 & A_1 \\ -1 & A_2 \end{pmatrix} \begin{pmatrix} e^{-t} & 0 \\ 0 & e^{\left(\frac{\bar{A}}{a} - 1\right)t} \end{pmatrix} \begin{pmatrix} A_2 & -A_1 \\ 1 & 1 \end{pmatrix} \begin{pmatrix} \int_0^t e^{-\tau} d\tau & 0 \\ 0 & \int_0^t e^{\left(\frac{\bar{A}}{a} - 1\right)\tau} d\tau \end{pmatrix} \times$$

$$\times \begin{pmatrix} A_2 & -A_1 \\ 1 & 1 \end{pmatrix} \begin{pmatrix} -\frac{\varepsilon(1-2a)}{2a} \\ -\frac{\varepsilon(1-2a)}{2a} \end{pmatrix}. \tag{6.22}$$

However, in the stochastic case, we had to use matrix exponentials and Ito's lemma to obtain expressions similar to (6.22). Moreover, we could not write the integrals of the Wiener processes in the closed form.

All possible equilibria in the deterministic version of the model under consideration are listed in [8].

Proposition 6.5 ([8]) *Let in deterministic dyad $A_1 > A_2$. Then there are 6 equilibria:*

(1) equilibrium in which $k_1 = k_2 = 0$;
(2) equilibrium in which $k_1 = k_2 = \varepsilon$;
(3) equilibrium in which $k_1 = \varepsilon, 0 < k_2 < \varepsilon$;
(4) equilibrium in which $0 < k_1 < \varepsilon, k_2 = 0$;
(5) equilibrium in which $k_1 = \varepsilon, k_2 = 0$;
(6) equilibrium in which both agents are active.

While equilibria 1, 2, 3, 5 are dynamically stable, equilibria 4 and 6 are unstable.

In the stochastic case, there are not so many equilibria. Equilibrium in this case is a point in the phase space to which a (stochastic) transition process converges as t tends to infinity. Thus, in our stochastic model, the very concept of unstable equilibrium is senseless.

To perform a qualitative analysis of the behavior of the solution, it is important to know in the case $A_1 + A_2 > 2a + \frac{(\alpha_1+\alpha_2)^2}{4a}$ whether the values of the integral terms in the right hand side of Eqs. (6.19) and (6.20) reach the values respectively $\frac{A_1(k_1^0+k_2^0)}{A_1+A_2}$ and $\frac{A_2(k_1^0+k_2^0)}{A_1+A_2}$ when $t \to \infty$, i.e. whether the random process

$$\int_0^t e^{\left(-\frac{\bar{A}}{a}+\frac{(\alpha_1+\alpha_2)^2}{8a^2}+1\right)\tau-\frac{\alpha_1+\alpha_2}{2a}W_\tau} d\tau$$

reaches the value

$$\frac{2a\bar{k}_0}{\varepsilon(1-2a)} = \frac{a(k_1^0+k_2^0)}{\varepsilon(1-2a)}.$$

Remark 6.6 Random density of

$$\int_0^\infty e^{\left(-\frac{\bar{A}}{a}+\frac{(\alpha_1+\alpha_2)^2}{8a^2}+1\right)t-\frac{\alpha_1+\alpha_2}{2a}W_t} dt$$

if $A_1 + A_2 > 2a + \frac{(\alpha_1+\alpha_2)^2}{4a}$ according to [1] is

$$f(k) = \frac{\left(\frac{(\alpha_1+\alpha_2)^2}{8a^2}\right)^{-\frac{4a(A_1+A_2-2a)}{(\alpha_1+\alpha_2)^2}+1}}{\Gamma\left(\frac{4a(A_1+A_2-2a)}{(\alpha_1+\alpha_2)^2}-1\right)} \cdot \frac{\exp\left(-\frac{8a^2}{(\alpha_1+\alpha_2)^2 k}\right)}{k^{\frac{4a(A_1+A_2-2a)}{(\alpha_1+\alpha_2)^2}}}.$$

Then the probability that

$$\frac{A_1\varepsilon(1-2a)}{a(A_1+A_2)}\int_0^t e^{\left(-\frac{\bar{A}}{a}+\frac{(\alpha_1+\alpha_2)^2}{8a^2}+1\right)\tau-\frac{\alpha_1+\alpha_2}{2a}W_\tau} d\tau < \frac{A_1(k_1^0+k_2^0)}{A_1+A_2}$$

which can be rewritten as

$$\frac{A_2 \varepsilon (1 - 2a)}{a(A_1 + A_2)} \int_0^t e^{\left(-\frac{\bar{A}}{a} + \frac{(\alpha_1 + \alpha_2)^2}{8a^2} + 1\right)\tau - \frac{\alpha_1 + \alpha_2}{2a} W_\tau} d\tau < \frac{A_2(k_1^0 + k_2^0)}{A_1 + A_2}$$

is equal to

$$\tilde{P} = \frac{\left(\frac{(\alpha_1 + \alpha_2)^2}{8a^2}\right)^{-\frac{4a(A_1 + A_2 - 2a)}{(\alpha_1 + \alpha_2)^2} + 1}}{\Gamma\left(\frac{4a(A_1 + A_2 - 2a)}{(\alpha_1 + \alpha_2)^2} - 1\right)} \cdot \int_0^{\frac{2a\bar{k}_0}{\varepsilon(1 - 2a)}} \frac{\exp\left(-\frac{8a^2}{(\alpha_1 + \alpha_2)^2 k}\right)}{k^{\frac{4a(A_1 + A_2 - 2a)}{(\alpha_1 + \alpha_2)^2}}} dk. \qquad (6.23)$$

Corollary 6.3 *In the stochastic case, the following cases take place.*

(1) If $A_1 + A_2 < 2a + \frac{(\alpha_1 + \alpha_2)^2}{4a}$ then according to the law of the iterated logarithm, we have

$$\lim_{t \to \infty} k_1(t) = 0 \quad a.s.$$

$$\lim_{t \to \infty} k_2(t) = 0 \quad a.s.$$

(2) If $A_1 + A_2 = 2a + \frac{(\alpha_1 + \alpha_2)^2}{4a}$ then the processes $k_1(t)$ and $k_2(t)$ will fluctuate between 0 and ε.

(3) If $A_1 + A_2 > 2a + \frac{(\alpha_1 + \alpha_2)^2}{4a}$ then with probability $\tilde{P}$ (see (6.23)) we get

$$\lim_{t \to \infty} k_1(t) = \varepsilon,$$

$$\lim_{t \to \infty} k_2(t) = \varepsilon.$$

and with probability $1 - \tilde{P}$ respectively

$$\lim_{t \to \infty} k_1(t) = 0,$$

$$\lim_{t \to \infty} k_2(t) = 0.$$

6.5 Conclusion

In this paper we develop and generalize the model proposed in [11, 12]. The contribution of this paper is in describing transition dynamics in the stochastic case where agent productivity has both deterministic and Brownian components. Previously, the transition dynamics between dynamically stable equilibria in the network were considered only in the deterministic case. It turns out that the boundaries of various scenarios of the agent's behavior (and the behavior itself) in the stochastic case are shifted compared to the deterministic case. We derive the explicit expressions for the dynamics of a single agent and dyad agents in the form of Brownian random processes (see Proposition 6.2 and Theorem 6.3), and provide a qualitative analysis of the solutions of stochastic equations and systems of stochastic equations (see Corollaries 6.2 and 6.3). We establish in which direction and with what probability the random process will evolve and to what state it will come in each case.

The next task is to study the transition dynamics in stochastic triangles with heterogeneous agents and in complete stochastic networks with agents having stochastic productivities. It might be also useful to consider the dynamics in networks with arbitrary correlation functions between the stochastic components of different parameters.

Appendix

The proof of Proposition 6.4.

Proof The system of differential equations in the deterministic case has the form

$$
\begin{cases}
\dot{k}_1 = \left(\frac{A_1}{2a} - 1\right) k_1 + \frac{A_1}{2a} k_2 - \frac{\varepsilon(1-2a)}{2a}, \\
\dot{k}_2 = \frac{A_2}{2a} k_1 + \left(\frac{A_2}{2a} - 1\right) k_2 - \frac{\varepsilon(1-2a)}{2a}.
\end{cases}
\tag{6.24}
$$

The characteristic equation for system (6.11) is as follows

$$
(\lambda + 1)^2 - \frac{A_1 + A_2}{2a}(\lambda + 1) = 0,
$$

therefore eigenvalues are

$$
\lambda_1 = -1; \quad \lambda_2 = -1 + \frac{\bar{A}}{a},
$$

where $\bar{A} = \frac{A_1 + A_2}{2}$. Obviously, we can choose as the eigenvectors of the matrix A the vectors

$$
e_1 = \begin{pmatrix} 1 \\ -1 \end{pmatrix}, \quad e_2 = \begin{pmatrix} A_1 \\ A_2 \end{pmatrix}.
$$

So the transition matrix is

$$
S = \begin{pmatrix} 1 & A_1 \\ -1 & A_2 \end{pmatrix},
$$

then

$$AS = SJ, \quad e^{tA} = Se^{tJ}S^{-1},$$

where

$$J = \begin{pmatrix} -1 & 0 \\ 0 & -1 + \frac{\bar{A}}{a} \end{pmatrix}, \quad e^{tJ} = \begin{pmatrix} e^{-t} & 0 \\ 0 & \exp\left(\left(\frac{\bar{A}}{a} - 1\right)t\right) \end{pmatrix},$$

$$S^{-1} = \frac{1}{A_1 + A_2} \begin{pmatrix} A_2 & -A_1 \\ 1 & 1 \end{pmatrix}.$$

The general solution of system (6.11) is as follows

$$\begin{pmatrix} k_1 \\ k_2 \end{pmatrix} = C_1 \begin{pmatrix} 1 \\ -1 \end{pmatrix} \exp(-t) + C_2 \begin{pmatrix} A_1 \\ A_2 \end{pmatrix} \exp\left(\left(\frac{\bar{A}}{a} - 1\right)t\right) + \begin{pmatrix} D_1 \\ D_2 \end{pmatrix}.$$

We find the constants D_1 and D_2 by solving the system of equations

$$\begin{cases} \left(\frac{A_1}{2a} - 1\right)k_1 + \frac{A_1}{2a}k_2 = \frac{\varepsilon(1-2a)}{2a}, \\ \frac{A_2}{2a}k_1 + \left(\frac{A_2}{2a} - 1\right)k_2 = \frac{\varepsilon(1-2a)}{2a}. \end{cases}$$

It is easy to verify that they are determined by the expression (6.18). We find the integration constants C_1 and C_2 from the initial conditions:

$$\begin{cases} k_1^0 = C_1 + A_1 C_2 + D_1, \\ k_2^0 = -C_1 + A_2 C_2 + D_2, \end{cases}$$

so

$$C_2 = \frac{\bar{k}^0 - \bar{D}}{\bar{A}},$$

where

$$\bar{k}^0 = \frac{k_1^0 + k_2^0}{2}, \quad \bar{D} = \frac{D_1 + D_2}{2} = \frac{\varepsilon(1 - 2a)}{2(\bar{A} - a)}, \quad A_2 D_1 - A_1 D_2 = \frac{\varepsilon(1 - 2a)(A_1 - A_2)}{2a}.$$

Then

$$C_1 = \frac{A_2 k_1^0 - A_1 k_2^0}{4\bar{A}} - \frac{\varepsilon(1 - 2a)(A_1 - A_2)}{8a\bar{A}}.$$

Thus, the solution is determined by expression (6.17).   ∎

The proof of Theorem 6.3.

Proof It is clear that the matrices A and α commute; therefore, for the matrix exponentials, the relation

$$e^{At} e^{\alpha W_t} = e^{At + \alpha W_t}$$

holds and we can solve the matrix equation (6.16) by multiplying from the left by the matrix exponent

$$e^{-At - \alpha W_t + \frac{\alpha^2}{2} t}.$$

Denote, as in the one-dimensional case, for brevity

$$\Psi = -At - \alpha W_t + \frac{\alpha^2}{2} t.$$

Then we have

$$d\left(e^{\Psi} k\right) = e^{\Psi} dk + de^{\Psi} k + de^{\Psi} dk =$$

$$= e^{\Psi} \left(Akdt + \alpha k dW_t + \bar{E} dt\right) + e^{\Psi} \left(-Adt - \alpha dW_t + \frac{\alpha^2}{2} dt + \frac{\alpha^2}{2} dt\right) k - e^{\Psi} k\alpha^2 dt =$$

$$= e^{\Psi} \bar{E} dt.$$

Thus, Eq. (6.16) takes the form

$$d\left(e^{-At - \alpha W_t + \frac{\alpha^2}{2} t} k\right) = e^{-At - \alpha W_t + \frac{\alpha^2}{2} t} \bar{E} dt,$$

therefore, the solution of matrix equation (6.8) can be written as

$$k(t) = e^{At + \alpha W_t - \frac{\alpha^2}{2} t} k_0 + e^{At + \alpha W_t - \frac{\alpha^2}{2} t} \left(\int_0^t e^{-A\tau - \alpha W_\tau + \frac{\alpha^2}{2} \tau} d\tau\right) \bar{E}. \qquad (6.25)$$

Notice, that

$$\alpha^2 = \frac{1}{2a} \begin{pmatrix} \alpha_1 & \alpha_1 \\ \alpha_2 & \alpha_2 \end{pmatrix} \cdot \frac{1}{2a} \begin{pmatrix} \alpha_1 & \alpha_1 \\ \alpha_2 & \alpha_2 \end{pmatrix} = \frac{1}{4a^2} \begin{pmatrix} \alpha_1^2 + \alpha_1\alpha_2 & \alpha_1^2 + \alpha_1\alpha_2 \\ \alpha_1\alpha_2 + \alpha_2^2 & \alpha_1\alpha_2 + \alpha_2^2 \end{pmatrix}.$$

The eigenvalues of the matrix

$$A - \frac{\alpha^2}{2} = \begin{pmatrix} \frac{A_1}{2a} - \frac{\alpha_1^2 + \alpha_1\alpha_2}{8q^2} - 1 & \frac{A_1}{2a} - \frac{\alpha_1^2 + \alpha_1\alpha_2}{8a^2} \\ \frac{A_2}{2a} - \frac{\alpha_1^2 + \alpha_1\alpha_2}{8a^2} & \frac{A_2}{2a} - \frac{\alpha_1^2 + \alpha_1\alpha_2}{8a^2} - 1 \end{pmatrix}.$$

are obviously $\lambda_1 = -1$ and $\lambda_2 = \frac{\bar{A}}{a} - \frac{(\alpha_1 + \alpha_2)^2}{8a^2} - 1$. As eigenvectors we can take

$$e_1 = \begin{pmatrix} 1 \\ -1 \end{pmatrix}$$

and

$$e_2 = \begin{pmatrix} A_1 \\ A_2 \end{pmatrix}$$

or in view of (6.14)

$$e_2 = \begin{pmatrix} \alpha_1 \\ \alpha_2 \end{pmatrix}.$$

The eigenvalues of the matrix α are $\lambda_1 = 0$ and $\lambda_2 = \alpha_1 + \alpha_2$, and obviously we can choose the same e_1 and e_2 as eigenvectors as for the matrix $A - \frac{\alpha^2}{2}$. Therefore, to reduce to the diagonal form of the matrices $\left(A - \frac{\alpha^2}{2}\right)t$ and αW_t we can use the same transition matrices

$$S = \begin{pmatrix} 1 & A_1 \\ -1 & A_2 \end{pmatrix}, \quad S^{-1} = \frac{1}{A_1 + A_2} \begin{pmatrix} 1 & A_1 \\ -1 & A_2 \end{pmatrix},$$

so we get

$$\left(A - \frac{\alpha^2}{2}\right)t + \alpha W_t = S(Jt + \Lambda W_t)S^{-1},$$

where

$$J = \begin{pmatrix} -1 & 0 \\ 0 & \frac{\bar{A}}{a} - \frac{(\alpha_1 + \alpha_2)^2}{8a^2} - 1 \end{pmatrix}, \quad \Lambda = \begin{pmatrix} 0 & 0 \\ 0 & \frac{\alpha_1 + \alpha_2}{2a} \end{pmatrix},$$

and correspondingly

$$\exp\left(\left(A - \frac{\alpha^2}{2}\right)t + \alpha W_t\right) =$$

$$= S \begin{pmatrix} \exp(-t) & 0 \\ 0 & \exp\left(\left(\frac{\bar{A}}{a} - \frac{(\alpha_1+\alpha_2)^2}{8a^2} - 1\right)t + \frac{\alpha_1+\alpha_2}{2a}W_t\right) \end{pmatrix} S^{-1}. \qquad (6.26)$$

Substituting (6.26) into (6.25) we obtain

$$\begin{pmatrix} k_1(t) \\ k_2(t) \end{pmatrix} = \frac{1}{A_1 + A_2} \begin{pmatrix} 1 & A_1 \\ -1 & A_2 \end{pmatrix} \times$$

$$\times \begin{pmatrix} \exp(-t) & 0 \\ 0 & \exp\left(\left(\frac{\bar{A}}{a} - \frac{(\alpha_1+\alpha_2)^2}{8a^2} - 1\right)t + \frac{\alpha_1+\alpha_2}{2a}W_t\right) \end{pmatrix} \begin{pmatrix} A_2 & -A_1 \\ 1 & 1 \end{pmatrix} \begin{pmatrix} k_1^0 \\ k_2^0 \end{pmatrix} -$$

$$- \frac{e(1-2a)}{2a} \cdot \frac{1}{A_1 + A_2} \begin{pmatrix} 1 & A_1 \\ -1 & A_2 \end{pmatrix} \times$$

$$\times \begin{pmatrix} \exp(-t) & 0 \\ 0 & \exp\left(\left(\frac{\bar{A}}{a} - \frac{(\alpha_1+\alpha_2)^2}{8a^2} - 1\right)t + \frac{\alpha_1+\alpha_2}{2a}W_t\right) \end{pmatrix} \times$$

$$\times \begin{pmatrix} \int_0^t \exp(\tau)d\tau & 0 \\ 0 & \int_0^t \exp\left(\left(-\frac{\bar{A}}{a} + \frac{(\alpha_1+\alpha_2)^2}{8a^2} + 1\right)\tau - \frac{\alpha_1+\alpha_2}{2a}W_\tau\right)d\tau \end{pmatrix} \times$$

$$\times \begin{pmatrix} A_2 & -A_1 \\ 1 & 1 \end{pmatrix} \begin{pmatrix} 1 \\ 1 \end{pmatrix}. \qquad (6.27)$$

Calculating expression (6.27) we get expressions (6.19)–(6.20).   ∎

References

1. Borodin, A.N., Salminen, P.: Handbook of Brownian Motion. Facts and Formulae. Birkhauser, Basel (1996)
2. Bramoullé, Y., Kranton, R.: Public goods in networks. J. Econ. Theory **135**, 478–494 (2007)
3. Galeotti, A., Goyal, S., Jackson, M.O., Vega-Redondo, F., Yariv, L.: Network games. Rev. Econ. Stud. **77**, 218–244 (2010)
4. Garmash, M.V., Kaneva, X.A.: Game equilibria and adjustment dynamics in full networks and in triangle with heterogeneous agents. Autom. Remote Control. (2020) (Math. Game Theory Appl. (2018) **10**(2), 3–26 in Russian, and accepted to print in 2020 year in Automation and Remote Control)

5. Granovetter, M.S.: The strength of weak ties. Am. J. Sociol. **78**, 1360–1380 (1973)
6. Jackson, M.O.: Social and Economic Networks. Princeton University, Princeton (2008)
7. Jackson, M.O., Zenou, Y.: Games on networks. In: Young, P., Zamir, S. (eds.) Handbook of Game Theory, vol. 4, pp. 95–163. Elsevier, Amsterdam (2014)
8. Kiselev, A.O., Yurchenko, N.I.: Game equilibria and transient dynamics in dyad with heterogeneous agents. Math. Game Theory Appl. **10**(1), 40–64 (2018, in Russian)
9. Lamperti, J.: Stochastic Processes. Springer, New York (1977)
10. Martemyanov, Y.P., Matveenko, V.D.: On the dependence of the growth rate on the elasticity of substitution in a network. Int. J. Process Manag. Benchmarking **4**(4), 475–492 (2014)
11. Matveenko, V.D., Korolev, A.V.: Network game with production and knowledge externalities. Contrib. Game Theory Manag. **8**, 199–222 (2015)
12. Matveenko, V., Korolev, A., Zhdanova, M.: Game equilibria and unification dynamics in networks with heterogeneous agents. Int. J. Eng. Bus. Manag. **9**, 1–17 (2017)
13. Matveenko, V., Garmash, M., Korolev, A.: Chapter 10: game equilibria and transition dynamics in networks with heterogeneous agents. In: Petrosyan, L.A., Mazalov, V.V., Zenkevich, N.A. (eds.) Frontiers of Dynamic Games, vol. 10, pp. 165–188. Springer, Birkhauser (2018)
14. Romer, P.M.: Increasing returns and long-run growth. J. Polit. Econ. **94**, 1002–1037 (1986)

Chapter 7
New Characteristic Function
for Cooperative Games with Hypergraph
Communication Structure

David A. Kosian and Leon A. Petrosyan

Abstract Cooperation in the games with hypergraph communication structure is considered. As usual in cooperative game theory, to define the allocation rule, the characteristic function is used. The communication possibilities are described by the hypergraph in which the nodes are players and hyperlinks are the communicating subgroups of players. The payoff of each player is influenced by actions of other players dependent from a distance between them on hypergraph. The new approach for constructing the characteristic function in the game is proposed. This approach does not require the use of maxmin operations which substantially simplifies the calculations. It is proved that the constructed characteristic function satisfies the convexity property. The results are shown in an example.

Keywords Cooperation · Charactetistic function · Hypergraph · Communication structure

7.1 Introduction

In classic cooperative games with transferable utility, it is assumed that all players have an opportunity to form a grand coalition and the total payoff from the cooperation can be distributed among the players.

The undirected graph can represent a communication structure. Initially, games with undirected graph communication structure were studied by Myerson [1]. After that, games with a communication structure have received attention in cooperative game theory. As a particular case, the hypergraph communication structure can be considered.

D. A. Kosian (✉) · L. A. Petrosyan
Saint Petersburg State University, St. Petersburg, Russia
e-mail: kosyan-david@mail.ru; l.petrosyan@spbu.ru

© The Editor(s) (if applicable) and The Author(s), under exclusive licence
to Springer Nature Switzerland AG 2020
L. A. Petrosyan et al. (eds.), *Frontiers of Dynamic Games*,
Static & Dynamic Game Theory: Foundations & Applications,
https://doi.org/10.1007/978-3-030-51941-4_7

Owen [2] studied games with a tree as a communication structure. Meessen [3] introduced the positional value for the games with graph communication structure.

The TU-games with hypergraph communication structure were studied by Nouweland, Borm and Tijs [4]. They characterized the Myerson value and the positional value for these games. The third value, which is called degree value for games with a hypergraph communication structure, was introduced in [5].

Hypergraph communication structure is an extension of the graph communication structure with hyperlinks containing more than two players. A hyperlink can be a model of groups of people or members in associations, sports teams etc. Also, people in social network chats can be modelled as hyperlinks. In this case, the cooperation between associations, clubs or companies can be described by transferable utility games with hypergraph communication structure.

Despite a large number of researches in the field of cooperative game theory, they are mostly focused on allocation rules while this paper focuses on the construction of characteristic functions for the games with hypergraph communication structure.

The paper is organized as follows. First, basic definitions and notations concerning the hypergraph communication structure are given. Then the definition of the game with specially defined discounted payoff function is provided. Next, the new characteristic function is introduced, and its convexity proved. Finally, the results are explained in an example.

7.2   Communication Structure

The hypergraph is a pair $(N, \mathscr{H})$, $\mathscr{H} \subseteq \{H \in 2^N \,||H| \geq 2\}$, N is a finite set and $\mathscr{H}$ is a given set of subsets of N. The cardinality of H should be greater or equal to two. If $|H| = 2$, for all i this structure will be a graph. $H \in \mathscr{H}$ is called a hyperlink. The elements of the set N are called vertexes.

The hypergraph $(N, \mathscr{H}')$ is called reduction of $(N, \mathscr{H})$ if it is obtained by removing all hyperlinks which are entirely contained in other hyperlinks. A hypergraph is called reduced if it coincides to its reduction, that is, it does not have a hyperlink inside other hyperlinks.

A simple cycle with length s in hypergraph $(N, \mathscr{H})$ is a sequence

$$(H_0, n_0, H_1, n_1, \ldots, H_{s-1}, n_{s-1}, H_s),$$

where $H_0, \ldots, H_{s-1}$—are different hyperlinks and hyperlink H_s coincides with H_0, $n_0, \ldots, n_{s-1}$—are different vertexes, and $n_i \in H_i \cap H_{i+1}$ for all $i = 0, \ldots, s - 1$.

An acyclic hypergraph is a hypergraph without cycles.

In the paper, we shall consider games with acyclic reduced hypergraph communication structure $(N, \mathscr{H})$.

The path between two hyperlinks H_i and H_j is a sequence $H_{k_1}, H_{k_2}, \ldots, H_{k_l}$ where:

1. $H_i = H_{k_1}, H_j = H_{k_l}$.
2. $\forall i : 1 \leq i \leq k-1, H_{l_i} \cap H_{l_{i+1}} \neq \emptyset$.

The path between the two hyperlinks is called "minimal" path if it contains a minimal number of hyperlinks.

The distance between two vertexes i, j is defined as the number of hyperlinks l in the minimal path between hyperlinks H_{k_1}, H_{k_l} where $i \in H_{k_1}, j \in H_{k_l}$.

Neighbours of vertex $i \in N$ with level 1 are the vertexes $j \in N_i^1$ that are at a distance 1 from i. Neighbours of vertex $i \in N$ with level k are the vertexes $j \in N_i^k$ that are at a distance k from i, etc. Finally, the vertexes which have no connection with i are denoted by N_i^{-1}. For any $i \in N$ the set of vertexes is split into neighbors with levels $k \in [1, \ldots, l_i, -1]$ denoted by N_i^k.

7.3 Definition of the Game

In the game setting, the set of vertexes N is identified with the set of players.

Thus $N := \{1, \ldots, n\}$ is considered as set of players. The communication structure is described by the acyclic reduced hypergraph. Each player is a vertex in this structure.

Denote by l_i the distance between player i and the farthest player.

Players i and j are called connected if there is a path between the hyperlinks in which they are included.

Define the strategy set $\mathfrak{U}_i$ of each player. In this paper, we suppose that the sets $\mathfrak{U}_i, i \in N$ are finite. Let $u_i \in \mathfrak{U}_i$ be a strategy of player i. After players have chosen their strategies, the strategy profile $u = (u_i, \ldots, u_n)$ is formed.

The payoff function of the player i is defined as:

$$K_i(u) = \sum_{m=1}^{l_i} \sum_{j \in N_i^m} \delta^{m-1} h_i^j(u_i, u_j), \quad \delta \in (0, 1), \quad h_i^j(u_i, u_j) \geq 0, i, j \in N.$$

7.4 Cooperation

Consider now the cooperative version of the game. It is supposed that the players agree to choose cooperative strategies $u_i^*, i \in N$ which maximize the sum of their payoffs:

$$\max_{\mathfrak{u}} \sum_{i \in N} K_i(u_i, \ldots, u_n) = \sum_{i \in N} K_i(u_1^*, \ldots, u_n^*).$$

Strategy profile $u^* = (u_i^*, \ldots, u_n^*)$ we shall call cooperative behaviour.

In the cooperative game theory, the important role takes the characteristic function defined over the subsets of the player set N.

This paper focused on the definition of characteristic function for the games with hypergraph communication structure. Characteristic function is a real-valued function $v : 2^N \to \mathbb{R}$ and $v(\emptyset) = 0$.

In the classical cooperative game theory the characteristic function is defined by Von Neumann and Morgenstern [6]:

$$v(S) = \max_{u_S} \min_{u_{N\setminus S}} \sum_{i \in S} K_i(u_1, \ldots, u_n), \quad S \subseteq N,$$

here $u_S = \{u_i\}, i \in S$ and $u_{N\setminus S} = \{u_i\}, i \in N \setminus S$.

In this paper, we propose another definition of characteristic function which does not require complicated maxmin computations. The approach is similar to one proposed in [7].

7.4.1 Characteristic Function

For each coalition $S \subseteq N$ define now the hypergraph $(S, \mathcal{H}_S)$ as a restriction of hypergraph $(N, \mathcal{H})$ over S, here $\mathcal{H}_S = \{H \setminus (N \setminus S) | H \in \mathcal{H}\}$.

If player i does not belong to coalition S, he does not interact with the players from this coalition. Also, the player does not act as a link between these players. Thus the player can be eliminated from the hypergraph. For any coalition S, a new communication structure is created by eliminating the players who are not in S. For a better understanding show this elimination by example.

Consider six player game with communication structure which is shown on the Fig. 7.1.

Construct the communication structure for coalition $S = 1, 2, 4, 5$. It is needed to eliminate players $3, 6$ (Fig. 7.2).

Fig. 7.1 Example of hypergraph

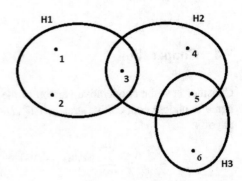

Fig. 7.2 New communication structure for coalition S

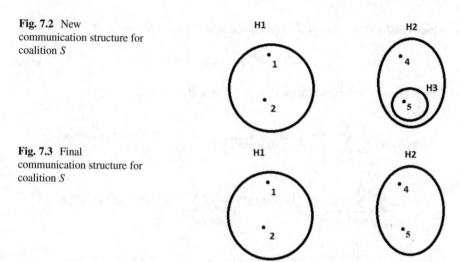

Fig. 7.3 Final communication structure for coalition S

The resulted hypergraph is not reduced since hyperlink H_3 is contained in hyperlink H_2, then it is necessary to construct the reduction of the hypergraph (Fig. 7.3).

Suppose that for any coalition S, the communication structure is constructed and denote by $S_i^k, i \in S$ the neighbours with a connection level k in S. Also let l_i^S be a distance between player i and the farthest player in S. Define the characteristic function in the following way:

$$v(\emptyset) = 0,$$

$$v(\{i\}) = 0,$$

$$v(S) = \sum_{i \in S} \sum_{m=1}^{l_i^S} \sum_{j \in S_i^m} \delta^{m-1} h_i^j(u_i^*, u_j^*), \qquad (7.1)$$

$$v(N) = \sum_{i \in N} \sum_{m=1}^{l_i} \sum_{j \in N_i^m} \delta^{m-1} h_i^j(u_i^*, u_j^*).$$

As follows from (7.1), the value of the characteristic function for each coalition additively depends on the pairwise interaction of the players in S when they use cooperative strategies. Also, from (7.1) it follows that for calculating $v(S)$ for each player $i \in S$, only payoffs from interaction with neighbours with a level $k \neq -1$ are taken into account. Because of the communication structure is acyclic reduced hypergraph the distance between players i and j in coalition S does not change or the path between does not exist.

Theorem 7.1 (Convexity) *For any two coalitions A, B follows that*

$$v(A \cup B) \geq v(B) + v(A) - v(A \cap B). \tag{7.2}$$

Proof Denote $A \cap B = C$ then $\bar{A} = A \setminus C$ and $\bar{B} = B \setminus C$.

$$v(A) = \sum_{i \in \bar{A}} \sum_{m=1}^{l_i^A} \sum_{j \in A_i^m} \delta^{m-1} h_i^j(u_i^*, u_j^*) + \sum_{i \in C} \sum_{m=1}^{l_i^A} \sum_{j \in A_i^m} \delta^{m-1} h_i^j(u_i^*, u_j^*) =$$

$$= \sum_{i \in \bar{A}} \sum_{m=1}^{l_i^{\bar{A}}} \sum_{j \in \bar{A}_i^m} \delta^{m-1} h_i^j(u_i^*, u_j^*) + \sum_{i \in \bar{A}} \sum_{m=1}^{l_i^C} \sum_{j \in C_i^m} \delta^{m-1} h_i^j(u_i^*, u_j^*) +$$

$$+ \sum_{i \in C} \sum_{m=1}^{l_i^C} \sum_{j \in C_i^m} \delta^{m-1} h_i^j(u_i^*, u_j^*) + \sum_{i \in C} \sum_{m=1}^{l_i^{\bar{A}}} \sum_{j \in \bar{A}_i^m} \delta^{m-1} h_i^j(u_i^*, u_j^*)$$

$$v(B) = \sum_{i \in \bar{B}} \sum_{m=1}^{l_i^B} \sum_{j \in B_i^m} \delta^{m-1} h_i^j(u_i^*, u_j^*) + \sum_{i \in C} \sum_{m=1}^{l_i^B} \sum_{j \in B_i^m} \delta^{m-1} h_i^j(u_i^*, u_j^*) =$$

$$= \sum_{i \in \bar{B}} \sum_{m=1}^{l_i^{\bar{B}}} \sum_{j \in \bar{B}_i^m} \delta^{m-1} h_i^j(u_i^*, u_j^*) + \sum_{i \in \bar{B}} \sum_{m=1}^{l_i^C} \sum_{j \in C_i^m} \delta^{m-1} h_i^j(u_i^*, u_j^*) +$$

$$+ \sum_{i \in C} \sum_{m=1}^{l_i^C} \sum_{j \in C_i^m} \delta^{m-1} h_i^j(u_i^*, u_j^*) + \sum_{i \in C} \sum_{m=1}^{l_i^{\bar{B}}} \sum_{j \in \bar{B}_i^m} \delta^{m-1} h_i^j(u_i^*, u_j^*)$$

$$v(A \cap B) = v(C) = \sum_{i \in C} \sum_{m=1}^{l_i^C} \sum_{j \in C_i^m} \delta^{m-1} h_i^j(u_i^*, u_j^*)$$

$$v(A \cup B) = \sum_{i \in \bar{A}} \sum_{m=1}^{l_i^{\bar{A}}} \sum_{j \in \bar{A}_i^m} \delta^{m-1} h_i^j(u_i^*, u_j^*) + \sum_{i \in \bar{A}} \sum_{m=1}^{l_i^C} \sum_{j \in C_i^m} \delta^{m-1} h_i^j(u_i^*, u_j^*) +$$

$$+ \sum_{i \in C} \sum_{m=1}^{l_i^C} \sum_{j \in C_i^m} \delta^{m-1} h_i^j(u_i^*, u_j^*) + \sum_{i \in C} \sum_{m=1}^{l_i^{\bar{A}}} \sum_{j \in \bar{A}_i^m} \delta^{m-1} h_i^j(u_i^*, u_j^*) +$$

$$+ \sum_{i \in \bar{B}} \sum_{m=1}^{l_i^{\bar{B}}} \sum_{j \in \bar{B}_i^m} \delta^{m-1} h_i^j(u_i^*, u_j^*) + \sum_{i \in \bar{B}} \sum_{m=1}^{l_i^C} \sum_{j \in C_i^m} \delta^{m-1} h_i^j(u_i^*, u_j^*) +$$

$$+\sum_{i\in C}\sum_{m=1}^{l_i^{\bar{B}}}\sum_{j\in\bar{B}_i^m}\delta^{m-1}h_i^j(u_i^*,u_j^*)+\sum_{i\in\bar{B}}\sum_{m=1}^{l_i^{\bar{A}}}\sum_{j\in\bar{A}_i^m}\delta^{m-1}h_i^j(u_i^*,u_j^*)+$$

$$+\sum_{i\in\bar{A}}\sum_{m=1}^{l_i^{\bar{B}}}\sum_{j\in\bar{B}_i^m}\delta^{m-1}h_i^j(u_i^*,u_j^*).$$

After substituting these expressions into (7.2) and moving all the components to the left, we get the following inequality:

$$v(A\cup B)-v(B)-v(A)+v(A\cap B)=$$

$$=\sum_{i\in\bar{B}}\sum_{m=1}^{l_i^{\bar{A}}}\sum_{j\in\bar{A}_i^m}\delta^{m-1}h_i^j(u_i^*,u_j^*)+\sum_{i\in\bar{A}}\sum_{m=1}^{l_i^{\bar{B}}}\sum_{j\in\bar{B}_i^m}\delta^{m-1}h_i^j(u_i^*,u_j^*)\geq 0. \qquad (7.3)$$

Because $h_i^j\geq 0$ then the (7.3) holds and the theorem is proved. ∎

The vector $\xi=(\xi_1,\ldots,\xi_n)$ which satisfies the conditions:

$$\xi_j\geq v(\{j\}),\ j\in N,$$

$$\sum_{j=1}^n\xi_j=v(N),$$

where v($\{j\}$)—is the value of characteristic function for coalition $S=\{j\}$ is called an imputation.

The set of nondominant imputations in the cooperative game (N,v) is called core. For the imputation $\alpha=(\alpha_1,\ldots,\alpha_n)$ to belong to the core, it is necessary and sufficient that:

$$v(S)\leq\sum_{i\in S}\alpha_i \qquad (7.4)$$

hold for all $S\subset N$.

Consider the set of neighbours with level k of player i. For any coalition $S, i\in S$ it holds that $S_i^k\subseteq N_i^k$. Denote by $C_i^k=N_i^k\setminus S_i^k$. Consider the following imputation $\hat{\alpha}=(\hat{\alpha}_1,\ldots,\hat{\alpha}_n)$ where

$$\hat{\alpha}_i=K_i(u^*)=\sum_{m=1}^{l_i}\sum_{j\in N_i^m}\delta^{m-1}h_i^j(u_i^*,u_j^*)$$

is the payoff of player i under cooperative strategy profile.

Lemma 7.1 *The imputation $\hat{\alpha}$ belongs to the core.*

Proof

$$v(S) = \sum_{i \in S} \sum_{m=1}^{l_i^S} \sum_{j \in S_i^m} \delta^{m-1} h_i^j (u_i^*, u_j^*),$$

$$\sum_{i \in S} \hat{\alpha}_i = \sum_{i \in S} K_i(u^*) = \sum_{i \in S} \sum_{m=1}^{l_i} \sum_{j \in N_i^m} \delta^{m-1} h_i^j (u_i^*, u_j^*) =$$

$$= \sum_{i \in S} \sum_{m=1}^{l_i^S} \sum_{j \in S_i^m} \delta^{m-1} h_i^j (u_i^*, u_j^*) + \sum_{i \in S} \sum_{m=1}^{l_i} \sum_{j \in C_i^m} \delta^{m-1} h_i^j (u_i^*, u_j^*)$$

After substituting these expressions into (7.4) and moving all the components to the left, we get the following inequality:

$$v(S) - \sum_{i \in S} \alpha_i = \sum_{i \in S} \sum_{m=1}^{l_i} \sum_{j \in C_i^m} \delta^{m-1} h_i^j (u_i^*, u_j^*) \geq 0. \tag{7.5}$$

Because $h_i^j \geq 0$ then (7.5) holds and the lemma is proved. ∎

From the theorem, it follows that the game is convex and the core of this game is not empty, and the Shapley value belongs to the core.

7.4.2 Example

Consider the cooperative game with player set $N = \{1, 2, 3, 4, 5, 6\}$ $\delta = 0.5$ and hypergraph $\mathcal{H} = \{H_1 = \{1, 2, 3\}, H_2 = \{3, 4, 5\}, H_3 = \{5, 6\}\}$ which is shown on the Fig. 7.1:

In this example, we consider the case when each player plays bimatrix game with players which have a connection with him. All players have the same sets of strategies $u_i = \{A, B\}$. A is the strategy to choose the first row/column and B to choose the second.

Payoffs of players 1 and 2, 1 and 3, 1 and 4 are represented in the form of following 2×2 matrices:

$$\begin{pmatrix} (5, 1) & (8, 6) \\ (6, 5) & (2, 3) \end{pmatrix} \begin{pmatrix} (1, 0) & (4, 7) \\ (4, 1) & (6, 2) \end{pmatrix} \begin{pmatrix} (2, 0) & (4, 14) \\ (12, 6) & (20, 8) \end{pmatrix}.$$

Payoffs of players 1 and 5, 1 and 6, 2 and 3:

$$\begin{pmatrix} (0,2) & (6,16) \\ (6,4) & (4,12) \end{pmatrix} \begin{pmatrix} (24,32) & (24,20) \\ (36,8) & (20,16) \end{pmatrix} \begin{pmatrix} (9,3) & (5,1) \\ (1,6) & (2,1) \end{pmatrix}.$$

For players 2 and 4, 2 and 5, 2 and 6:

$$\begin{pmatrix} (10,2) & (4,0) \\ (10,14) & (10,12) \end{pmatrix} \begin{pmatrix} (0,8) & (12,2) \\ (8,12) & (4,6) \end{pmatrix} \begin{pmatrix} (24,4) & (16,8) \\ (20,4) & (8,16) \end{pmatrix}.$$

Payoffs of players 3 and 4, 3 and 5, 3 and 6:

$$\begin{pmatrix} (4,3) & (1,3) \\ (5,6) & (4,2) \end{pmatrix} \begin{pmatrix} (0,5) & (3,0) \\ (5,3) & (5,6) \end{pmatrix} \begin{pmatrix} (0,2) & (8,4) \\ (2,6) & (6,0) \end{pmatrix}.$$

Payoffs of players 4 and 5, 4 and 6, 5 and 6:

$$\begin{pmatrix} (0,2) & (2,2) \\ (5,0) & (6,5) \end{pmatrix} \begin{pmatrix} (2,2) & (12,8) \\ (12,4) & (2,6) \end{pmatrix} \begin{pmatrix} (3,5) & (2,3) \\ (3,4) & (3,6) \end{pmatrix}.$$

The sets N_i^k have the form:

$$N_1^1 = \{2,3\}, \ N_1^2 = \{4,5\}, \ N_1^3 = \{6\},$$

$$N_2^1 = \{1,3\}, \ N_2^2 = \{4,5\}, \ N_2^3 = \{6\},$$

$$N_3^1 = \{1,2,4,5\}, \ N_3^2 = \{6\},$$

$$N_4^1 = \{3,5\}, \ N_4^2 = \{1,2,6\},$$

$$N_5^1 = \{3,4,6\}, \ N_5^2 = \{1,2\},$$

$$N_6^1 = \{5\}, \ N_6^2 = \{3,4\}, \ N_6^3 = \{1,2\}.$$

The sum of the players' payoffs is equal to:

$$\sum_{i \in N} K_i(u_i^*, \dots, u_n^*) = \sum_{i \in N} \sum_{m=1}^{l_i} \sum_{j \in N_i^m} \delta^{m-1} h_i^j(u_i^*, u_j^*) =$$

$$= h_1^2(u_1^*, u_2^*) + h_1^3(u_1^*, u_3^*) + \delta h_1^4(u_1^*, u_4^*) + \delta h_1^5(u_1^*, u_5^*) + \delta^2 h_1^6(u_1^*, u_6^*) + h_2^1(u_2^*, u_1^*) +$$

$$+ h_2^3(u_2^*, u_3^*) + \delta h_2^4(u_2^*, u_4^*) + \delta h_2^5(u_2^*, u_5^*) + \delta^2 h_2^6(u_2^*, u_6^*) + h_3^1(u_3^*, u_1^*) + h_3^2(u_3^*, u_2^*) +$$

$$+h_3^4(u_3^*, u_4^*) + h_3^5(u_3^*, u_5^*) + \delta h_3^6(u_3^*, u_6^*) + h_4^3(u_4^*, u_3^*) + h_4^5(u_4^*, u_5^*) + \delta h_4^1(u_4^*, u_1^*) +$$

$$+\delta h_4^2(u_4^*, u_2^*) + \delta h_4^6(u_4^*, u_6^*) + h_5^3(u_5^*, u_3^*) + h_5^4(u_5^*, u_4^*) + h_5^6(u_5^*, u_6^*) + \delta h_5^1(u_5^*, u_1^*) +$$

$$+\delta h_5^2(u_5^*, u_2^*) + h_6^5(u_6^*, u_5^*) + \delta h_6^3(u_6^*, u_3^*) + \delta h_6^4(u_6^*, u_4^*) + \delta^2 h_6^1(u_6^*, u_1^*) + \delta^2 h_6^2(u_6^*, u_2^*).$$

The cooperative strategy profile is $u^* = \{A, B, B, B, B, A\}$ and the corresponding joint payoff is equal to 131.

$$h_1^2(A, B) = 8 \quad h_1^3(A, B) = 4 \quad \delta h_1^4(A, B) = 2 \quad \delta h_1^5(A, B) = 3$$

$$\delta^2 h_1^6(A, A) = 6 \quad h_2^1(B, A) = 6 \quad h_2^3(B, B) = 2 \quad \delta h_2^4(B, B) = 5$$

$$\delta h_2^5(B, B) = 2 \quad \delta^2 h_2^6(B, A) = 5 \quad h_3^1(B, A) = 7 \quad h_3^2(B, B) = 1$$

$$h_3^4(B, B) = 4 \quad h_3^5(B, B) = 5 \quad \delta h_3^6(B, A) = 1 \quad h_4^3(B, B) = 2$$

$$h_4^5(B, B) = 6 \quad \delta h_4^1(B, A) = 7 \quad \delta h_4^2(B, B) = 6 \quad \delta h_4^6(B, A) = 6$$

$$h_5^3(B, B) = 6 \quad h_5^4(B, B) = 5 \quad h_5^6(B, A) = 3 \quad \delta h_5^1(B, A) = 8$$

$$\delta h_5^2(B, B) = 3 \quad h_6^5(A, B) = 4 \quad \delta h_6^3(A, B) = 3 \quad \delta h_6^4(A, B) = 2$$

$$\delta^2 h_6^1(A, A) = 8 \quad \delta^2 h_6^2(A, B) = 1$$

The characteristic function has the form (see Table 7.1).
Consider the imputation $\hat{\alpha} = (\hat{\alpha}_1, \ldots, \hat{\alpha}_6)$.

$$\hat{\alpha}_1 = 23; \quad \hat{\alpha}_2 = 20; \quad \hat{\alpha}_3 = 18;$$

$$\hat{\alpha}_4 = 27; \quad \hat{\alpha}_5 = 25; \quad \hat{\alpha}_6 = 18.$$

From Table 7.1, it is seen that conditions (7.4) hold for all $S \subset N$ and the imputation $\hat{\alpha} = (\hat{\alpha}_1, \ldots, \hat{\alpha}_6)$ belongs to the core.
Compute now the Shapley value $\phi(v) = (\phi_1(v), \ldots, \phi_6(v))$ for this example

$$\phi_i(v) = \sum_{S \subseteq N, i \in S} \frac{(|S| - 1)!(|N| - |S|)!}{|N|!} (v(S) - v(S \setminus \{i\})).$$

We get:

$$\phi_1(v) = 22, 6; \quad \phi_2(v) = 15, 4(3); \quad \phi_3(v) = 33, 7(6)$$

$$\phi_4(v) = 17, 9(3); \quad \phi_5(v) = 28, 7(6); \quad \phi_6(v) = 12, 5.$$

Table 7.1 Values of characteristic function for all coalitions

S	v(S)	S	v(S)	S	v(S)	S	v(S)
Ø	0	3,4	6	2,3,4	20	1,3,4,5	59
1	0	3,5	11	2,3,5	19	1,3,4,6	26
2	0	3,6	0	2,3,6	3	1,3,5,6	56
3	0	4,5	11	2,4,5	11	1,4,5,6	26
4	0	4,6	0	2,4,6	0	2,3,4,5	47
5	0	5,6	7	2,5,6	7	2,3,4,6	20
6	0	1,2,3	28	3,4,5	28	2,3,5,6	36
1,2	14	1,2,4	14	3,4,6	6	2,4,5,6	26
1,3	11	1,2,5	14	3,5,6	22	3,4,5,6	47
1,4	0	1,2,6	14	4,5,6	26	1,2,3,4,5	92
1,5	0	1,3,4	26	1,2,3,4	54	1,2,3,4,6	54
1,6	0	1,3,5	31	1,2,3,5	55	1,2,3,5,6	86
2,3	3	1,3,6	11	1,2,3,6	28	1,2,4,5,6	40
2,4	0	1,4,5	11	1,2,4,5	25	1,3,4,5,6	92
2,5	0	1,4,6	0	1,2,4,6	14	2,3,4,5,6	72
2,6	0	1,5,6	7	1,2,5,6	21	1,2,3,4,5,6	131

7.5 Conclusion

In the paper games with hypergraph communication structure were considered. The discounted payoff function, which depends on the distance between players, was proposed. The coalition-building process was described. The characteristic function based on cooperative strategy profiles was introduced. This form of characteristic function substantially simplifies the computation of the characteristic function and based on this function, different solution concepts. It was proved that the newly introduced characteristic function is convex, which implies that the Shapley Value belongs to the core, and the core is not empty. An example illustrated the results with the Shapley Value as a solution.

Acknowledgments This work was supported by Russian Science Foundation under grants No.17-11-01079.

References

1. Myerson, R.B.: Graphs and cooperation in games. Math. Oper. Res. **2**, 225–229 (1977)
2. Owen, G.: Values of graph-restricted games. SIAM J. Algebraic Discrete Methods **7**, 210–220 (1986)
3. Meessen, R.: Communication games, Master's thesis. Department of Mathematics, University of Nijmegen, The Netherlands (1988, in Dutch)

4. van den Nouweland, A., Borm, P., Tijs, S.: Allocation rules for hypergraph communication situations. Int. J. Game Theory **20**, 255–268 (1992)
5. Shan, E., Zhang, G., Shan, X.: The degree value for games with communication structure. Int. J. Game Theory **47**, 857–871 (2018)
6. Von Neumann J., Morgenstern O.: Theory of Games and Economic Behavior. Princeton University, Princeton (1994)
7. Bulgakova, M.A., Petrosyan, L.A.: About one multistage non-antagonistic network game. In: Vestnik of Saint Petersburg University, Applied Mathematics. Computer Science. Control Processes, vol. 15(4), pp. 603–615 (2019, in Russian)

Chapter 8
Minimax Generalized Solutions
of Hamilton-Jacobi Equations
in Dynamic Bimatrix Games

Nikolay A. Krasovskii and Alexander M. Tarasyev

Abstract The paper deals with the dynamic bimatrix game whose dynamics describes the motion of flows corresponding to the control signals. A specific feature of the model is connected with increasing the dimension of the matrix game to 2×3 payoff matrices. This increase of the dimension is necessary as an initial step in solving the dynamic version of the well-known matrix game "rock-paper-scissors", which is an algorithmically complex problem. For the considered 2×3 dynamic bimatrix game the sets of acceptable situations for players are constructed in the static setting. Basing on these constructions, an algorithm is developed for finding the value function for the first player in the antagonistic setting of a differential game on an infinite time interval. The stability properties of the value function are verified using the conjugate derivatives apparatus in the framework of the theory of generalized minimax (viscosity) solutions of the Hamilton-Jacobi equations.

Keywords Dynamic bimatrix games · Value functions · Generalized minimax solutions of the Hamilton-Jacobi equations · Optimal control strategies

N. A. Krasovskii (✉)
Krasovskii Institute of Mathematics and Mechanics UrB RAS, Yekaterinburg, Russia
e-mail: n.a.krasovskii@imm.uran.ru

A. M. Tarasyev
Krasovskii Institute of Mathematics and Mechanics UrB RAS, Yekaterinburg, Russia

Ural Federal University named after the first President of Russia B.N. Yeltsin, Yekaterinburg, Russia
e-mail: tam@imm.uran.ru; a.m.tarasyev@urfu.ru

L. A. Petrosyan et al. (eds.), *Frontiers of Dynamic Games*, Static & Dynamic Game Theory: Foundations & Applications, https://doi.org/10.1007/978-3-030-51941-4_8

8.1   Introduction

The paper is devoted to the study of a dynamic bimatrix game, the dynamics of which is associated with evolutionary changes [3, 12–14], investment processes [8, 11], decomposition algorithms of the auction type [9], and Kolmogorov's differential equations for probabilities [5]. The novelty of this work is in increasing the dimension of the bimatrix game to 2×3 payoff matrices. Such increase in the dimension seriously complicates the construction of solutions in a dynamic bimatrix game. Besides, such a transition is a step towards the consideration of the well-known algorithmically complex game "rock-paper-scissors" [2], especially in the dynamic setting.

At the first step of solving the problem, acceptable situations of the players are constructed in a static setting. Saddle points in antagonistic games and Nash equilibrium points [20] are found on the prism of possible game situations.

At the second step, a differential game of the antagonistic type is considered on an infinite time horizon. The problem is to construct the value function of such a game as a generalized minimax (viscosity) solution of the Hamilton-Jacobi equations [1, 15–17]. For this purpose, on the basis of a generalized method of characteristics, we construct domains on the prism of possible situations in which the value function is described by smooth components. The continuous pasting of these smooth components is verified on the boundaries of the domains. Continuous pasting is not smooth on all boundaries of the domains. It is worth to remind that at smooth pasting points the value function satisfies the Hamilton-Jacobi equation. At the points of non-smooth pasting, it is necessary to check differential inequalities that provide stability properties of the value function. In the paper, such a verification is implemented basing on the apparatus of conjugated derivatives [16].

Let us note that basing on the value function, one can construct guaranteeing positional strategies [6, 7] of players and use them to generate equilibrium trajectories in a dynamic bimatrix game based on the approach proposed in [4] for obtaining Nash solutions in the dynamic setting. The papers [18, 19] are devoted to numerical methods and approximation schemes for the construction of attainability sets and value functions.

In the future studies, it is planned to construct equilibrium trajectories based on guaranteeing strategies generated by value functions as generalized minimax solutions of the Hamilton-Jacobi equations.

8.2   Players' Payoff Functions

The paper is devoted to the construction of the value function in the dynamic bimatrix game on the infinite time horizon $t \in [t_0, +\infty)$. In this section we consider a static statement of 2×3 bimatrix game for each period of time t. Let us assume, that payoffs of the first player is described by the matrix $A = a_{ij}$, and payoffs of the second player is described by the matrix $B = b_{ij}$:

$$A = \begin{pmatrix} a_{11} \ a_{12} \ a_{13} \\ a_{21} \ a_{22} \ a_{23} \end{pmatrix}, \quad B = \begin{pmatrix} b_{11} \ b_{12} \ b_{13} \\ b_{21} \ b_{22} \ b_{23} \end{pmatrix}.$$

The structure of matrices A and B means that the first player has two possible strategies and the second player has three possible strategies in the static game. Let us introduce notations. By the symbol x, $0 \leq x \leq 1$ we denote the probability that the first player chooses the first strategy and, respectively, $(1 - x)$ is the probability that he holds to the second strategy. The symbol y_1, $0 \leq y_1 \leq 1$ stands for the probability that the second player chooses the first strategy, the symbol y_2, $0 \leq y_2 \leq (1 - y_1)$ denotes the probability that the second player selects the second strategy, and, respectively $(1 - y_1 - y_2)$ is the probability that he plays the third strategy. Thus we consider the game on the rectangular prism of possible situations:

$$P = \{(x, y_1, y_2) : \ 0 \leq x \leq 1, \ 0 \leq y_1 \leq 1, \ 0 \leq y_2 \leq (1 - y_1)\}. \qquad (8.1)$$

The payoff functions of players in the time period t, $t \in [t_0, +\infty)$ is determined as mathematical expectations of payoffs, given by corresponding matrices A and B in the bimatrix game, and can be interpreted as "local" interest of players.

Payoff function of the first player is given as follows:

$$g_A(x, y_1, y_2) = XAY^T =$$

$$= (x, 1 - x) \times \begin{pmatrix} a_{11} \ a_{12} \ a_{13} \\ a_{21} \ a_{22} \ a_{23} \end{pmatrix} \times \begin{pmatrix} y_1 \\ y_2 \\ 1 - y_1 - y_2 \end{pmatrix} = \qquad (8.2)$$

$$= a_{11}xy_1 + a_{21}(1 - x)y_1 + a_{12}xy_2 + a_{22}(1 - x)y_2 +$$
$$+ a_{13}x(1 - y_1 - y_2) + a_{23}(1 - x)(1 - y_1 - y_2) =$$
$$= xy_1 C_A^1 + xy_2 C_A^2 - x\alpha_A - y_1\beta_A - y_2\gamma_A + a_{23},$$

where the game situation (x, y_1, y_2) depends on time period t, $x = x(t)$, $y_1 = y_1(t)$, $y_2 = y_2(t)$, and parameters C_A^1, C_A^2, α_A, β_A, γ_A are determined according to the classical theory of bimatrix games (see [20])

$$C_A^1 = a_{11} - a_{21} - a_{13} + a_{23};$$
$$C_A^2 = a_{12} - a_{22} - a_{13} + a_{23};$$
$$\alpha_A = a_{23} - a_{13};$$
$$\beta_A = a_{23} - a_{21};$$
$$\gamma_A = a_{23} - a_{22}.$$

Payoff function of the second player is determined analogously:

$$g_B(x, y_1, y_2) = XBY^T =$$

$$= (x, 1 - x) \times \begin{pmatrix} b_{11} & b_{12} & b_{13} \\ b_{21} & b_{22} & b_{23} \end{pmatrix} \times \begin{pmatrix} y_1 \\ y_2 \\ 1 - y_1 - y_2 \end{pmatrix} =$$

$$= xy_1 C_B^1 + xy_2 C_B^2 - x\alpha_B - y_1\beta_B - y_2\gamma_B + b_{23},$$

where

$$C_B^1 = b_{11} - b_{21} - b_{13} + b_{23};$$
$$C_A^2 = b_{12} - b_{22} - b_{13} + b_{23};$$
$$\alpha_B = b_{23} - b_{13};$$
$$\beta_B = b_{23} - b_{21};$$
$$\gamma_B = b_{23} - b_{22}.$$

8.3 Sets of Acceptable Situations for Players in Antagonistic Games

In this section we consider sets of acceptable situations for players in the antagonistic games. For this we analyze cases of maximum and minimum values of players' payoff functions on the prism P (8.1).

8.3.1 Antagonistic Game with Payoff Matrix A

For definiteness and convenience of graphical illustrations, we consider the following case for the matrix A:

$$C_A^1 > 0, \quad C_A^2 > 0, \quad \alpha_A > 0, \quad 0 < \frac{\alpha_A}{C_A^1} < 1, \quad 0 < \frac{\alpha_A}{C_A^2} < 1.$$

The set of acceptable situations for the first player in the antagonistic game with the matrix A is determined as follows:

$$\max_x \{x[y_1 C_A^1 + y_2 C_A^2 - \alpha_A]\} - y_1\beta_A - y_2\gamma_A + a_{23}.$$

Situations for the realization of the maximum are the following:

1. $(y_1 C_A^1 + y_2 C_A^2 - \alpha_A) > 0$:
   $0 \leq y_1 \leq 1, \quad 0 \leq y_2 \leq (1 - y_1), \quad x = 1.$

2. $(y_1 C_A^1 + y_2 C_A^2 - \alpha_A) = 0$:

$0 \le y_1 \le 1, \quad 0 \le y_2 \le (1 - y_1), \quad 0 \le x \le 1.$

3. $(y_1 C_A^1 + y_2 C_A^2 - \alpha_A) < 0$:

$0 \le y_1 \le 1, \quad 0 \le y_2 \le (1 - y_1), \quad x = 0.$

The set of acceptable situations for the second player in the antagonistic game with the matrix A is determined as follows:

$$\min_{(y_1, y_2)} \left\{ y_1 (C_A^1 x - \beta_A) + y_2 (C_A^2 x - \gamma_A) \right\} - x \alpha_A + a_{23}.$$

Situations for the realization of the minimum are the following:

1. $(C_A^1 x - \beta_A) > 0$ and $(C_A^2 x - \gamma_A) > 0$.

We obtain: $x > \frac{\beta_A}{C_A^1}$ and $x > \frac{\gamma_A}{C_A^2}$.

Then the minimum is reached under the following conditions:

$y_1 = 0, \quad y_2 = 0, \quad x > a = \max \left\{ \frac{\beta_A}{C_A^1}, \frac{\gamma_A}{C_A^2} \right\}.$

2. $(C_A^1 x - \beta_A) = 0$ and $(C_A^2 x - \gamma_A) > 0$.

We obtain: $x = \frac{\beta_A}{C_A^1}$ and $x > \frac{\gamma_A}{C_A^2}$. Let $\frac{\beta_A}{C_A^1} \ge \frac{\gamma_A}{C_A^2}$.

Then the minimum is reached under the following conditions:

$0 \le y_1 \le 1, \quad y_2 = 0, \quad x = \frac{\beta_A}{C_A^1}.$

3. $(C_A^1 x - \beta_A) < 0$ and $(C_A^2 x - \gamma_A) > 0$.

We obtain: $x < \frac{\beta_A}{C_A^1}$ and $x > \frac{\gamma_A}{C_A^2}$.

Then the minimum is reached under the following conditions:

$y_1 = 1, \quad y_2 = 0, \quad x \in \left(\frac{\gamma_A}{C_A^2}, \frac{\beta_A}{C_A^1} \right).$

4. $(C_A^1 x - \beta_A) < 0$ and $(C_A^2 x - \gamma_A) = 0$.

We obtain: $x < \frac{\beta_A}{C_A^1}$ and $x = \frac{\gamma_A}{C_A^2}$.

Then the minimum is reached under the following conditions:

$y_1 = 1, \quad y_2 = 0, \quad x = \frac{\gamma_A}{C_A^2}.$

5. $(C_A^1 x - \beta_A) < 0$ and $(C_A^2 x - \gamma_A) < 0$.

We obtain: $x < \frac{\beta_A}{C_A^1}$ and $x < \frac{\gamma_A}{C_A^2}$. For the definiteness, let us assume that $(C_A^1 x - \beta_A) < (C_A^2 x - \gamma_A)$. These inequalities are not burdensome and a change in the signs of these inequalities leads to similar calculations with a change in the signs of the parameters.

Then the minimum is reached under the following conditions:

$y_1 = 1, \quad y_2 = 0, \quad x < \frac{\gamma_A}{C_A^2}.$

On Fig. 8.1 we depict the planes of acceptable situations of the first player for the matrix A by the pink (shaded) color. The red (bold) polygon line with numbered segments represents the corresponding acceptable situations of the second player for the matrix A. The green circle on the polygon line has an abscissa $\frac{\gamma_A}{C_A^2}$. The saddle

Fig. 8.1 Antagonistic game
with payoff matrix A

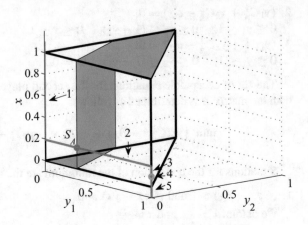

point S_A is located on the intersection of acceptable sets of players and is depicted
by the pink circle.

The saddle point S_A has the following coordinates:

$$x = \frac{\beta_A}{C_A^1}, \qquad y_1 = \frac{\alpha_A}{C_A^1}, \qquad y_2 = 0.$$

8.3.2 Antagonistic Game with Payoff Matrix B

Without loss of generality let us consider the following case for the matrix B:

$$C_B^1 < 0, \quad C_B^2 < 0, \quad \alpha_B < 0, \quad 0 < \frac{\alpha_B}{C_B^1} < 1, \quad 0 < \frac{\alpha_B}{C_B^2} < 1.$$

The set of acceptable situations for the first player in the antagonistic game with
the matrix B is determined as follows:

$$\min_x \left\{ x[y_1 C_B^1 + y_2 C_B^2 - \alpha_B] \right\} - y_1 \beta_B - y_2 \gamma_B + b_{23}.$$

Situations of the realization of the minimum are the following:

1. $(y_1 C_B^1 + y_2 C_B^2 - \alpha_B) > 0$:
   $0 \le y_1 \le 1, \quad 0 \le y_2 \le (1 - y_1), \quad x = 0.$
2. $(y_1 C_B^1 + y_2 C_B^2 - \alpha_B) = 0$:
   $0 \le y_1 \le 1, \quad 0 \le y_2 \le (1 - y_1), \quad 0 \le x \le 1.$
3. $(y_1 C_B^1 + y_2 C_B^2 - \alpha_B) < 0$:
   $0 \le y_1 \le 1, \quad 0 \le y_2 \le (1 - y_1), \quad x = 1.$

The set of acceptable situations for the second player in the antagonistic game with the matrix B is determined as follows:

$$\max_{(y_1, y_2)} \left\{ y_1(C_B^1 x - \beta_B) + y_2(C_B^2 x - \gamma_B) \right\} - x\alpha_B + b_{23}.$$

Situations of the realization of the maximum are the following:

1. $(C_B^1 x - \beta_B) > 0$ and $(C_B^2 x - \gamma_B) > 0$.

   We obtain: $x < \frac{\beta_B}{C_B^1}$ and $x < \frac{\gamma_B}{C_B^2}$. Let us assume for the definiteness that: $\frac{\beta_B}{C_B^1} < \frac{\gamma_B}{C_B^2}$. These inequalities are not burdensome and a change in the signs of these inequalities leads to similar calculations with a change in the signs of the parameters.

   We get three cases:

   1a. $(C_B^1 x - \beta_B) > (C_B^2 x - \gamma_B)$.

   Then the maximum is reached under the following conditions:
   $y_1 = 1, \quad y_2 = 0, \quad x < \frac{\beta_B}{C_B^1}$.

   1b. $(C_B^1 x - \beta_B) = (C_B^2 x - \gamma_B)$.

   Then the maximum is reached under the following conditions:
   $0 \leq y_1 \leq 1, \quad 0 \leq y_2 \leq (1 - y_1), \quad x < \frac{\beta_B}{C_B^1}$.

   1c. $(C_B^1 x - \beta_B) < (C_B^2 x - \gamma_B)$.

   Then the maximum is reached under the following conditions:
   $y_1 = 0, \quad y_2 = 1, \quad x < \frac{\beta_B}{C_B^1}$.

2. $(C_B^1 x - \beta_B) \leq 0$ and $(C_B^2 x - \gamma_B) > 0$.

   We obtain: $x \geq \frac{\beta_B}{C_B^1}$ and $x < \frac{\gamma_B}{C_B^2}$.

   Then the maximum is reached under the following conditions:
   $y_1 = 0, \quad y_2 = 1, \quad x \in \left(\frac{\beta_B}{C_B^1}, \frac{\gamma_B}{C_B^2} \right)$.

3. $(C_B^1 x - \beta_B) < 0$ and $(C_B^2 x - \gamma_B) = 0$.

   We obtain: $x > \frac{\beta_B}{C_B^1}$ and $x = \frac{\gamma_B}{C_B^2}$.

   Then the maximum is reached on the interval:
   $y_1 = 0, \quad 0 \leq y_2 \leq 1, \quad x = \frac{\gamma_B}{C_B^2}$.

4. $(C_B^1 x - \beta_B) < 0$ and $(C_B^2 x - \gamma_B) < 0$.

   We obtain: $x > \frac{\beta_B}{C_B^1}$ and $x > \frac{\gamma_B}{C_B^2}$.

   Then the maximum is reached under the following conditions:
   $y_1 = 0, \quad y_2 = 0, \quad x > \frac{\gamma_B}{C_B^2}$.

   The saddle point S_B has the following coordinates:

$$x = \frac{\gamma_B}{C_B^2}, \quad y_1 = 0, \quad y_2 = \frac{\alpha_B}{C_B^2}.$$

Fig. 8.2 Nash equilibrium

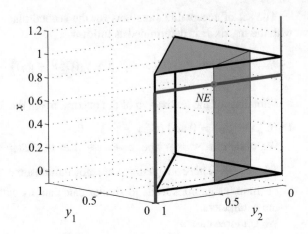

8.3.3 Nash Equilibrium in the Static Game

In order to construct the Nash equilibria in the static game we combine the acceptable situations of the first player in the game with the matrix A and the acceptable situations of the second player in the game with the matrix B. The intersection points of these acceptable sets generate the Nash equilibria. In the considered case we obtain the unique Nash equilibrium with coordinates:

$$x^N = \frac{\gamma_B}{C_B^2}, \quad y_1^N = 0, \quad y_2^N = \frac{\alpha_A}{C_A^2}.$$

On Fig. 8.2 we depict by the pink circle the Nash Equilibrium point N_E at the intersection of situations of realizations of maximum for both players.

8.4 Game Dynamics

In this section we consider game dynamics and construct Hamilton-Jacobi equation.

The system of differential equations is considered, which sets the dynamics of behavior of two players:

$$\begin{cases} \dot{x}(t) = -x(t) + u(t), & x(t_0) = x^0, \\ \dot{y}_1(t) = -y_1(t) + v_1(t), & y_1(t_0) = y_1^0, \\ \dot{y}_2(t) = -y_2(t) + v_2(t), & y_2(t_0) = y_2^0, \end{cases} \tag{8.3}$$

$$0 \le u(t) \le 1, \ 0 \le v_1(t) \le 1, \ 0 \le v_2(t) \le 1 - v_1(t).$$

Here the parameter $x = x(t)$, $0 \leq x \leq 1$ means the probability that the first player holds to the first strategy (respectively, $(1 - x)$ is the probability that he holds to the second strategy). Parameter $y_1 = y_1(t)$, $0 \leq y_1 \leq 1$ means the probability of choosing the first strategy by the second player, parameter $y_2 = y_2(t)$, $0 \leq y_2 \leq (1 - y_1)$ is the probability that he holds to the second strategy. And, respectively, $(1 - y_1 - y_2)$ is the probability that he plays the third strategy.

Control parameters $u = u(t)$, $v_1 = v_1(t)$ and $v_2 = v_2(t)$ satisfy conditions $0 \leq u \leq 1, 0 \leq v_1 \leq 1, 0 \leq v_2 \leq (1 - v_1)$ and can be interpreted as signals that recommend changing strategies by players.

For example, the value $u = 0$ corresponds to the signal for the first player: "change the first strategy to the second". The value $u = 1$ corresponds to the signal for the first player: "change the second strategy to the first". The value $u = x$ corresponds to the signal for the first player: "keep the previous strategy".

The value $v_1 = 1$ corresponds to the signal for the second player: "hold to the first strategy". The value $v_2 = 1$ corresponds to the signal for the second player: "hold to the second strategy". The value $(1 - v_1 - v_2) = 1$ corresponds to the signal for the second player "hold to the third strategy".

Let us note, that the basis for the dynamics (8.3) and its properties were considered in [12]. In this dynamics Kolmogorov's differential equations are generalized under the assumption that the coefficients of incoming and outgoing flows are not set a priori and can be constructed in the controlled process on the feedback principle.

"Global" interests J_A^∞ of the first player are determined as multivalued (two-digit) functions formed by lower and upper limits of "local" payoffs (8.2)

$$
\begin{aligned}
J_A^\infty &= [J_A^-, J_A^+], \\
J_A^- &= J_A^-(x(\cdot), y_1(\cdot), y_2(\cdot)) = \liminf_{t \to \infty} g_A(x(t), y_1(t), y_2(t)), \\
J_A^+ &= J_A^+(x(\cdot), y_1(\cdot), y_2(\cdot)) = \limsup_{t \to \infty} g_A(x(t), y_1(t), y_2(t)),
\end{aligned}
\tag{8.4}
$$

calculated for the trajectories $(x(\cdot), y_1(\cdot), y_2(\cdot))$ of the system (8.3). For the second player "global" interests J_B^∞ are determined symmetrically.

8.5 The Value Function

In the paper [11] it is proved, that in the game with functionals (8.4), despite the multi-valuedness of these functionals, there exists an single-valued value function $w_A = w_A(x, y_1, y_2)$, which for each initial position $x = x^0$, $y_1 = y_1^0$, $y_2 = y_2^0$ sets the minimax (maximin) value of functionals (8.4), determined in the class of positional strategies of players [6, 7].

Let us note, that the function w_A at points of smoothness should satisfy the Hamilton-Jacobi equation:

$$-\frac{\partial w_A}{\partial x}x - \frac{\partial w_A}{\partial y_1}y_1 - \frac{\partial w_A}{\partial y_2}y_2+$$
$$+\max\left\{0, \frac{\partial w_A}{\partial x}\right\} + \min\left\{0, \frac{\partial w_A}{\partial y_1}, \frac{\partial w_A}{\partial y_2}\right\} = 0. \tag{8.5}$$

The Hamilton-Jacobi equation (8.5) is associated with the Hamiltonian of the following form:

$$H(x, y_1, y_2, s_1, s_2, s_3) = -s_1x - s_2y_1 - s_3y_2 + \max\{0, s_1\} + \min\{0, s_2, s_3\}. \tag{8.6}$$

Here vector $s = (s_1, s_2, s_3) \in R^3$ is a conjugate vector, which at points of differentiability of the value function coincides with the gradient vector:

$$s_1 = \frac{\partial w_A}{\partial x}, \quad s_2 = \frac{\partial w_A}{\partial y_1}, \quad s_3 = \frac{\partial w_A}{\partial y_2}. \tag{8.7}$$

In addition, the property of u-stability [6, 7] should be fulfilled for all initial values (x, y_1, y_2):

$$w_A(x, y_1, y_2) \le g_A(x, y_1, y_2), \ 0 \le x \le 1, \ 0 \le y_1 \le 1, \ 0 \le y_2 \le (1 - y_1). \tag{8.8}$$

The property of v-stability is valid for the points (x, y_1, y_2), that satisfy the strict inequality:

$$w_A(x, y_1, y_2) < g_A(x, y_1, y_2).$$

Let us remind, that the properties of u-stability and v-stability at points of nondifferentiability of the value function can be effectively tested by means of the conjugate derivatives technique [16].

Let us provide the definitions of conjugate derivatives. The definition of the upper conjugate derivative is presented as follows:

$$D^*w_A(x, y_1, y_2)|(s) = \sup_{h \in R^3} \left(\langle s, h \rangle - \partial_- w_A(x, y_1, y_2)|(h)\right), \ s = (s_1, s_2, s_3) \in R^3,$$

where the derivative in the direction $h = (h_1, h_2, h_3)$ is given by the formula:

$$\partial_- w_A(x, y_1, y_2)|(h_1, h_2, h_3) =$$
$$= \liminf_{\delta \downarrow 0} \frac{w_A(x + \delta h_1, y_1 + \delta h_2, y_2 + \delta h_3) - w_A(x, y_1, y_2)}{\delta}.$$

The definition of the lower conjugate derivative is given as follows:

$$D_* w_A(x, y_1, y_2)|(s) = \inf_{h \in R^3} \left(\langle s, h \rangle - \partial_+ w_A(x, y_1, y_2)|(h) \right), \quad s = (s_1, s_2, s_3) \in R^3,$$
(8.9)

where the derivative in the direction $h = (h_1, h_2, h_3)$ is given by the formula:

$$\partial_+ w_A(x, y_1, y_2)|(h_1, h_2, h_3) =$$
$$= \limsup_{\delta \downarrow 0} \frac{w_A(x + \delta h_1, y_1 + \delta h_2, y_2 + \delta h_3) - w_A(x, y_1, y_2)}{\delta}.$$
(8.10)

The property of u-stability at all interior points of the prism (8.1) is expressed by the inequality:

$$D_* w_A(x, y_1, y_2)|(s) \leq H(x, y_1, y_2, s),$$
$$\left(0 < x < 1, \ 0 < y_1 < 1, \ 0 < y_2 < (1 - y_1), \ s = (s_1, s_2, s_3) \in R^3 \right),$$
(8.11)

and the property of v-stability at interior points of the prism (8.1), where the value function is strictly less than the payoff function, is determined by the inequality:

$$D^* w_A(x, y_1, y_2)|(s) \geq H(x, y_1, y_2, s),$$
$$\left(0 < x < 1, \ 0 < y_1 < 1, \ 0 < y_2 < (1 - y_1), \ w_A(x, y_1, y_2) < g_A(x, y_1, y_2), \right.$$
$$\left. s = (s_1, s_2, s_3) \in R^3 \right).$$
(8.12)

8.6 Smooth Components of the Value Function

In this section, we provide the description for the structure of the value function, which is presented by four components of smooth functions, that satisfy Hamilton-Jacobi equation (8.5). The structure of the value function presumes, that smooth components are pasted together continuously on the junction surfaces. Wherein, such pasting can be smooth, and then the Hamilton-Jacobi equation is fulfilled automatically at the junction points. The non-smooth (but continuous) pasting of the components is also possible, and in that case, it is necessary to verify the properties of u- and v-stability, for example, in the form of the inequalities for the conjugate derivatives [16].

Smooth components are calculated basing on the generalized method of characteristics [11, 15], and its construction is implemented by the following scheme.

Step 1. On the first step, we construct smooth components of the value function with fixed time moment T of the game termination. For that, the value function is determined along the characteristics of Hamilton-Jacobi equations, which are straight lines directed to the vertices of the prism of possible game situations and are generated by marginal values of controls u, v_1, and v_2.

Step 2. On the second step, continuous pasting of these smooth components is implemented, and the conditions of u- and v-stability are checked at the points of nonsmoothness on the basis of the technique of differential inequalities for Hamilton-Jacobi equations [15, 16].

Step 3. On the third step, we calculate lower envelopes of these terminal value functions by the time parameter T of the game termination. To this end, we calculate derivatives of these functions by the time parameter T, equate these derivatives to zero, exclude parameter T, and obtain stationary smooth components for the value function on infinite horizon.

Step 4. On the fourth step, we check that these smooth components satisfy the stationary Hamilton-Jacobi equation.

Step 5. On the fifth step, using smooth components we paste the continuous function, for which we verify properties of u- and v-stability basing on the techniques of conjugate derivatives [16].

A more detailed description of the algorithm in the 2×2-dimensional case is provided in the papers [10, 11], and here this algorithm is generalized for the 2×3-dimensional case.

As a result of the described calculations, we obtain the following four smooth functions, as the elements of the "mosaic":

$$\psi_A^1 = a_{23} - \frac{\left(C_A^1 \alpha_A x + \beta_A (C_A^1 y_1 + C_A^2 y_2)\right)^2}{4(C_A^1)^2 x (C_A^1 y_1 + C_A^2 y_2)},$$

$$\psi_A^2 = a_{11} - \frac{\left((C_A^1 - \alpha_A)(1 - x) + (C_A^1 - \beta_A)\left(1 - y_1 - \frac{C_A^2}{C_A^1} y_2\right)\right)^2}{4 C_A^1 (1 - x)\left(1 - y_1 - \frac{C_A^2}{C_A^1} y_2\right)},$$

$$\psi_A^3 = x y_1 C_A^1 + x y_2 C_A^2 - x \alpha_A - y_1 \beta_A - y_2 \gamma_A + a_{23},$$

$$\psi_A^4 = \frac{a_{23} C_A^1 - \alpha_A \beta_A}{C_A^1} = v_A.$$

Smooth functions ψ_A^i, $i = 1, \ldots, 4$ are pasted together along the next surfaces K_A^j, $j = 1, \ldots, 5$.

$$K_A^1 = \{0 \leq x \leq 1, \ 0 \leq y_1 \leq 1, \ 0 \leq y_2 \leq (1 - y_1), \ x = \xi_A^1(y_1, y_2) = \frac{\beta_A}{C_A^1}\},$$

$$K_A^2 = \{0 \leq x \leq 1, \ 0 \leq y_1 \leq 1, \ 0 \leq y_2 \leq (1 - y_1),$$

$$x = \xi_A^2(y_1, y_2) = \frac{\beta_A}{\alpha_A} y_1 + \frac{C_A^2 \beta_A}{C_A^1 \alpha_A} y_2, \ C_A^1 y_1 + C_A^2 y_2 - \alpha_A \geq 0\},$$

$$K_A^3 = \{0 \leq x \leq 1, \ 0 \leq y_1 \leq 1, \ 0 \leq y_2 \leq (1 - y_1),$$

$$x = \xi_A^3(y_1, y_2) = \frac{(C_A^1 - \beta_A)}{(C_A^1 - \alpha_A)} y_1 + \frac{C_A^2(C_A^1 - \beta_A)}{C_A^1(C_A^1 - \alpha_A)} y_2 - \frac{(\alpha_A - \beta_A)}{(C_A^1 - \alpha_A)},$$

$$C_A^1 y_1 + C_A^2 y_2 - \alpha_A \leq 0\}.$$

On the surface K_A^1 two smooth components ψ_A^3 and ψ_A^4 are merged. The surface K_A^2 connects two smooth components ψ_A^1 and ψ_A^4. And two smooth components ψ_A^2 and ψ_A^4 are joined on the surface K_A^3.

Let us note, that continuity of junction for surfaces K_A^1, K_A^2, K_A^3 of the corresponding components can be checked directly.

On the surface K_A^4 we have the junction of two smooth components ψ_A^1 and ψ_A^3:

$$K_A^4 = \{0 \leq x \leq 1, \ 0 \leq y_1 \leq 1, \ 0 \leq y_2 \leq (1 - y_1),$$

$$x = \xi_A^4(y_1, y_2) = \frac{\beta_A(C_A^1 y_1 + C_A^2 y_2)}{C_A^1(2(C_A^1 y_1 + C_A^2 y_2) - \alpha_A)},$$

$$C_A^1 y_1 + C_A^2 y_2 - \alpha_A \leq 0\}.$$

In order to obtain the continuous junction on the surface K_A^4 it is necessary to introduce the additional condition, which is not restrictive:

$$\left(\beta_A \frac{C_A^2}{C_A^1} - \gamma_A\right) = 0. \tag{8.13}$$

On the last fifth surface K_A^5 the smooth components ψ_A^2 and ψ_A^3 are pasted together continuously:

$$K_A^5 = \{0 \leq x \leq 1, \ 0 \leq y_1 \leq 1, \ 0 \leq y_2 \leq (1 - y_1),$$

$$x = \xi_A^5(y_1, y_2) = 1 - \frac{(C_A^1 - \beta_A)(C_A^1 - (C_A^1 y_1 + C_A^2 y_2))}{C_A^1(C_A^1 + \alpha_A - 2(C_A^1 y_1 + C_A^2 y_2))},$$

$$C_A^1 y_1 + C_A^2 y_2 - \alpha_A \geq 0\}.$$

Let us turn to the analytical description of the value function w_A.

In the considered case the value function $(x, y_1, y_2) \rightarrow w_A(x, y_1, y_2)$ is determined as follows:

$$w_A(x, y_1, y_2) = \psi_A^i(x, y_1, y_2), \quad \text{if} \ (x, y_1, y_2) \in E_A^i, \quad i = 1, \ldots, 4.$$
$$\tag{8.14}$$

Here domains E_A^i, $i = 1, \ldots, 4$, are determined as follows:

$$E_A^{11} = \{0 \le x \le 1,\ 0 \le y_1 \le 1,\ 0 \le y_2 \le (1 - y_1),$$
$$\xi_A^4(y_1, y_2) \le x \le 1,\ C_A^1 y_1 + C_A^2 y_2 - \alpha_A \le 0\},$$
$$E_A^{12} = \{0 \le x \le 1,\ 0 \le y_1 \le 1,\ 0 \le y_2 \le (1 - y_1),$$
$$\xi_A^2(y_1, y_2) \le x \le 1,\ C_A^1 y_1 + C_A^2 y_2 - \alpha_A \ge 0\},$$
$$E_A^1 = E_A^{11} \cup E_A^{12},$$
$$E_A^{21} = \{0 \le x \le 1,\ 0 \le y_1 \le 1,\ 0 \le y_2 \le (1 - y_1),$$
$$0 \le x \le \xi_A^3(y_1, y_2),\ C_A^1 y_1 + C_A^2 y_2 - \alpha_A \le 0\},$$
$$E_A^{22} = \{0 \le x \le 1,\ 0 \le y_1 \le 1,\ 0 \le y_2 \le (1 - y_1),$$
$$0 \le x \le \xi_A^5(y_1, y_2),\ C_A^1 y_1 + C_A^2 y_2 - \alpha_A \ge 0\},$$
$$E_A^2 = E_A^{21} \cup E_A^{22},$$
$$E_A^{31} = \{0 \le x \le 1,\ 0 \le y_1 \le 1,\ 0 \le y_2 \le (1 - y_1),$$
$$\xi_A^1(y_1, y_2) \le x \le \xi_A^4(y_1, y_2),\ C_A^1 y_1 + C_A^2 y_2 - \alpha_A \le 0\},$$
$$E_A^{32} = \{0 \le x \le 1,\ 0 \le y_1 \le 1,\ 0 \le y_2 \le (1 - y_1),$$
$$\xi_A^5(y_1, y_2) \le x \le \xi_A^1(y_1, y_2),\ C_A^1 y_1 + C_A^2 y_2 - \alpha_A \ge 0\},$$
$$E_A^3 = E_A^{31} \cup E_A^{32},$$
$$E_A^{41} = \{0 \le x \le 1,\ 0 \le y_1 \le 1,\ 0 \le y_2 \le (1 - y_1),$$
$$\xi_A^1(y_1, y_2) \le x \le \xi_A^2(y_1, y_2),\ C_A^1 y_1 + C_A^2 y_2 - \alpha_A \ge 0\},$$
$$E_A^{42} = \{0 \le x \le 1,\ 0 \le y_1 \le 1,\ 0 \le y_2 \le (1 - y_1),$$
$$\xi_A^3(y_1, y_2) \le x \le \xi_A^1(y_1, y_2),\ C_A^1 y_1 + C_A^2 y_2 - \alpha_A \le 0\},$$
$$E_A^4 = E_A^{41} \cup E_A^{42}. \tag{8.15}$$

8.7 Feedback Strategy of the First Player

The feedback (positional) strategy of the first player is generated by the structure of the value function w_A. Let us note that the feedback strategy has a discontinuous character on the surfaces K_A^2 and K_A^3 on which one can observe the change of control signals. More precisely, the structure of the feedback strategy looks like:

$$u = u(x, y_1, y_2) = \begin{cases} 1, & \text{if } (x, y_1, y_2) \in E_A^2 \cup E_A^{32} \cup E_A^{41}, \\ 0, & \text{if } (x, y_1, y_2) \in E_A^1 \cup E_A^{31} \cup E_A^{42}. \end{cases} \tag{8.16}$$

8.8 Properties of u- and v-Stability

In this section, we prove that for the function w_A the necessary and sufficient conditions are fulfilled. Namely, we need to prove the validity of the differential inequalities (8.11) and (8.12) and the boundedness condition (8.8). These necessary and sufficient conditions imply that the function w_A is the generalized minimax solution of the Hamilton-Jacobi equation (8.5), and, hence, coincides with the value function.

Let us start with the boundedness condition. It is obviously fulfilled since the functions ψ_A^i, $i = 1, \ldots, 4$ are the lower envelopes of the terminal solution $w_1(T, t, x, y_1, y_2)$ and, hence,

$$\psi_A^i(x, y_1, y_2) \le g_A(x, y_1, y_2), \quad i = 1, \cdots, 4,$$
$$0 \le x \le 1, \ 0 \le y_1 \le 1, \ 0 \le y_2 \le (1 - y_1).$$

Let us check, that differential inequalities (8.11), (8.12) are fulfilled for the function w_A. It is not difficult to prove, that functions $\psi_A^i, i = 1, 2, 4$, satisfy the Hamilton-Jacobi equation (8.5) at internal points of the domains $E_A^i, i = 1, 2, 4$.

As to the function ψ_A^3, one can verify that it coincides with the boundary function g_A and, therefore, for it is necessary to check only the inequality, which expresses the property of u-stability:

$$-\frac{\partial \psi_A^3}{\partial x} x - \frac{\partial \psi_A^3}{\partial y_1} y_1 - \frac{\partial \psi_A^3}{\partial y_2} y_2 +$$
$$+ \max \left\{ 0, \frac{\partial \psi_A^3}{\partial x} \right\} + \min \left\{ 0, \frac{\partial \psi_A^3}{\partial y_1}, \frac{\partial \psi_A^3}{\partial y_2} \right\} \ge 0$$

at the internal points of the domain E_A^3. This inequality is verified by direct calculations for the function ψ_A^3.

It remains to check the differential inequalities (8.11), (8.12) on the surfaces K_A^j, $j = 1, \ldots, 5$.

At the points of the surface K_A^2 the functions ψ_A^1 and ψ_A^4 are continuously pasted together. Let us show that this pasting is also smooth.

Let us calculate partial derivatives of these functions:

$$\frac{\partial \psi_A^1}{\partial x} = \frac{(\beta_A^2(C_A^1 y_1 + C_A^2 y_2)^2 - (C_A^1 \alpha_A x)^2)}{4(C_A^1 x)^2(C_A^1 y_1 + C_A^2 y_2)},$$

$$\frac{\partial \psi_A^1}{\partial y_1} = \frac{((C_A^1 \alpha_A x)^2 - \beta_A^2(C_A^1 y_1 + C_A^2 y_2)^2)}{4 C_A^1 x (C_A^1 y_1 + C_A^2 y_2)^2},$$

$$\frac{\partial \psi_A^1}{\partial y_2} = \frac{C_A^2((C_A^1 \alpha_A x)^2 - \beta_A^2(C_A^1 y_1 + C_A^2 y_2)^2)}{4(C_A^1)^2 x(C_A^1 y_1 + C_A^2 y_2)^2},$$

$$\frac{\partial \psi_A^4}{\partial x} = 0, \qquad \frac{\partial \psi_A^4}{\partial y_1} = 0, \qquad \frac{\partial \psi_A^4}{\partial y_2} = 0.$$

One can see that these derivatives are equal to zero on the surface K_A^2:

$$\frac{\partial \psi_A^1}{\partial x} |_{K_A^2} = \frac{\partial \psi_A^4}{\partial x} |_{K_A^2} = 0, \quad \frac{\partial \psi_A^1}{\partial y_1} |_{K_A^2} = \frac{\partial \psi_A^4}{\partial y_1} |_{K_A^2} = 0, \quad \frac{\partial \psi_A^1}{\partial y_2} |_{K_A^2} = \frac{\partial \psi_A^4}{\partial y_2} |_{K_A^2} = 0.$$

In other words, functions ψ_A^1 and ψ_A^4 are smoothly pasted together here.

Analogously one can prove smooth pasting of the functions ψ_A^2 and ψ_A^4 on the surface K_A^3.

Let us consider the surface K_A^4, where the functions ψ_A^1 and ψ_A^3 are merged. We have already calculated values of partial derivatives $\partial \psi_A^1/\partial x$, $\partial \psi_A^1/\partial y_1$, $\partial \psi_A^1/\partial y_2$. Let us estimate partial derivatives $\partial \psi_A^3/\partial x$, $\partial \psi_A^3/\partial y_1$, $\partial \psi_A^3/\partial y_2$:

$$\frac{\partial \psi_A^3}{\partial x} = C_A^1 y_1 + C_A^2 y_2 - \alpha_A,$$

$$\frac{\partial \psi_A^3}{\partial y_1} = C_A^1 x - \beta_A,$$

$$\frac{\partial \psi_A^3}{\partial y_2} = C_A^2 x - \gamma_A.$$

One can verify that pasting is smooth here. Really, for partial derivatives, calculated on the surface K_A^4 we have:

$$\frac{\partial \psi_A^1}{\partial x} |_{K_A^4} = \frac{\partial \psi_A^3}{\partial x} |_{K_A^4} = C_A^1 y_1 + C_A^2 y_2 - \alpha_A,$$

$$\frac{\partial \psi_A^1}{\partial y_1} |_{K_A^4} = \frac{\partial \psi_A^3}{\partial y_1} |_{K_A^4} = \frac{\beta_A(C_A^1 y_1 + C_A^2 y_2 - \alpha_A)}{(\alpha_A - 2(C_A^1 y_1 + C_A^2 y_2))},$$

$$\frac{\partial \psi_A^1}{\partial y_2} |_{K_A^4} = \frac{\partial \psi_A^3}{\partial y_2} |_{K_A^4} = \frac{C_A^2 \beta_A(C_A^1 y_1 + C_A^2 y_2 - \alpha_A)}{C_A^1(\alpha_A - 2(C_A^1 y_1 + C_A^2 y_2))}.$$

Analogously one can check the smoothness of the function w_A on the surface K_A^5.

Along the last surface K_A^1 functions ψ_A^3 and ψ_A^4 are pasted together. We have already calculated partial derivatives of these functions. Values of these derivatives on the surface K_A^1 are determined by relations:

$$\frac{\partial \psi_A^3}{\partial x}|_{K_A^1} = C_A^1 y_1 + C_A^2 y_2 - \alpha_A, \qquad \frac{\partial \psi_A^4}{\partial x}|_{K_A^1} = 0,$$

$$\frac{\partial \psi_A^3}{\partial y_1}|_{K_A^1} = (C_A^1 x - \beta_A)|_{K_A^1} = \frac{\partial \psi_A^4}{\partial y_1}|_{K_A^1} = 0,$$

$$\frac{\partial \psi_A^3}{\partial y_2}|_{K_A^1} = (C_A^2 x - \gamma_A)|_{K_A^1} = \frac{C_A^2}{C_A^1}\beta_A - \gamma_A = \frac{\partial \psi_A^4}{\partial y_2}|_{K_A^1} = 0.$$

The last equality is fulfilled under the condition (8.13).

Let us note that in the case of fulfillment of relations $w_A = \psi_A^3 = \psi_A^4 = g_A$ on the surface K_A^1 it is necessary to check only the u-stability condition (8.11). The pasting on this surface is continuous but nonsmooth. In some neighbourhood of the surface K_A^1 this pasting is based on the minimum operation for the functions ψ_A^3 and ψ_A^4:

$$w_A(x, y_1, y_2) = \min\{\psi_A^3(x, y_1, y_2), \psi_A^4(x, y_1, y_2)\}.$$

For the calculation of directional derivatives for the function w_A at the points $(x, y_1, y_2) \in K_A^1$ we obtain relations (see (8.9), (8.10)), which are generated by the minimum operation of two linear functions of directions h:

$$\partial w_A(x, y_1, y_2)|(h_1, h_2, h_3) =$$
$$= \min\{(C_A^1 y_1 + C_A^2 y_2 - \alpha_A) \cdot h_1 + 0 \cdot h_2 + 0 \cdot h_3, \ 0 \cdot h_1 + 0 \cdot h_2 + 0 \cdot h_3\} =$$
$$= \min\{(C_A^1 y_1 + C_A^2 y_2 - \alpha_A) \cdot h_1, \ 0\}, \quad h = (h_1, h_2, h_3) \in R^3.$$

The lower conjugate derivative of the function w_A at points $(x, y_1, y_2) \in K_A^1$ in directions $s = (s_1, s_2, s_3) \in R^3$ is determined by the following formula:

$$D_* w_A(x, y_1, y_2)|(s_1, s_2, s_3) =$$

$$= \begin{cases} 0, & \\ \quad \text{if } s_1 = \lambda(C_A^1 y_1 + C_A^2 y_2 - \alpha_A) + (1 - \lambda) \cdot 0 \text{ and } s_2 = 0, \ s_3 = 0, & (8.17) \\ -\infty, & \\ \quad \text{otherwise}, & \end{cases}$$

here the parameter λ, $0 \leq \lambda \leq 1$, serves for the construction of the convex hull of gradient vectors for the functions ψ_A^3 and ψ_A^4.

Taking into account relations (8.6) and (8.7) for the Hamiltonian and the conjugate vectors, we obtain that for the points $(x, y_1, y_2) \in K_A^1$ and conjugate vectors $s = (s_1, s_2, s_3)$, $s_1 = \lambda(C_A^1 y_1 + C_A^2 y_2 - \alpha_A)$, $s_2 = 0$, $s_3 = 0$ the Hamiltonian $H(x, y_1, y_2, s_1, s_2, s_3)$ is determined by the following formula:

$$H(x, y_1, y_2, s_1, 0, 0) = -s_1 x + \max\{0, s_1\} =$$
$$= \begin{cases} -s_1 x, & \text{if } s_1 \leq 0, \\ s_1(1 - x), & \text{otherwise.} \end{cases} \tag{8.18}$$

It is obvious that for these values the Hamiltonian (8.18) does not exceed the lower conjugate derivative (8.17). Hence, the property of u-stability expressed by the inequality (8.11) on the surface K_A^1 is proved.

The verification of conditions (8.11), (8.12) for the function w_A at each point of the prism (8.1) of the acceptable situations is completed. Thus, we have proved that the function w_A (8.14)–(8.15) satisfies the properties of u- and v-stability in the form of conjugate derivatives for the Hamilton-Jacobi equations and, hence, it is the value function of the game with functional (8.4) on the infinite horizon.

8.9 Example of the Value Function Construction

Let us consider as an example payoff matrices of two players. We present matrices A, B and their main "game" parameters in the following form:

$$A = \begin{pmatrix} 100 & 58 & 5 \\ 18 & 28 & 40 \end{pmatrix}, \qquad B = \begin{pmatrix} -50 & 30 & 40 \\ 100 & 55 & 0 \end{pmatrix},$$

$$C_A^1 = a_{11} - a_{21} - a_{13} + a_{23} = 117,$$
$$C_A^2 = a_{12} - a_{22} - a_{13} + a_{23} = 65,$$
$$\alpha_A = a_{23} - a_{13} = 35,$$
$$\beta_A = a_{23} - a_{21} = 22,$$
$$\gamma_A = a_{23} - a_{22} = 12,$$
$$x_A = \frac{\beta_A}{C_A^1} = 0.19, \quad y_1^A = \frac{\alpha_A}{C_A^1} = 0.3, \quad y_2^A = \frac{\alpha_A}{C_A^2} = 0.54.$$

Let us note, that the condition (8.13) is valid for the matrix A.
Saddle point S_A has the following coordinates:

$$S_A = (x_A; y_1^A; 0) = (\frac{\beta_A}{C_A^1}; \frac{\alpha_A}{C_A^1}; 0) = (0.19; 0.3; 0).$$

$$C_B^1 = b_{11} - b_{21} - b_{13} + b_{23} = -190;$$
$$C_B^2 = b_{12} - b_{22} - b_{13} + b_{23} = -65;$$
$$\alpha_B = b_{23} - b_{13} = -40;$$
$$\beta_B = b_{23} - b_{21} = -100;$$
$$\gamma_B = b_{23} - b_{22} = -55;$$
$$x_B = \frac{\gamma_B}{C_B^2} = 0.85; \quad y_1^B = \frac{\alpha_B}{C_B^1} = 0.21; \quad y_2^B = \frac{\alpha_B}{C_B^2} = 0.62.$$

Saddle point S_B has the following coordinates:

$$S_B(x_B; 0; y_2^B) = (\frac{\gamma_B}{C_B^2}; 0; \frac{\alpha_B}{C_B^2}) = (0.85; 0; 0.62).$$

The point of the static Nash equilibrium N_E has the following coordinates:

$$N_E(x_B; 0; y_2^A) = (\frac{\gamma_B}{C_B^2}; 0; \frac{\alpha_A}{C_A^2}) = (0.85; 0; 0.54).$$

On Fig. 8.3 we present surfaces of pasting of smooth components for the value function in the dynamic game with the matrix A. The surface K_A^4 is depicted in dark blue color and the surface K_A^5 is depicted in dark red. The surface K_A^2 is shown in violet. The surface K_A^3 is presented in orange. The set of acceptable situation of the first player is given in grey color. The set of acceptable situation for the second player in the antagonistic game with the matrix A is presented by the red polygon line. The saddle point S_A lies on the intersection of the set of acceptable situations of players for the matrix A.

All surfaces which determine the components of the value function are generated from the saddle point S_A. The feedback strategy of the first player in the dynamic game is also connected with the saddle point S_A. It has the discontinuous character,

Fig. 8.3 Value function for the matrix A

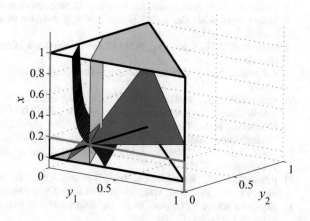

which is given by the switching surfaces K_A^2, K_A^3 for control signals according to the structure (8.16) generated by the value function (8.14)–(8.15).

8.10 Conclusion

For the considered 2×3 dynamic bimatrix game we construct the sets of acceptable situations for players in the static setting. Basing on these constructions, we develop an algorithm for finding the value function for the first player in the antagonistic setting of a differential game on an infinite time interval. The stability properties of the value function are verified using the conjugate derivatives apparatus in the framework of the theory of generalized minimax (viscosity) solutions of Hamilton-Jacobi equations. The illustration example is presented for construction of the value function in the 2×3 dynamic bimatrix game. The next step would be the development of the proposed methodology for dynamic bimatrix games of higher dimensions, i.e. for a dynamic version of the well-known matrix game "rock-paper-scissors".

Acknowledgments The first author, Nikolay A. Krasovskii, is supported by the Russian Science Foundation (Project No. 19-11-00105).

References

1. Crandall, M.G., Lions, P.L.: Viscosity solutions of Hamilton-Jacobi equations. Trans. Am. Math. Soc. **277**(4), 1–42 (1983)
2. Hofbauer, J., Bednarik, P.: Discretized best-response dynamics for the rock-paper-scissors game. J. Dyn. Games **4**(1), 75–86 (2017). http://dx.doi.org/10.3934/jdg.2017005
3. Hofbauer, J., Sigmund, K.: The Theory of Evolution and Dynamic Systems. Cambridge Univ. Press, Cambridge (1988)
4. Kleimenov, A.F.: Nonantagonistic Positional Differential Games. Nauka, Yekaterinburg (1993)
5. Kolmogorov, A.N.: On analytical methods in probability theory. Uspekhi Matematicheskih Nauk **5**, 5–41 (1938)
6. Krasovskii, A.N., Krasovskii, N.N.: Control Under Lack of Information. Birkhauser, Boston etc. (1995)
7. Krasovskii, N.N., Subbotin, A.I.: Game-Theoretical Control Problems. Springer, New York (1988)
8. Krasovskii, A.A., Taras'ev, A.M.: Dynamic optimization of investments in the economic growth models. Autom. Remote Control **68**(10), 1765–1777 (2007)
9. Krasovskii, N.A., Tarasyev, A.M.: Decomposition algorithm of searching equilibria in a dynamic game. Math. Game Theory Appl. **3**(4), 49–88 (2011)
10. Krasovskii, N.A., Tarasyev, A.M.: Equilibrium Solutions in Dynamical Games. UrGAU, Yekaterinburg (2015)
11. Krasovskii, N.A., Kryazhimskiy, A.V., Tarasyev, A.M.: Hamilton–Jacobi equations in evolutionary games. Proc. Inst. Math. Mech. UrB RAS **20**(3), 114–131 (2014)
12. Kryazhimskii, A.V., Osipov, Yu.S.: On differential–evolutionary games. Proc. Steklov Inst. Math. **211**, 234–261 (1995)

13. Mazalov, V.V., Rettieva, A.N.: Asymmetry in a cooperative bioresource management problem. Large Scale Syst. Control **55**, 280–325 (2015)
14. Petrosjan, L.A., Zenkevich, N.A.: Conditions for sustainable cooperation. Autom. Remote Control **76**(10), 1984–1904 (2015). http://dx.doi.org/10.1134/S0005117915100148
15. Subbotin, A.I.: Generalized Solutions of First Order PDEs. The Dynamical Optimization Perspective. Birkhauser, Boston (1995)
16. Subbotin, A.I., Tarasyev, A.M.: Conjugate derivatives of the value function of a differetnial game. Doklady AN SSSR **283**(3), 559–564 (1985)
17. Subbotin, A.I., Tarasyev, A.M.: Stability properties of the value function of a differential game and viscosity solutions of Hamilton-Jacobi equations. Probl. Control Inf. Theory **15**, 451–463 (1986)
18. Taras'ev, A.M.: Approximation schemes for constructing minimax solutions of Hamilton-Jacobi equations. J. Appl. Math. Mech. **58**(2), 207–221 (1994) http://dx.doi.org/10.1016/0021-8928(94)90049-3
19. Ushakov, V.N., Lebedev, P.D.: Algorithms for the construction of an optimal cover for sets in three-dimensional Euclidean space. Proc. Steklov Inst. Math. **293**(1), 225–237 (2016) http://dx.doi.org/10.1134/S0081543816050205
20. Vorobyev, N.N.: Game Theory for Economists and System Scientists. Nauka, Moscow (1985)

Chapter 9
Analysis of Economic Behaviour in Evolutionary Model of Tax Control Under Information Diffusion

Suriya Kumacheva, Elena Gubar, Ekaterina Zhitkova, and Galina Tomilina

Abstract Nowadays many economic fields are dependent on information flows. Information serves as a strategic tool in economics, business and many social processes. In particular, spreading information among taxpayers can be used as a control parameter in tax control. The probability of auditing which encourages tax payments used to be considered as a unique tool to stimulate tax collection. The current study represents a combined approach where tools of evolutionary game theory and network modeling are applied to the analysis of agents' economic behavior. Information of possible tax auditing is disseminated across the population of taxpayers and is supposed to be a main factor influencing their decision on whether to evade or not. According to the previous research, interactions and dissemination of information or rumors among taxpayers in long-term period can be formulated as an evolutionary process. It is assumed that agents tend to spread information or rumors over their own contact network of neighbors and colleagues rather than over randomly chosen agents. Thus, the network model of social interaction is constructed. Information spreading in the network of various topology (e.g. grid, random network, etc.) is considered as an evolutionary process where agents' behavior is described by the stochastic imitation dynamics and their interaction is described by different modifications of the bimatrix games which generate evolutionary dynamics. Scenario analysis is supported by the series of experiments. Numerous simulations help visualize the process of information spreading across different types of network, imitation protocols and players payoffs. The results show that information flow helps encourage tax payments in the population.

Keywords Evolutionary games · Tax control · Information diffusion · Economic behavior · Network modeling

S. Kumacheva (✉) · E. Gubar · E. Zhitkova · G. Tomilina
Saint Petersburg State University, St. Petersburg, Russia
e-mail: s.kumacheva@spbu.ru; e.gubar@spbu.ru; e.zhitkova@spbu.ru; g.tomilina@yandex.ru

L. A. Petrosyan et al. (eds.), *Frontiers of Dynamic Games*,
Static & Dynamic Game Theory: Foundations & Applications,
https://doi.org/10.1007/978-3-030-51941-4_9

9.1   Introduction

Nowadays various types of tax noncompliance [1–3, 7, 24, 27, 32–34] represent a very significant problem for many economies and fiscal policies. Over the last decades mathematical science has been solving this problem using different approaches and taking into consideration the related fields of knowledge. Mathematical modeling considers the problem of assessing tax risks and modeling tax control from a position of institutional economics, decision and game theory, probabilistic modeling, methods of modern applied statistics, data mining and data analysis methods and many others. Following the tradition of combining different approaches, this work has been designed as the intersection of game-theoretic approach for principal-agent models, evolutionary dynamics and network modeling.

Earlier mathematical models of tax control had a number of common features. Among those features there is a hierarchical structure. The games that imply such models are often based on the "principle-agent" scheme, which considers risk neutral agents as taxpayers. Usually risk neutral agents do not possess such characteristics as conscientiousness and honesty, or, conversely, an excessive tendency to risk and, possibly, fraud, and they make rational decisions taking into account the economic envelopment. In many mathematical models of tax control, the solution of the problem is always formulated in the form of some optimality criteria. For example, in [24] the solution is mentioned as the agent's optimal reporting rule. In [7] and [27] similar results were titled as an optimal contract or an optimal scheme. In [32–34] the solution was formulated in the form of the threshold rule.

However, the considered mathematical models have some natural limitations. In practice, tax authorities have a strong limitation of the budget. Thus, the application of the majority of so-called "ideal" solutions to real-life cases is a very expensive procedure. Due to this fact, reaching the optimal proportion of the audited taxpayers is extremely rarely achieved. This fact can be considered as Problem One which will be studied (discussed) further. This problem encourages tax authority to find additional ways to stimulate taxpayers to fair payments subject to the limitation of resources. As Problem Two we choose the complete information in various mathematical models. However, the information sources or mechanism from which taxpayers receive information about upcoming tax audits are not usually discussed. The formulation of both problems leads to the fact that information about future tax audits can be disseminated among taxpayers, encouraging them to avoid of high penalties and pay taxes in accordance with their actual income level.

The network modeling is one of the actively developing approaches for simulating and visualizing the dynamic processes in social trends. To investigate tax noncompliance, this technique was successfully used in [1, 14]. In the current work the information spreading process plays a major role in changing the ratio of tax evaders and non-evaders in society. Various papers [9, 15, 22] consider the phenomenon of information dissemination as an epidemic or evolutionary processes. According to the previous research, we considered information spreading in terms of epidemic processes [12] by applying the modification of Susceptible-Infected-

Susceptible (SIS) and Susceptible-Infected-Recovered (SIR) models. They have been transformed further into the opinion dynamic models by using the tools of the Markov process [13]. In [10, 11] this problem was considered as a combination of network modeling and evolutionary game theory. The current work modifies all previous results and generalizes the recent approach.

We consider a large but finite population of taxpayers (hereafter—economic agents) structured as an indirect network. Unlike the models, which consider taxpayers with different propensity to risk [13, 18–20], the current study focuses solely on the behavior of risk neutral agents, since they have the greatest impact on the final state of the system. Generally, taxpayers with fixed propensity to risk, such as risk-avoiding and risk-loving, behave the way determined by their risk-status and rarely change their opinion. Whereas the risk-neutral taxpayers are independent and rational in their choices, they are inclined to change the taxation strategy according to the economic environment. To analyze how economic behavior of rational taxpayers changes depending on the received information, we conduct imitation analysis applied to the network models with different topology (e.g. grid, random network, etc.). The social interaction is defined by imitation dynamics [6, 28, 35], which are based on the classes of the bimatrix games with known structures, i.e. the Prisoner's Dilemma, the Stag Hunt game, the Hawk-Dove game. We consider a set of scenarios describing the decision making process with different initial data, imitation rules and definitions of payoff functions.

A series of repeated experiments were carried out to analyze the economic behavior of agents in the simulated networks. The specially designed software runs 10^2 times to evaluate the statistical stability of the received stochastic results for analyzing the tendencies in taxpayers behavior.

The paper has the following structure. Section 9.1 represents the overview of static and dynamic models. Section 9.2 contains behavioral models of risk-neutral agents. Section 9.3 discusses the network model of annual process of tax auditing. Section 9.4 contains numerical simulations and their results. Section 9.5 concludes the paper.

9.2 The Model of Risk-Neutral Agents

Based on the previous research, we assume that the majority of agents consists of the risk-neutral taxpayers. The assumption of the rationality of agents behavior arises from [5] which can be considered as the basic model for the current paper. The usage of two possible levels of income ξ considered—H (high) and L (low)—among the members of population leads to the simplification of the computation. Generally, the gradation of the taxpayers' income can be expanded, see [4, 17] or replaced by a continuous variable as in [5, 27, 34]. This approach has been applied in [7, 33] and such simplification does not impact on the essence of the results. Thereby, we consider a model where $\xi, \eta \in \{L, H\}$, where $0 < L < H, \eta \leq \xi$, here ξ is a true income and η is a declared income. Thus, the total taxable population of size

n consists of two subgroups defined by numbers n_H and n_L, which are satisfied to condition $n_L + n_H = n$.

The assessment of efficiency of the fiscal system can be evaluated by its total revenue. First, it is necessary to analyze possible behavior and the corresponding profit function of taxpayers. Only three options are possible:

$$u\,(L(L)) = (1 - \theta) \cdot L; \tag{9.1}$$

$$u\,(H(H)) = (1 - \theta) \cdot H; \tag{9.2}$$

$$u\,(L(H)) = H - \theta L - P_L(\theta + \pi)(H - L), \tag{9.3}$$

where P_L is the probability of audit agents, who declared $\eta = L$, constants θ and π are tax and penalty rates correspondingly, c is the unit cost of auditing.

Following the results of [5], which are similar to the "Threshold rule" obtained for the class of taxation models, we formulate a theorem which will be used further as one of the main prerequisites for the current study.

Theorem 9.1 *The optimal value P^* of the auditing probability is*

$$P^* = \frac{\theta}{\theta + \pi}. \tag{9.4}$$

For each risk neutral taxpayer the optimal strategy is

$$\eta^*(\xi) = \begin{cases} H, & P_L \geq P^*; \\ L, & P_L < P^*. \end{cases} \tag{9.5}$$

The proof of this theorem was also given in [5].

Considering a classical taxation model, where an information about possible tax auditing is excluded, we assume that the total taxable population evades due to their risk neutrality (according to the Theorem 9.1). From (9.5) it follows that our assumption corresponds to the pessimistic case (where the tax auditing is absent and everyone evades) represented in [5]. In this case only agents with true income level L pay and therefore the total tax revenue TTR_0 is

$$TTR_0 = n_L \theta L + n_H \,(\theta\, L + P_L\,(\theta + \pi)(H - L)) - n\, P_L\, c. \tag{9.6}$$

Previously, in Sect. 9.1 the necessity of elimination the problem of incomplete information in the game-theoretical models and method of stimulation the honest payments in the population was discussed. The following scheme was offered, if the model admits a possibility of disseminating information over the taxpayers then an injection of information occurs in the form "$P_L \geq P^*$", referring to the optimistic case of equality (9.5). It means that at the initial time moment the number of taxpayers $n_{inf}^0 = n_{nev}^0$ was informed about the grows of probability of tax auditing. Informed taxpayers become spreaders in population and according to their rationality, formalized in the Theorem 9.1, they decide to remain honest taxpayers.

In each time moment $t \in [0, T]$ the relation between the number of evaders n_{ev} and payers n_{nev} will change continuously:

$$n_H = n_{ev}(t) + n_{nev}(t),$$

or, which is the same, $v_{nev}(t) + v_{ev}(t) = 1$ (where v_{ev}, v_{nev} are the portions of evaders and payers correspondingly).

Thus, the system has reached a steady state at time moment T, since the information has flown through the population. The corresponding total tax revenue TTR_T is defined as follows:

$$TTR_T = n_L \theta L + n_H \left(v_{nev}^T \theta H + v_{ev}^T (\theta L + P_L(\theta + \pi)(H - L)) \right) - n(P_L c + v_{inf}^0 c_{inf}). \tag{9.7}$$

Variable v_{nev}^T defines the portion of honest taxpayers at the moment $t = T$, v_{ev}^T corresponds with the portion of evaders at the moment $t = T$, value v_{inf}^0 is the fraction of the initially informed agents at initial time moment $t_0 = 0$ ($v_{inf}^0 = v_{inf}(t_0)$) and c_{inf} is the unit cost of the initial injection. It is assumed that information spreading is more cost-effective than the audit: $c_{inf} << c$.

9.3   Network Model

In the previous research, the process of information spreading about the upcoming tax audit in a taxpayer population was formulated in terms of the epidemic models [12] or the Markov process [13], and also as the simplest evolutionary processes [10, 11, 19]. Based on [20], this paper contributes to the study of evolutionary processes in the network extending the previous results. In this paragraph we formulate some basic assumptions of the network evolutionary model.

Generally, a population of agents with different behaviors can be divided into several subpopulations according to a number of behavioral types. As opposed to the original formulation of the evolutionary game, where evolution of the population is described through the set of random meetings in the well-mixed population, here we suppose that the social connections of each taxpayer can be represented by networks of different topology [20, 22, 25]. Therefore, the inter-agent interactions are only feasible among the connected taxpayers. Hence, the evolution process occurs over the links between nodes in the network. The evolution process describes the transformation of Uninformed taxpayers into Informed ones according to the stochastic imitation rules [6, 21, 28, 35]. Following the recent papers [13, 19], the current study discusses three methods of selecting neighbors in terms of the imitation dynamics [6]. The first method describes a random choice: the agent who receives information is randomly selected from a set of agents connected with the active one. The second rule is formulated as a principle of the most influential neighbor: information is spread by the agents with the largest number of direct connections with their neighbors. If there are several agents with an equally large numbers of contacts, the choice between them is made at random. The third method is based on the neighbor with the highest income: information is spread by the agents with the highest income.

Thus, we consider the indirect network (N, K), where $N = \{1, \ldots, n_H\}$ is a set of economic agents with high level of income and $K \subset N \times N$ is an edge set (each edge in K represents two-players symmetric game Γ between the connected taxpayers). It is assumed that the taxpayers choose strategies from a binary set $X = \{ev, nev\}$ and receive payoffs according to the 2×2 matrix of payoffs. Each instant time moment agents use a single strategy against all opponents, and thus the games occur simultaneously. The strategy state is $x(t) = (x_1(t), \ldots, x_{n_H}(t))^T$, where $x_i(t) \in X$ is a strategy of taxpayer, who use strategy i, $i = \overline{1, n_H}$, at time moment t. Aggregated payoff of agent i will be defined as in [25, 26]:

$$u_i = \omega_i \sum_{j \in M_i} a_{x_i(t),\, x_j(t)}, \tag{9.8}$$

where $a_{x_i(t),\, x_j(t)}$ is a component of payoff matrix, $M_i := \{j \in L : \{i, j\} \in K\}$ is a set of neighbors for taxpayer i, weighted coefficient $\omega_i = 1$ for cumulative payoffs and $\omega_i = \frac{1}{|M_i|}$ for average payoffs. Vector of payoffs of the entire population is $u(t) = (u_1(t), \ldots, u_{n_H}(t))^T$. In current paper the payoff matrix has one of the structures represented further.

9.3.1 Payoff Matrices

The instant interactions between taxpayers are defined through the two-players symmetric bimatrix game $\Gamma(A, B)$, where a payoff matrix of the first player is A and a payoff matrix of the second player is symmetric $B = A^T$ [28, 35]. In this particular case each taxpayer can choose between two behaviors $X = \{ev, nev\}$, where nev is the behavior "not evade", ev is the behavior "to evade". Following [20] we continue to use the payoff matrix with the known structure to simplify the prediction of taxpayers behavior. Here we use the modification of tree classical bimatrix games such as the Prisoner's Dilemma, the Stag Hunt game, the Hawk-Dove game. Since the structure of these bimatrix games is well-known as well as the structure of its equilibrium sets, then they help estimate the impact of the network structure and the imitation rules.

Modifying the Prisoner's Dilemma game [23] we obtain the following payoff matrix:

	nev	ev
nev	$\bar{u} + SW, (\bar{u} + SW)$	$(\bar{u} - SW, u(L(H)))$
ev	$(u(L(H)), \bar{u} - SW)$	$(\bar{u}, \bar{u})$

where nev is the strategy "to cooperate", which in our case corresponds to the behavior "to pay taxes" or "not evade", ev is the strategy "to defect" and describes the behavior "to evade". Payoff $\bar{u} = 1/2u(L(L)) + 1/2u(H(H))$ defines the average profit of a "mean" agent, variable SW reflects a social welfare, obtained in the participation in social consolidation. The payoff matrix defines all possible meetings between taxpayers with different attitude to tax payments (i.e. the meeting between honest taxpayers or two evaders, etc.).

Hereafter in the next two games we follow the same technique.

The modified version of the Stag Hunt game [30] is represented below:

	nev	ev
nev	$(\bar{u} + SW, \bar{u} + SW)$	$(0, \bar{u} - SW)$
ev	$(\bar{u} - SW, 0)$	$(\bar{u} - SW, \bar{u} - SW)$

Here the strategy nev, which is corresponding to social strategy "to hunt a stag" in classical form of the game, means "to pay taxes/not evade" in our framework, similarly the strategy ev, which describes the individual behavior "to hunt a hare" in original game, in our interpretation recommends taxpayer "to evade".

Finally, the following matrix is the modified case of the Hawk-Dove game:

	ev	nev
ev	$\left(\dfrac{u(L(H)) - (\theta + \pi)(H - L)}{2}, \dfrac{u(L(H)) - (\theta + \pi)(H - L)}{2} \right)$	$(\bar{u} + SW, 0)$
nev	$(0, \bar{u} + SW)$	$\left(\dfrac{\bar{u} + SW}{2}, \dfrac{\bar{u} + SW}{2} \right)$

The strategy ev, which originally corresponds "to be a Hawk" and dictates to an agent the usage of an aggressive behavior, in our case, leads taxpayer "to evade". The strategy nev fits to the behavior of a Dove and imposes to remain passive, that in our case can be considered as the behavior "to pay taxes/not evade". Additionally, we assume that the condition $u(L(H)) << (\theta + \pi)(H - L)$ should be satisfied and it works for the large values of the parameters θ and π or if the value of the difference $(H - L)$ is large.

9.3.2 Behavioral Dynamics

In evolutionary game agents of population adopt their behavior to the better performing strategies. Many studies, for example, [8] and [29], focus on the dynamics with a strategy updated rules, which dictate agents to choose from their self-inclusive neighborhood an exemplary player who received the greatest payoff. If this payoff is not greater than the payoff of the updating player, then the player keeps his/her own strategy.

Population changes according to the rule, which is a function of the strategies and payoffs of neighboring agents:

$$x_i(t+1) = f(\{x_j(t), u_j(t) : j \in M_i \cup \{i\}\}), \ t > 0. \tag{9.9}$$

This rule corresponds to a strategy of each player at time $t + 1$, which depends on its neighborhood players' information at the moment t, including their strategies and payoffs.

Remark 9.1 The assumption for the update rule is that it is payoff monotonic [35]. It means that if there exists a better performing strategy in the neighborhood of agent i, then the agent will switch to that strategy with a probability $\sigma > 0$.

This rule appoints a method of adaptation agents behavior to the changes in his/her environment, which means that taxpayer can switch on the another strategy if at least one neighbor has the better payoff. As the basis of the dynamics we use the proportional imitation rule [28, 29, 35], in which each agent chooses a neighbor randomly and if this neighbor receives a higher payoff by using a different strategy, then the agent will switch with a probability proportional to the payoff difference. In [25] the proportional imitation rule was presented as

$$p\left(x_i(t+1) = x_j(t)\right) := \left[\frac{\lambda}{|M_i|}(u_j(t) - u_i(t))\right]_0^1 \tag{9.10}$$

for each agent $i \in K$, where $j \in M_i$ is a uniformly randomly chosen neighbor, $\lambda > 0$ is an arbitrary rate constant, and the notation $[z]_0^1$ indicates $\max(0, \min(1, z))$.

Remark 9.2 Imitation rule can be considered as a revision protocol, which is a map $\rho \colon R^n \times X \to R_+^{n \times n}$. The scalar $\rho_{ij}(u_i(x), x)$ is called the conditional switch rate from strategy $i \in X$ to strategy $j \in X$ given payoff vector $u(x)$ and population state x.

This proportional imitation rule is a widely studied model with important property that the dynamics in well-mixed population can be approximated by the replicator dynamics.

Thereby we can determine the following term:

Definition 9.1 A network evolutionary game is a system $((N, K), \Gamma)$ which includes

- a network (graph) (N, K);
- a game G such that if $(i, j) \in K$, then i and j play Γ iteratively with strategies $x_i(t)$ and $x_j(t)$ respectively

and takes into account an information based strategy updating rule, which can be expressed as (9.9).

In contrast to previous research [6, 28, 35], in the current work, we define three modifications of imitation rules, each of them is based on either a choice made by an exemplary agent, or taxpayers' payoffs, etc. The main assumption of the current study is the application of these imitation rules only to the subgroup of risk-neutral taxpayers, as far as, in our hypothesis, this group is the most influenceable on the tax collections in the entire population. Proportions of risk-loving and risk-averse taxpayers are fixed at the initial time moment.

The strategy of player i at time $t + 1$, $x_i(t + 1)$, is selected as the best strategy from strategies of neighborhood players $j \in N_i$ at time t according to the chosen imitation rule.

- **Rule 1.** *A random neighbor.* When a taxpayer i receives an opportunity to revise his/her strategy then he/she chooses an exemplary agent at random with equal probability to all connected neighbors.
- **Rule 2.** *Neighbor with the highest payoff.* When agent i receives an opportunity to revise his/her strategy then he/she considers current payoffs of all taxpayers and chooses an agent (or a set of agents) with maximum payoff. If there are several agents with the maximum payoff then an exemplary agent is chosen at random among this subset.
- **Rule 3.** *The most influenceable neighbor.* Firstly, taxpayer i estimates a number of connections of all the nearest neighbors and selects an exemplary agent from the set of agents with the maximum number of links. If there are several agents with the maximum links, then an opponent is selected at each iteration of the dynamic process at random among the subset of the most influenceable agents.

9.4 Numerical Simulations and Experimental Results

In this section we represent the numerical simulations which illustrate the scenario analysis of taxpayers behavior. Each series of experiments corresponds to the specific network configuration such as grid, strongly and weakly connected random graphs or different imitation rules or the instant games. Each agent evaluates his/her own profit, which is determined by the matrix of the instant game and

Table 9.1 Two modeled groups and average income

Group	Income interval (rub./month)	Average income (rub.)	Proportion of population (%)
L	Less 25,000	$L = 12,500$	51
H	More 25,000	$H = 50,000$	49

information about the neighboring agents' choice, following to the next calculating algorithms:

- Cumulative: the computed sum of the profits from each interaction;
- Average: the sum of the profits from interactions divided by their number.

The values of parameters are fixed throughout experiments: a tax rate is $\theta = 13\%$, penalty rate is $\pi = 13\%$, optimal value of the probability of audit is $P^* = 0.5$, actual value of the probability of audit for agents who declared L is $P_L = 0.1$, unit cost of auditing is $c = 7455$ (rub.), as a unit cost of information injection we consider $c_{inf} = 10\%c = 745.5$ (rub.).

Two possible taxpayers' income levels L and H are calculated as the mean values of the uniform and Pareto distributions [16] by using the empirical data based on the distribution of the income among the population of Russian Federation in 2018 [31]. The proportions of the entire population (see Table 9.1) correspond with the results from [19].

The stopping point of the iteration process can be defined by the condition

$$\sqrt{\sum_{i=1}^{n}(x_i(t) - x_i(t+1))} \leq 10^{-2},$$ where $x_i(t)$ is the i-th agent's profit at iteration t

for each $t \in [0, T]$.

To illustrate the application of the model of information spreading, we run different experiments 10^2 times for each initial distribution of honest taxpayers and evaders. This allows us to discuss the statistically stable trends received after conducting a scenario analysis of the taxpayers' behavior, taking into account their rationality. These initial distributions are represented by the networks with different topology and are examined in the series of numerical simulations.

The network consists of 25 nodes, which correspond to the number of the taxpayers with high level of income $n_H = 25$, and helps to visualize the evolutionary process and use to illustrate the main results. The specially designed software allows to simulate a process with a larger number of nodes and receives data for statistical analysis. For this modeled population different agents' strategies are depicted as colored dots—agents who use the strategy "to pay taxes" are drown as yellow dots, and agents who use the strategy "to evade" are drown as blue dots correspondingly. We study the dynamic process in the population using the following modifications of the network:

- strongly connected network, where the probability of link formation is 1/10;
- weakly connected network, where the probability of link formation is 1/3;
- grid.

Fig. 9.1 Series 1.
Example 9.1. The Prisoner's
Dilemma; initial state:
$(n_{nev}, n_{ev}) = (5, 20)$

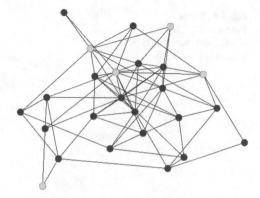

Fig. 9.2 Series 1.
Example 9.1. The Prisoner's
Dilemma; final state:
$(n_{nev}, n_{ev}) = (0, 25)$

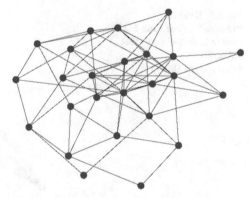

Fig. 9.3 Series 1.
Example 9.2. The
Hawk-Dove game; initial
state: $(n_{nev}, n_{ev}) = (5, 20)$

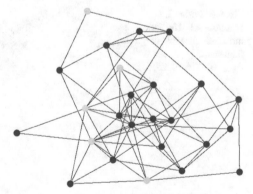

Each experiment is designed for the same number of nodes. Thus, according to the
formula (9.6) the initial value of total tax revenue $TTR_0 = 69219.75$ is constant in
each of the examples.

Figures 9.1, 9.2, 9.3, 9.4, 9.5, 9.6, 9.7, 9.8, 9.9, 9.10, 9.11, 9.12, 9.13, 9.14
demonstrate the evolution of the proportions of taxable population during the
iteration process.

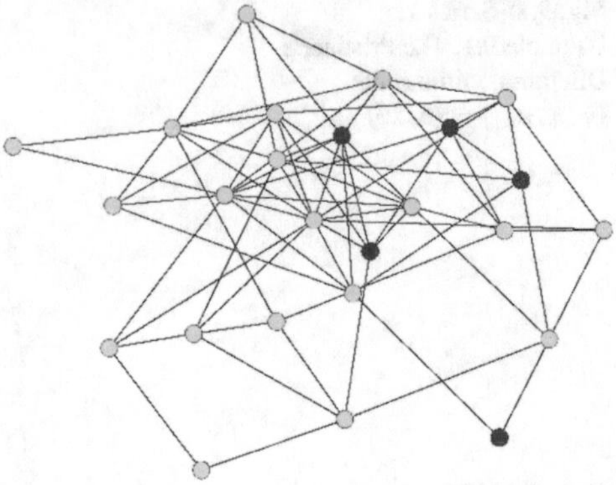

Fig. 9.4 Series 1.
Example 9.2. The
Hawk-Dove game; final state:
$(n_{nev}, n_{ev}) = (20, 5)$

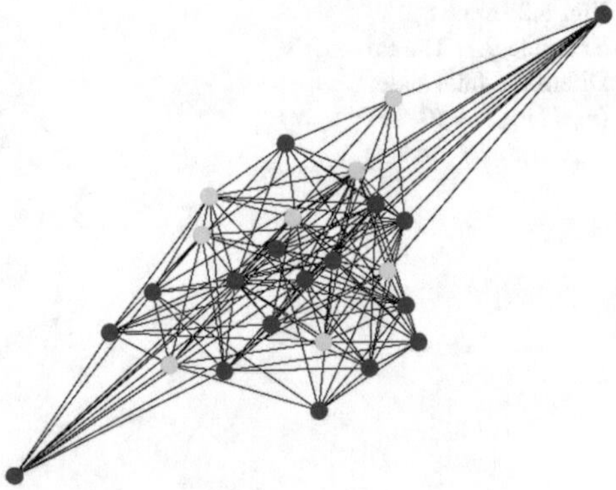

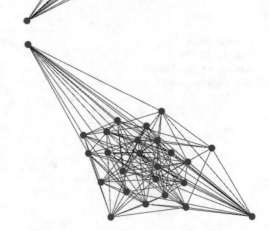

Fig. 9.5 Series 2.
Example 9.1. The Rule 3. The
most influenceable neighbor;
initial state:
$(n_{nev}, n_{ev}) = (8, 17)$

Fig. 9.6 Series 2.
Example 9.1. The Rule 3. The
most influenceable neighbor;
final state:
$(n_{nev}, n_{ev}) = (0, 25)$

9.4.1 Series 1

Here the series of numerical simulation implemented with the following combination of parameters: the network topology is a weakly connected graph and the
imitation rule is a random neighbor. Figures 9.1, 9.2, 9.3, 9.4 represent the initial
and final states of the system subject to an average method of computing of the
agent's profit. Initial state of system is $(n_{nev}, n_{ev}) = (5, 20)$.

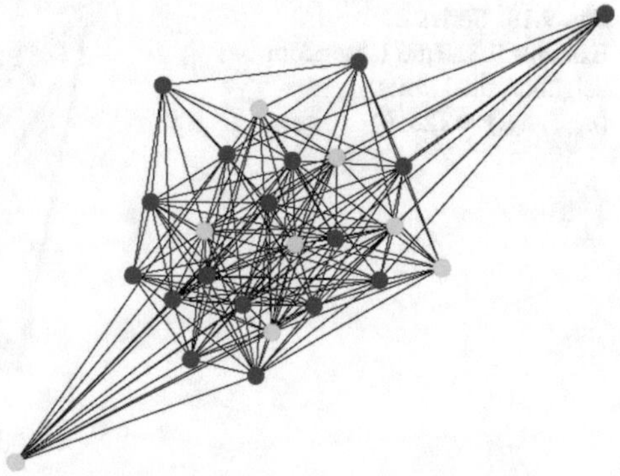

Fig. 9.7 Series 2.
Example 9.2. Rule 2.
Neighbor with the highest
payoff; initial state:
$(n_{nev}, n_{ev}) = (8, 17)$

Fig. 9.8 Series 2.
Example 9.2. Rule 2.
Neighbor with the highest
payoff; final state:
$(n_{nev}, n_{ev}) = (25, 0)$

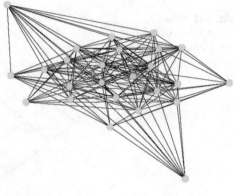

Fig. 9.9 Series 2.
Example 9.3. Rule 1. Random
neighbor; initial state:
$(n_{nev}, n_{ev}) = (8, 17)$

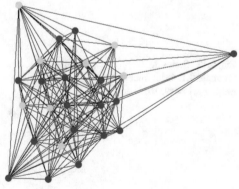

Example 9.1 This example uses the Prisoner's Dilemma as an instant.

The system reaches its steady state at the moment $t = T$ and the final value of the total tax revenue computed by the formula (9.6) is $TTR_T = 65518.25$ rubles. Comparing this value with the initial value TTR_0 we can see that in this experiment the total tax revenue decreases.This result does not contradict the previous research as far as the result of the information spreading over the population is the decreasing of the portion of non-evaders.

Fig. 9.10 Series 2.
Example 9.3. Rule 1. Random
neighbor; final state:
$(n_{nev}, n_{ev}) = (25, 0)$

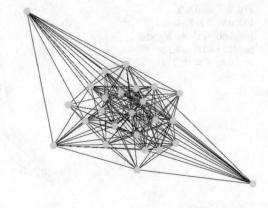

Fig. 9.11 Series 3.
Example 9.1. Grid; initial
state: $(n_{nev}, n_{ev}) = (13, 12)$

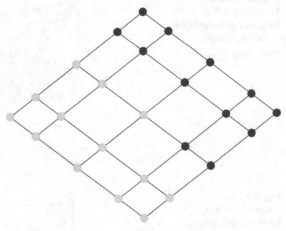

Fig. 9.12 Series 3.
Example 9.1. Grid; final state:
$(n_{nev}, n_{ev}) = (9, 16)$

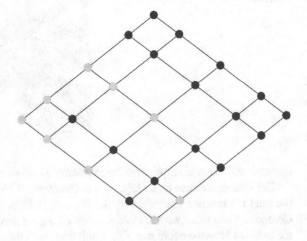

Fig. 9.13 Series 3.
Example 9.2. Weakly
connected network; initial
state: $(n_{nev}, n_{ev}) = (13, 12)$

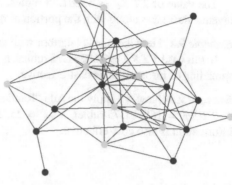

Fig. 9.14 Series 3.
Example 9.2. Weakly
connected network; final
state: $(n_{nev}, n_{ev}) = (15, 10)$

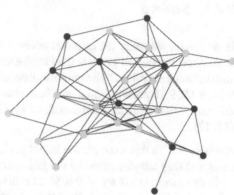

Example 9.2 An instant game we use is the Hawk-Dove game with the same parameters as in the previous example.

The total tax revenue is $TTR_T = 143518.25$ rubles at the final moment. Therefore, in this case spreading of information is cost-effective unlike to the previous experiment. This situation is caused by the number of honest taxpayers having increased from $n_{nev} = 5$ to $n_{nev} = 20$.

9.4.2 Series 2

In this series of experiments the following parameters are constant: the network is a strongly connected graph, the instant game is the Stag Hunt game, the payoff is cumulative. Figures 9.5, 9.6, 9.7, 9.8, 9.9, 9.10 present initial and final states of the system for application of the different imitation rules. Initial state of system for each example is $(n_{nev}, n_{ev}) = (8, 17)$.

Example 9.1 In this example the rule "The most influenceable neighbor" is applied.

The value of $TTR_T = 63281.75$ rubles, therefore, in this example the total tax revenue decreases along with the portion of non-evaders.

Example 9.2 Here the rule "Neighbor with the highest payoff" is applied.

In this case $TTR_T = 160781.75$ rubles, n_{nev} is increasing, thus, for this example spreading of information is cost-effective.

Example 9.3 Now we apply the rule "Random neighbor".

$TTR_T = 160781.75$ rubles, $n_{nev} = 25$. It means that in this case spreading of information is also cost-effective.

9.4.3 Series 3

In this series we study the different network modifications for the following initial conditions: the instant game is the Hawk-Dove game, the type of payoff is average and the imitation rule is "The most influenceable neighbor". Figures 9.11, 9.12, 9.13, 9.14, 9.15, 9.16 present initial and final states of the system for application of the different imitation rules. Initial state of system for each example is $(n_{nev}, n_{ev}) = (13, 12)$.

Example 9.1 In this example the initial parameters are the same as in the previous cases and they are visualized by the grid network.

In this example $TTR_T = 94654.25$ rubles and we can see that the spreading of information is cost-effective. Here the value of n_{nev} increases.

Example 9.2 In this example we consider weakly connected graph as a network.

The example shows that the spreading of information is also cost-effective as far as $TTR_T = 118054.25$ rubles and n_{nev} increases.

Example 9.3 Here an example with strongly connected network is represented.

In this example n_{nev} increases, $TTR_T = 106354.25$ rubles, thus, spreading of information is also cost-effective.

Fig. 9.15 Series 3.
Example 9.3. Strongly
connected network; initial
state: $(n_{nev}, n_{ev}) = (13, 12)$

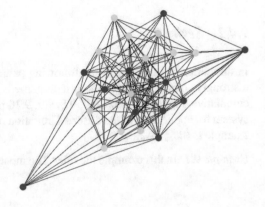

Fig. 9.16 Series 3.
Example 9.3. Strongly
connected network; final
state: $(n_{nev}, n_{ev}) = (12, 13)$

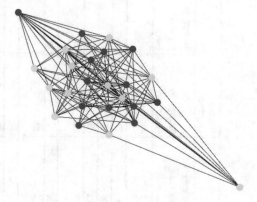

9.4.4   Results on Numerical Simulations

Based on the aggregated data (see Table 9.2) the following tendencies have been revealed. If the Prisoner's Dilemma is used as an instant game, then the agents' behavior depends on the network topology or the imitation rule. Any stochastic element in the structure of scenario (for example, if the network is a random graph, or if the imitation protocol is the "Random neighbor") leads to the fact that the steady state of the evolutionary game is the equilibrium (ev, ev), which means that both agent prefer to evade. If the dynamic process occurs in the grid then the steady state is the equilibrium (nev, nev) and it means that agents incline to prefer social behavior and pay taxes honestly.

The choice of the Stag Hunt game as an instant game simulated evolutionary process and provides the following results. In the majority experiments agents choose honest behavior and pay their taxes, the equilibrium profile (nev, nev) corresponds to this case. Agents are inclined to use the strategy "to evade" only in few experiments. Moreover, imitation rule "Neighbor with the highest payoff" leads to the mixed strategy profile as a steady state, where honest agents prevail regardless of the network topology.

If the instant game is the Hawk-Dove game, then we obtain a steady state as an equilibrium profile, which depends on the network parameters. Thus, if information is propagated in random networks, then two pure asymmetric equilibriums (ev, nev) or (nev, ev) are possible as a steady state. The mixed equilibrium profile can be used as a steady state only if the topology of the network is grid. In this case we can say that in contrast to the powerful influence of network topology, an imitation rule does not impact on the propagation process.

Table 9.2 Aggregated data of the scenario analysis

Topology of the network	Types of payoff matrices	Payoff	Total number of experiments	Most influential neighbour		Neighbour with the highest payoff		Random neighbour	
				Strategy "to evade" (ev)	Strategy "to pay taxes" (nev)	Strategy "to evade" (ev)	Strategy "to pay taxes" (nev)	Strategy "to evade" (ev)	Strategy "to pay taxes" (nev)
Strongly connected network	The Prisoner's Dilemma	Average	24	24	0	24	0	23	1
		Cumulative	24	24	0	21	3	23	1
	The Stag Hunt game	Average	24	11	13	4	20	7	17
		Cumulative	24	17	7	5	19	7	17
	The Hawk-Dove game	Average	24	23	1	0	24	1	23
		Cumulative	24	24	0	0	24	1	23
Weakly connected network	The Prisoner's Dilemma	Average	24	21	3	20	4	19	5
		Cumulative	24	21	3	17	7	19	5
	The Stag Hunt game	Average	24	12	12	4	20	12	12
		Cumulative	24	7	17	8	16	14	10
	The Hawk-Dove game	Average	24	6	18	0	24	3	21
		Cumulative	24	6	18	4	20	4	20
Grid	The Prisoner's Dilemma	Average	24	11	13	4	20	22	2
		Cumulative	24	11	13	4	20	14	10
	The Stag Hunt game	Average	24	10	14	2	22	6	18
		Cumulative	24	8	16	3	21	6	18
	The Hawk-Dove game	Average	24	12	12	0	24	4	20
		Cumulative	24	12	12	12	12	9	15

9.5   Conclusions

In the current paper, the model of tax control, which includes the information dissemination about the future tax auditing have been applied to the analysis of economic agents' behavior. The process of information spreading was formulated as an evolutionary process on the networks with different topology, the dynamics of the process was described by the specially designed imitation rules. The scenario analysis of the taxpayers behavior has shown that the propagation of information about possible tax audit has a positive effect on the tax collecting. The total amount of honest taxpayers increases which leads to the growth of the total revenue of fiscal system. Moreover, if the structure of payoff matrix and the imitation rule are known it is simpler to analyse how information dissemination impacts on the effectiveness of tax control.

References

1. Andrei, A.L. Comer, K., Koehler M.: An agent-based model of network effects on tax compliance and evasion. J. Econ. Psychol. **40**, 119–133 (2014)
2. Antocia, A., Russua, P., Zarrib, L.: Tax evasion in a behaviorally heterogeneous society: An evolutionary analysis. Econ. Modell. **10**(42), 106–115 (2014)
3. Antunes, L., Balsa J., Urbano, P., Moniz, L., Roseta-Palma, C.: Tax compliance in a simulated heterogeneous multi-agent society. In: Sichman, J.S., Antunes, L. (eds.) Multi-Agent-Based Simulation VI. MABS 2005. LNCS, vol. 3891, pp. 147–161. Springer, Berlin, Heidelberg (2006)
4. Boure, V., Kumacheva, S.: A model of audit with using of statistical information about taxpayers' income. Vestnik SPbGU **10**(1–2), 140–145 (2005) (in Russian)
5. Boure, V., Kumacheva, S.: A game theory model of tax auditing using statistical information about taxpayers. Vestnik SPbGU **10**(4), 16–24 (2010) (in Russian)
6. Carrera, E.J.S., Gubar, E., Oleynik, A.F.: Network structures and poverty traps. Dyn. Games Appl. **9**(1), 236–253 (2019)
7. Chander, P., Wilde, L.L.: A general characterization of optimal income tax enforcement. Rev. Econ. Stud. **65**, 165–183 (1998)
8. Cheng, D., He, F., Qi, H., Xu, T.: Modeling, analysis and control of networked evolutionary games. IEEE Trans. Autom. Control **60**(9), 2402–2415 (2015)
9. Goffman, W., Newill, V.A.: Generalization of epidemic theory: An application to the transmission of ideas. Nature **204**(4955), 225–228 (1964)
10. Gubar, E., Kumacheva, S., Zhitkova, E., Kurnosykh, Z.: Evolutionary behavior of taxpayers in the model of information dissemination. In: Proceedings of the Constructive Nonsmooth Analysis and Related Topics (Dedicated to the Memory of V.F. Demyanov, CNSA 2017), pp. 1–4. IEEE Conference Publications, St. Petersburg (2017)
11. Gubar, E., Kumacheva, S., Zhitkova, E., Kurnosykh, Z., Skovorodina, T.: Modelling of information spreading in the population of taxpayers: evolutionary approach. Contrib. Game Theory Manag. **10**, 100–128 (2017)
12. Gubar, E.A., Kumacheva, S.Sh., Zhitkova, E.M., Porokhnyavaya, O.Yu.: Propagation of information over the network of taxpayers in the model of tax auditing. In: Proceedings of the 2015 International Conference on Stability and Control Processes in Memory of V.I. Zubov (SCP 2015), pp. 244–247. IEEE Conference Publications, St. Petersburg (2015)

13. Gubar, E., Kumacheva, S., Zhitkova, E., Tomilina, G.: Modeling of the impact of information on tax audits on the risk statuses and evasions of individuals. Vestnik SPbGU **15**(2), 245–258 (2019) (in Russian)
14. Hashimzade, N., Myles, G.D. Page F., Rablen M.D.: Social networks and occupational choice: The endogenous formation of attitudes and beliefs about tax compliance. J. Econ. Psychol. **40**, 134–146 (2014)
15. Kandhway, K., Kuri, J.: Optimal control of information epidemics modeled as Maki Thompson rumors. Commun. Nonlinear Sci. Numer. Simul. (2014, Preprint)
16. Kendall, M.G., Stuart, A.: Distribution Theory (in Russian). Nauka, Moscow (1966)
17. Kumacheva, S.Sh.: Tax auditing using statistical information about taxpayers. Contrib. Game Theory Manag. **5**, 156–167 (2012)
18. Kumacheva, S.Sh., Gubar, E.A.: Evolutionary model of tax auditing. Contrib. Game Theory Manag. **8**, 164–175 (2015)
19. Kumacheva, S., Gubar, E., Zhitkova, E., Tomilina, G.: Evolution of risk-statuses in one model of tax control. In: Petrosyan, L., Mazalov, V., Zenkevich, N. (eds.) Frontiers of Dynamic Games. Static & Dynamic Game Theory: Foundations & Applications. pp. 121–138. Birkhauser, Cham (2018)
20. Kumacheva, S., Gubar, E., Zhitkova, E., Tomilina, G.: Modeling the behaviour of economic agents as a response to information on tax audits. In: Agarwal, N., Sakalauskas, L., Weber, G.-W. (eds.) Modeling and Simulation of Social-Behavioral Phenomena in Creative Societies. First International EURO Mini Conference (MSBC 2019) Proceedings, pp. 96–111. Springer International Publishing (2019)
21. Luthi, L., Tomassini, M., Pestelacci, E.: Evolutionary games on networks and payoff invariance under replicator dynamics. Biosystems **96**(3), 213–222 (2009)
22. Nekovee, A.M., Moreno Y., Bianconi G., Marsili M.: Theory of rumor spreading in complex social networks. Physica A **374**, 457–470 (2007)
23. Owen, G.: Game Theory. Saunders Company, Philadelphia: London: Toronto (1968)
24. Reinganum J.R., Wilde L.L.: Income tax compliance in a principal-agent framework. J. Public Econ. **26**, 1–18 (1985)
25. Riehl, J. R., Cao, M.: Control of stochastic evolutionary games on networks. In: 5th IFAC Workshop on Distributed Estimation and Control in Networked Systems, pp. 458–462, Philadelphia, PA, United States (2015)
26. Riehl, J. R., Cao M.: Towards optimal control of evolutionary games on networks. IEEE Trans. Autom. Control **62**(1), 458–462 (2017)
27. Sanchez I., Sobel J.: Hierarchical design and enforcement of income tax policies. J. Public Econ. **50**, 345–369 (1993)
28. Sandholm, W.H.: Population Games and Evolutionary Dynamics. MIT Press, Cambridge, MA (2010)
29. Schlad, K.H.: Why imitate, and if so, how? A boundedly rational approach to multi-armed bandits. J. Econ. Theory **7**(1), 130–156 (1998)
30. Skyrms, B.: The Stag Hunt and the Evolution of Social Structure. Cambridge University Press, Cambridge (2003)
31. The web-site of the Russian Federation State Statistics Service, http://www.gks.ru/
32. Vasin, A., Morozov, V.: The Game Theory and Models of Mathematical Economics. MAKS-press (in Russian), Moscow (2005)
33. Vasina, P.A.: Otimal tax enforcement with imperfect taxpayers and inspectors. Comput. Math. Model. **14**(3), 309–318 (2003)
34. Vasin, A.A., Panova, E.I.: Tax collection and corruption in Fiscal Bodies. Final Report on EERC Project, 31 p. (1999)
35. Weibull, J.: Evolutionary Game Theory. MIT Press, Cambridge, MA (1995)

Chapter 10
Cooperation Enforcing in Multistage Multicriteria Game: New Algorithm and Its Implementation

Denis Kuzyutin, Ivan Lipko, Yaroslavna Pankratova, and Igor Tantlevskij

Abstract To enforce the long-term cooperation in a multistage multicriteria game we use the imputation distribution procedure (IDP) based approach. We mainly focus on such useful properties of the IDP like "reward immediately after the move" assumption, time consistency inequality, efficiency and non-negativity constraint. To overcome the problem of negative payments along the optimal cooperative trajectory the novel refined A-incremental IDP is designed. We establish the properties of the proposed A-incremental payment schedule and provide an illustrative example to clarify how the algorithm works.

Keywords Dynamic game · Multistage game · Multicriteria game · Cooperative solution · Shapley value · Time consistency · Imputation distribution procedure

10.1 Introduction

The theory of multicriteria games (multiobjective games or the games with vector payoffs) develops at the overlap of classical game theory and multiple criteria decision analysis. It can be used to model various real-world decision-making problems where several objectives (or criteria) have to be taken into account (see, e.g., [1, 2, 14, 26] a player aims at simultaneously increasing production, obtaining large quote for the use of a common resource, saving costs of water purification, saving health care costs, etc. Starting from [29], much research has

D. Kuzyutin (✉)
Saint Petersburg State University, St. Petersburg, Russia

National Research University Higher School of Economics (HSE), St. Petersburg, Russia
e-mail: d.kuzyutin@spbu.ru

I. Lipko · Y. Pankratova · I. Tantlevskij
Saint Petersburg State University, St. Petersburg, Russia
e-mail: st052710@student.spbu.ru; y.pankratova@spbu.ru; tantigor@bk.ru

L. A. Petrosyan et al. (eds.), *Frontiers of Dynamic Games*,
Static & Dynamic Game Theory: Foundations & Applications,
https://doi.org/10.1007/978-3-030-51941-4_10

been done on non-cooperative multicriteria games (see, e.g., [8, 12, 24, 31]). Different cooperative solutions for static and dynamic multicriteria games were examined in [9–11, 13, 22].

This paper is mainly focused on the dynamic aspects of cooperation enforcing in an n-person multistage multicriteria games in extensive form (see, e.g., [6, 7, 20]) with perfect information. In order to achieve and implement a long-term cooperative agreement in a multicriteria dynamic game the players have to solve the following problems. First, when players seek to reach the maximal total vector payoff of the grand coalition, they face the problem of choosing a unique Pareto optimal payoffs vector. In the dynamic setting it is necessary that a specific method the players agreed to accept in order to choose a particular Pareto optimal solution not only takes into account the relative importance of the criteria, but also satisfies time consistency [5, 6, 9, 17–20, 25, 27], i.e., a fragment of the optimal cooperative trajectory in the subgame should remain optimal in this subgame. In the paper, we assume that the players employ the refined leximin (RL) algorithm, introduced in [11], to select a unique Pareto optimal solution for each multicriteria optimization problem they face. This approach allows constructing time consistent cooperative trajectory and vector-valued characteristic function. Another appropriate method— the rule of the minimal sum of relative deviations from the ideal payoffs vector—was suggested in [13].

After choosing the cooperative trajectory it is necessary to construct a vector-valued characteristic function. For instance, when analyzing the Example 10.1, we employ a friendly computable ζ-characteristic function introduced in [3] as well as the RL-algorithm in order to choose a particular Pareto efficient solution for the auxiliary vector optimization problems. To determine the optimal payoff allocation we adopt the vector analogue of the Shapley value [9, 22, 28]. Such an approach is based on the assumption that the payoff can be transferred between the players within the same criterion. It is worth noting that the main measurable criteria used in multicriteria resource management problems usually satisfy this component-wise transferable utility property.

Lastly, to guarantee the sustainability of the achieved long-term cooperative agreement the players are expected to use an appropriate imputation distribution procedure (IDP), i.e. a payoff allocation rule that determines the actual current payments to every player along the optimal cooperative trajectory. The IDP based approach was extensively studied for single-criterion differential and multistage games (see, e.g. [15, 16, 18, 20, 21]) and was extended to multicriteria multistage games in [9, 10]. The detailed review of useful properties the IDP may satisfy for multistage multicriteria games is presented in [9–13].

In particular, two novel properties an acceptable payment schedule for the multistage game should satisfy which take into account the sequence of the players' actions along the optimal cooperative trajectory were suggested in [12]. Firstly, a player which moves at position x according to the cooperative scenario expects to receive some reward for the "correct" move immediately after this move, while the other players (which are inactive at x) should get zero current payments. Furthermore, if the position x is the last player i's node along the cooperative trajectory this player should get the rest of her optimal payoff right after her last move. These properties were formalised in the so-called Reward Immediately after the Move (RIM) assumption (see [12] for details).

In this paper we mainly focus on the RIM assumption, efficiency and non-negativity constraint as well as time consistency property. The first so-called "incremental" IDP was suggested in [18] to ensure time consistency of the solution in differential single-criterion game, then this simple IDP was extended to different classes of dynamic games. The A-incremental IDP that satisfies RIM assumption, efficiency constraint and time consistency for multicriteria multistage game was designed in [12]. However, as it is demonstrated in the paper, the A-incremental IDP, as well as the classical incremental IDP may imply negative current payments to some players at some nodes (see [4, 9, 20] for details). One approach how to overcome this negative feature of the incremental IDP—the refined payment schedule for multicriteria games—was constructed in [9]. Another regularisation method for single-criterion multistage game was proposed in [4]. In this paper we provide a refinement of the A-incremental imputation distribution procedure for multicriteria multistage game. This "refined A-incremental IDP" is proved to satisfy the RIM assumption, non-negativity constraint, efficiency condition and time consistency inequality.

Hence, the main contribution of this paper is twofold:

- we reveal one possible disadvantage of the A-incremental payment schedule, namely that it may imply negative current payments to the players. To overcome this drawback we design the novel A-refined imputation distribution procedure which satisfies a number of usefull properties (in particular, non-negativity).
- we provide the step-by-step algorithm how to implement this novel allocation rule. Then we compare the implementation of the simple A-incremental IDP and the refined A-incremental IDP for given 3-person bicriteria multistage game.

The rest of the paper is organized as follows: The class of r-criteria multistage n-person games in extensive form with perfect information is formalized in Sect. 10.2. The optimal cooperative trajectory and vector-valued characteristic function are constructed in Sect. 10.3 using the refined leximin algorithm. We provide an illustrative example of the 3-person bicriteria multistage game here. Different useful properties of imputation distribution procedure are formulated in Sect. 10.4. In Sect. 10.5, we discuss the implementation of the A-incremental IDP and reveal the problem of negative payments. We provide a refined A-incremental IDP and the algorithm of its implementation in Sect. 10.6 and a brief conclusion in Sect. 10.7.

10.2 Multistage Game with Vector Payoffs

We consider a finite multistage r-criteria game in extensive form with perfect information following [7, 9, 20]. First we define the following notations that will be used throughout the paper:

- $N = \{1, \ldots, n\}$ is the finite set of players;
- K is the game tree with the root x_0 and the set of all nodes P;
- $S(x)$ is the set of all direct successors (descendants) of the node x and $S^{-1}(y)$ is the unique predecessor (parent) of the node $y \neq x_0$ such that $y \in S(S^{-1}(y))$;
- P_i is the set of all player i's decision nodes, $P_i \cap P_j = \varnothing$ for $i \neq j$, and $P_{n+1} = \{y^j\}_{j=1}^m$ is the set of all terminal nodes, $S(y^j) = \varnothing \ \forall y^j \in P_{n+1}, \cup_{i=1}^{n+1} P_i = P$;
- $\omega = (x_0, \ldots, x_{t-1}, x_t, \ldots, x_T)$ is the trajectory (or path) in the game tree, $x_{t-1} = S^{-1}(x_t), 1 \leqslant t \leqslant T; x_T = y^j \in P_{n+1}$, the lower index t in x_t denotes the number of the node within the trajectory ω and can be interpreted as the "time index", T is an ordinal number of the last node of the trajectory ω;
- $h_i(x) = (h_{i/1}(x), \ldots, h_{i/r}(x))$ is the i-th player's vector payoff at the node $x \in P \backslash \{x_0\}$.

We assume that

$$h_{i/k}(x) \geq 0; \ \forall i \in N; \ k = 1, \ldots, r; \ x \in P \backslash \{x_0\}.$$

Let us use $MG^P(n, r)$ to denote the class of all finite multistage n-person r-criteria games in an extensive form with perfect information. Since we will focus on the games with perfect information we restrict ourselves to the class of pure strategies (see, e.g., [7, 20]). The pure strategy $u_i(\cdot)$ of player i is a function with domain P_i that specifies for every node $x \in P_i$ the next node $u_i(x) \in S(x)$ which the player i should choose at x. Let U_i denote the (finite) set of all i-th player's pure strategies, $U = \prod_{i \in N} U_i$. Every strategy profile $u = (u_1, \ldots, u_n) \in U$ generates the trajectory $\omega(u) = (x_0, \ldots, x_t, x_{t+1}, \ldots, x_T) = (x_0, x_1(u), \ldots, x_t(u), x_{t+1}(u), \ldots, x_T(u))$, where $x_{t+1} = u_j(x_t) \in S(x_t)$ if $x_t \in P_j, 0 \leq t \leq T - 1, x_T \in P_{n+1}$, and, respectively, a collection of all players' vector payoffs.

Denote by

$$H_i(u) = (H_{i/1}(u), \ldots, H_{i/r}(u)) = \tilde{h}_i(\omega(u)) = \sum_{\tau=1}^{T} h_i(x_\tau(u)),$$

the value of player i's vector payoff function, given by the strategy profile $u = (u_1, \ldots, u_n)$.

In the multistage multicriteria game Γ^{x_0} defined above every intermediate node $x_t \in P \backslash P_{n+1}$ generates a subgame Γ^{x_t} with the subgame tree K^{x_t} and the subroot x_t as well as a factor-game with the factor-game tree $K^D = \{x_t\} \cup (K \backslash K^{x_t})$ (see, for instance [20]). Decomposition of the original extensive game Γ^{x_0} at node x_t into the subgame Γ^{x_t} and the factor-game Γ^D generates the corresponding decomposition of pure strategies.

Let $P_i^{x_t}(P_i^D), i = 1, \ldots, n$ denote the restriction of P_i on the subtree $K^{x_t}(K^D)$, and $u_i^{x_t}(u_i^D), i = 1, \ldots, n$, denote the restriction of the player i's pure strategy $u_i(\cdot)$ in Γ^{x_0} on $P_i^{x_t}(P_i^D)$. The strategy profile $u^{x_t} = (u_1^{x_t}, \ldots, u_n^{x_t})$ generates the trajectory $\omega^{x_t}(u^{x_t}) = (x_t, x_{t+1}, \ldots, x_T) = (x_t, x_{t+1}(u^{x_t}), \ldots, x_T(u^{x_t}))$ and, respectively, a collection of all player's vector payoffs in the subgame. Denote by

$$H_i^{x_t}(u^{x_t}) = \tilde{h}_i^{x_t}(\omega^{x_t}(u^{x_t})) = \sum_{\tau=t+1}^{T} h_i(x_\tau(u^{x_t})), \tag{10.1}$$

the value of player i's vector payoff function in the subgame Γ^{x_t}, and by $U_i^{x_t}$ the set of all player i's pure strategies in Γ^{x_t}, $U^{x_t} = \prod_{i \in N} U_i^{x_t}$. Note that

$$H_i(u) = \tilde{h}_i(\omega(u)) = \sum_{\tau=1}^{T} h_i(x_\tau(u)) = \sum_{\tau=1}^{t} h_i(x_\tau(u)) +$$

$$\sum_{\tau=t+1}^{T} h_i(x_\tau(u^{x_t})) = \tilde{h}_i(\underline{\omega}^{x_t}(u)) + \tilde{h}_i^{x_t}(\omega^{x_t}(u^{x_t})), \tag{10.2}$$

where $\underline{\omega}^{x_t}(u) = (x_0, x_1, \ldots, x_{t-1}, x_t)$ denotes a part of trajectory $\omega(u)$ before the subgame Γ^{x_t} starts.

Remark 10.1 Since $P_i = P_i^{x_t} \cup P_i^D$ while $P_i^{x_t} \cap P_i^D = \varnothing$, one can compose the player i's pure strategy $W_i = (u_i^D, v_i^{x_t}) \in U_i$ in the original game Γ^{x_0} from his strategies $v_i^{x_t} \in U_i^{x_t}$ and $u_i^D \in U_i^D$ in the subgame Γ^{x_t} and factor-game Γ^D respectively [20].

Let $a, b \in R^m$; we use the following vector inequalities: $a \geqq b$ if $a_k \geqslant b_k, \forall k = 1, \ldots, m$; $a > b$ if $a_k > b_k, \forall k = 1, \ldots, m$; $a \geq b$, if $a \geqq b$ and $a \neq b$. The last vector inequality implies that vector b is Pareto dominated by a.

10.3 Designing a Cooperative Solution

If the players agree to cooperate in multicriteria game Γ^{x_0}, they maximize w.r.t. the binary relation $\geq$ the total vector payoff $\sum_{i=1}^{n} H_i(u)$. Denote by $PO(\Gamma^{x_0})$ the set of all Pareto optimal strategy profiles from U, i.e.:

$$u \in PO(\Gamma^{x_0}) \; if \; \nexists v \in U : \sum_{i \in N} H_i(v) \geq \sum_{i \in N} H_i(u)$$

The set $PO(\Gamma^{x_0})$ is known to be nonempty (see, e.g., [23]) and in general it contains multiple strategy profiles. Since the set $PO(\Gamma^{x_0})$ may contain more than one strategy profile, the players face the problem how to select a unique Pareto optimal cooperative strategy profile $\bar{u} \in PO(\Gamma^{x_0})$ and corresponding optimal cooperative trajectory $\bar{\omega} = \bar{\omega}(\bar{u}) = (\bar{x}_0, \bar{x}_1, \ldots, \bar{x}_T)$. In a dynamic game it is essential that a specific method the players agreed to employ in order to choose a particular Pareto optimal solution has to satisfy time consistency, that is, a fragment $\bar{\omega}^{x_t}(\bar{u}^{x_t}) = (\bar{x}_t, \bar{x}_{t+1}, \ldots, \bar{x}_T)$ of the optimal trajectory $\bar{\omega}$ in the subgame $\Gamma^{\bar{x}_t} \in G(\bar{u})$ should remain optimal trajectory for this subgame.

We employ the so-called Refined Leximin algorithm, introduced in [11] to find optimal cooperative trajectory in Example 10.1. This approach looks reasonable for the special case when the criteria have significantly different importance, and all the players rank the criteria in the same order. In other circumstances, the players may employ other appropriate methods to choose a unique Pareto optimal solutions (an example of such methods—the rule of minimal sum of relative deviations from the ideal payoffs vector—was suggested in [13]). Note that the main result of the paper—Proposition 10.1—does not depend on the particular time consistent rule which the players have agreed to use in order to choose a unique Pareto optimal solution.

Let us briefly remind the main idea of the RL algorithm and the notations (the reader could find the comprehensive specification of this algorithm in [11, 12]). Suppose that all the criteria are ordered in accordance with their relative importance for the players, namely let criterion 1 be the most important for every player $i \in N$, the next to be the 2-nd criterion, and so on, and the last criterion r be the least important one. When choosing the optimal cooperative trajectory the players are expected to maximise the total vector payoff primarily on the first criterion, i.e.

$$\max_{u \in U} \sum_{i \in N} H_{i/1}(u) = \sum_{i \in N} H_{i/1}(\bar{u}) = \bar{H}_1.$$

If there exists a unique trajectory $\bar{\omega} = \omega(\bar{u})$ satisfying this condition then this trajectory is called the optimal cooperative trajectory while $\bar{u})$ is the optimal cooperative strategy profile.

If there are several trajectories $\omega(u)$ with $\sum_{i \in N} H_{i/1}(u) = \bar{H}_1$ the players should choose such trajectory from this set $\overline{PO}_1(\Gamma^{x_0})$ that

$$\max_{u \in \overline{PO}_1(\Gamma^{x_0})} \sum_{i \in N} H_{i/2}(u) = \bar{H}_2,$$

and so on. Lastly, if there are several trajectories $\omega \in \{\omega(u), u \in \overline{PO}_r(\Gamma^{x_0})\}$, the players should choose the trajectory from this set with minimal number j of the terminal node y^j.

We will suppose henceforth that the players have agreed to use the RL algorithm in order to choose the *optimal cooperative strategy profile* $\bar{u} \in PO(\Gamma^{x_0})$ and the corresponding *optimal cooperative trajectory* $\bar{\omega} = \omega(\bar{u}) = (\bar{x}_0, \ldots, \bar{x}_T)$.

Let

$$Max_{u \in U}^L \sum_{i \in N} H_i(u) = \sum_{i \in N} H_i(\bar{u}) \tag{10.3}$$

denote the maximal (in the sense of the RL algorithm) total vector payoff. Note that the Pareto optimal cooperative trajectory $\bar{\omega} = \omega(\bar{u}) = (\bar{x}_0, \ldots, \bar{x}_T)$ based on the RL algorithm was proved to satisfy time consistency [11].

Let us use the following example to demonstrate how the players choose the cooperative trajectory and then to explore and compare the A-incremental IDP and the refined A-incremental payment schedule.

Example 10.1 (A 3-Player Bicriteria Multistage Game) The game tree K is shown in Fig. 10.1. Let $n = 3$, $r = 2$, $P_1 = \{\bar{x}_0, \bar{x}_2, \bar{x}_4, \bar{x}_6\}$, $P_2 = \{\bar{x}_1, \bar{x}_5\}$, $P_3 = \{\bar{x}_3\}$, $P_{n+1} = \{z_1, \ldots, z_9\}$,

$$h(x_t) = \begin{pmatrix} h_{1/1}(x_t) & h_{2/1}(x_t) & h_{3/1}(x_t) \\ h_{1/2}(x_t) & h_{2/2}(x_t) & h_{3/2}(x_t) \end{pmatrix},$$

i.e. the columns correspond to the players while the rows correspond to the criteria. The players' payoffs at all nodes $x \in P \setminus \{x_0\}$ are:

$$h(x_1) = \begin{pmatrix} 0 & 6 & 0 \\ 0 & 0 & 12 \end{pmatrix}, \quad h(x_2) = \begin{pmatrix} 6 & 0 & 0 \\ 0 & 0 & 12 \end{pmatrix}, \quad h(x_3) = \begin{pmatrix} 0 & 6 & 0 \\ 0 & 0 & 12 \end{pmatrix},$$

$$h(x_4) = \begin{pmatrix} 6 & 0 & 0 \\ 0 & 12 & 0 \end{pmatrix}, \quad h(x_5) = \begin{pmatrix} 0 & 6 & 0 \\ 0 & 0 & 12 \end{pmatrix}, \quad h(x_6) = \begin{pmatrix} 6 & 0 & 0 \\ 0 & 0 & 12 \end{pmatrix},$$

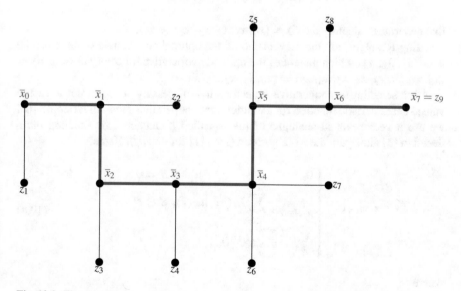

Fig. 10.1 The game tree

$$h(x_7) = \begin{pmatrix} 30\ 30\ 30 \\ 60\ 30\ 30 \end{pmatrix}, \ h(z_1) = \begin{pmatrix} 18\ 0\ 0 \\ 18\ 0\ 0 \end{pmatrix}, \ h(z_2) = \begin{pmatrix} 0\ 18\ 0 \\ 0\ 18\ 0 \end{pmatrix},$$

$$h(z_3) = \begin{pmatrix} 18\ 0\ 0 \\ 18\ 0\ 0 \end{pmatrix}, \ h(z_4) = \begin{pmatrix} 0\ 0\ 18 \\ 0\ 0\ 18 \end{pmatrix}, \ h(z_5) = \begin{pmatrix} 0\ \ 18\ 0 \\ 60\ 18\ 0 \end{pmatrix},$$

$$h(z_6) = \begin{pmatrix} 18\ 0\ 0 \\ 18\ 0\ 0 \end{pmatrix}, \ h(z_7) = \begin{pmatrix} 90\ \ 6\ 0 \\ 162\ 0\ 6 \end{pmatrix}, \ h(z_8) = \begin{pmatrix} 18\ \ 0\ 0 \\ 150\ 0\ 0 \end{pmatrix},$$

There are three pure strategy Pareto optimal strategy profiles in $PO(\Gamma^{x_0})$:

$$\bar{u}_1(x_0) = x_1, \ \bar{u}_2(x_1) = x_2, \ \bar{u}_1(x_2) = x_3, \ \bar{u}_3(x_3) = x_4, \ \bar{u}_1(x_4) = x_5,$$

$$\bar{u}_2(x_5) = x_6, \ \bar{u}_1(x_6) = x_7$$

that generates trajectory $\bar{\omega}(\bar{u}) = (x_0, x_1, x_2, x_3, x_4, x_5, x_6, x_7)$,

$$u_1'(x_0) = x_1, \ u_2'(x_1) = x_2, \ u_1'(x_2) = x_3, \ u_3'(x_3) = x_4, \ u_1'(x_4) = z_7,$$

that generates trajectory $\bar{\omega}(u') = (x_0, x_1, x_2, x_3, x_4, z_7)$ and

$$u_1''(x_0) = x_1, \ u_2''(x_1) = x_2, \ u_1''(x_2) = x_3, \ u_3''(x_3) = x_4, \ u_1''(x_4) = x_5,$$

$$u_2'(x_5) = x_6, \ u_1''(x_6) = z_8$$

that generates trajectory $\bar{\omega}(u'') = (x_0, x_1, x_2, x_3, x_4, x_5, z_8)$.

Using RL algorithm the players choose the optimal cooperative strategy profile $\bar{u} = (\bar{u}_1, \bar{u}_2, \bar{u}_3)$ which generates the optimal cooperative trajectory $\bar{\omega} = \omega(\bar{u}) = (x_0, x_1, x_2, x_3, x_4, x_5, x_6, x_7) = (\bar{x}_0, \bar{x}_1, \bar{x}_2, \bar{x}_3, \bar{x}_4, \bar{x}_5, \bar{x}_6, \bar{x}_7)$.

After selecting a cooperative trajectory it is necessary to construct a vector-valued characteristic function for a multicriteria cooperative game. In Example 10.1 we use a vector-valued analogue of the so-called ζ-characteristic function introduced in [3] and again the RL algorithm (see [11] for details). Namely:

$$V^{x_0}(S) = \begin{cases} 0, & S = \emptyset \\ \underset{u_j, j \in N \backslash S}{Min^L} \sum_{i \in S} H_i(\bar{u}_S, u_{N \backslash S}), & S \subset N, \\ \underset{u \in U}{Max^L} \sum_{i \in N} H_i(u), & S = N \end{cases} \qquad (10.4)$$

where

$$\underset{u_j, j \in N \backslash S}{Min^L} \sum_{i \in S} H_i(\bar{u}_S, u_{N \backslash S}) = - \underset{u_j, j \in N \backslash S}{Max^L} \left(- \sum_{i \in S} H_i(\bar{u}_S, u_{N \backslash S}) \right).$$

Let $\Gamma^{x_0}(N, V^{x_0})$ denote multicriteria game $\Gamma^{x_0} \in MG^P(n, r)$ with characteristic function V^{x_0}. It is worth noting that one can use other approaches to construct characteristic function (CF) for multicriteria game, say classical α-CF or δ-CF [21], but as it was mentioned in [3] the ζ-characteristic function is much more easy to compute (it is essential especially for multicriteria case). Note that the main result of the paper—Proposition 10.1—does not depend on the specific method which the players employ to calculate the vector-valued characteristic function.

We assumed that the players adopt a single-valued cooperative solution φ^{x_0} (for instance, the vector analogue of the Shapley value [9, 28]) for the cooperative game $\Gamma^{x_0}(N, V^{x_0})$ which satisfies the efficiency property

$$\sum_{i=1}^{n} \varphi_i^{x_0} = V^{x_0}(N) = \sum_{\tau=1}^{T} \sum_{i=1}^{n} h_i(\bar{x}_\tau), \tag{10.5}$$

and the individual rationality property

$$\varphi_i^{x_0} \geq V^{x_0}(\{i\}), \ i = 1, \ldots, n. \tag{10.6}$$

Denote by $\Gamma^{\bar{x}_t}(N, V^{\bar{x}_t})$, $\bar{x}_t \in \bar{\omega}(\bar{u})$, $t = 0, \ldots, T - 1$ a subgame along the optimal cooperative trajectory with the characteristic function $V^{\bar{x}_t}$ which can be computed in the subgame using (10.4). Note that $V^{\bar{x}_t}(N) = \sum_{\tau=t+1}^{T} \sum_{i \in N} h_i(\bar{x}_\tau)$.

In addition, we assume that the same properties (10.5) and (10.6) are valid for the cooperative solutions $\varphi^{\bar{x}_t}$ at each subgame $\Gamma^{\bar{x}_t}(N, V^{\bar{x}_t})$, $t = 0, \ldots, T - 1$.

10.4 Imputation Distribution Procedure and Its Properties

Let $\beta = \{\beta_{i/k}(\bar{x}_\tau)\}, i = 1, \ldots, n; k = 1, \ldots, r; \tau = 1, \ldots, T$ denote the Imputation Distribution Procedure—IDP [9, 18, 20, 25] or the payment schedule. The IDP-based approach implies that the players have agreed to accumulate the cooperative vector payoff $\sum_{i \in N} H_i(\bar{u}) = V^{x_0}(N)$, obtained using the initial payoffs $h_i(\bar{x}_\tau)$, and then allocate this summary payoff between the players along the optimal cooperative trajectory $\bar{\omega}(\bar{u})$. Then $\beta_{i/k}(\bar{x}_\tau)$ corresponds to the actual current payment which the player i receives at $\bar{x}_\tau$ w.r.t. criterion k (instead of $h_{i/k}(\bar{x}_\tau)$) according to the IDP β.

From now on we suppose that the IDP β should satisfy the following assumption (see [12] for details):

Assumption RIM (Reward Immediately After the Move) If $\bar{x}_t \in P_i, t = 0, \ldots, T - 1$, then $\beta_j(\bar{x}_{t+1}) = 0$ for all $j \in N\backslash\{i\}$, i.e. the only player who can receive nonzero current payment at node $\bar{x}_{t+1}$ is the player i which moves at the previous node $\bar{x}_t = S^{-1}(\bar{x}_{t+1})$.

For given player $i \in N$ let $(y_1^i, y_2^i, \ldots, y_{T(i)}^i)$ denote the ordered set of all the positions from the set $P_i \cap \bar{\omega}$ along the optimal trajectory $\bar{\omega}$, where nodes $\{y_\tau^i\}$ are listed in order of their location in $\bar{\omega}$. Namely,

$$y_1^i = \bar{x}_{t^i(1)}, \ y_2^i = \bar{x}_{t^i(2)}, \ldots, y_{T(i)}^i = \bar{x}_{t^i(T(i))};$$

and for all $y_\lambda^i = \bar{x}_{t^i(\lambda)}$ and $y_m^i = \bar{x}_{t^i(m)}$, we have $\lambda < m$ if and only if $t(\lambda) < t^i(m)$.

Below, we introduce a number of useful properties an acceptable IDP may satisfy (see [9–11]). Note that we need to modify known definitions of efficiency and time consistency to take assumption RIM into account.

To simplify the notations, henceforth we will omit superscript i in $t^i(\lambda)$, $\lambda = 1, \ldots, T(i)$, i.e. we will write $\beta_i(\bar{x}_{t(\lambda)+1})$ instead of $\beta_i(\bar{x}_{t^i(\lambda)+1})$, e.t.c.

Definition 10.1 ([12]) The imputation distribution procedure $\beta = \{\beta_{i/k}(\bar{x}_t)\}$ satisfies the *efficiency condition* if

$$\sum_{t=1}^{T} \beta_i(\bar{x}_t) = \sum_{\lambda=1}^{T(i)} \beta_i(\bar{x}_{t(\lambda)+1}) = \varphi_i^{\bar{x}_0}, \ i = 1, \ldots, n. \tag{10.7}$$

Indeed, if (10.7) holds then the payment schedule for every player can be considered as a rule for the step-by-step allocation of the player i's optimal payoff.

Definition 10.2 The IDP $\beta = \{\beta_{i/k}(\bar{x}_t)\}$ meets the *time consistency (TC) inequality* if for every player $i \in N$ such that $|T(i)| \geqslant 2$, for all $\tau = 1, \ldots, T(i) - 1$ it holds that

$$\sum_{\lambda=1}^{\tau} \beta_i(\bar{x}_{t(\lambda)+1}) + \varphi_i^{\bar{x}_{t(\tau)+1}} \geq \varphi_i^{\bar{x}_0}. \tag{10.8}$$

The vector inequality (10.8) implies that every player has an incentive to continue cooperation at every subgame along the cooperative trajectory.

Definition 10.3 ([9]) The imputation distribution procedure $\beta = \{\beta_{i/k}(\bar{x}_t)\}$ satisfies the *balance condition* if $\forall t = 0, \ldots, T; \forall k = 1, \ldots, r$ it holds that

$$\sum_{\tau=1}^{t} \sum_{i=1}^{n} \beta_{i/k}(\bar{x}_\tau) \leq \sum_{\tau=1}^{t} \sum_{i=1}^{n} h_{i/k}(\bar{x}_\tau) \tag{10.9}$$

Note that (10.9) is always satisfied for $t = T$ due to the efficiency condition (10.7) and (10.5). If β does not satisfy (10.9) at some intermediate node $\bar{x}_t$, we will suppose that the players may borrow the required amount on account of future earnings. For the sake of simplicity we assume that an interest-free loan is available for the grand coalition N while recognising that in general case the enforcing of a cooperative agreement may require extra costs (see [9]).

Definition 10.4 ([9]) The IDP β satisfies the *non-negativity constraint* if

$$\beta_{i/k}(\bar{x}_t) \geqslant 0, \ i = 1, \ldots, n; \ k = 1, \ldots, r; \ t = 1, \ldots, T.$$

Note that there could be different payment schedules that may or may not satisfy the properties listed above (several IDP for multicriteria games are examined in [9–11]). The A-incremental IDP that satisfies RIM assumption, efficiency constraint and time consistency (equation) for multicriteria multistage game was suggested in [12]:

Definition 10.5 The *A-incremental imputation distribution procedure* $\beta = \{\beta_{i/k}(\bar{x}_t)\}, t = 0, \ldots, T; i \in N$ is formulated as follows

(c1) $\beta_{i/k}(\bar{x}_0) = 0, i = 1, \ldots, n; k = 1, \ldots r;$

(c2) if $\bar{x}_t \in P_i, t = 0, \ldots, T - 1$, then $\beta_j(\bar{x}_{t+1}) = 0$ for all $j \in N\backslash\{i\};$

(c3) if $\bar{x}_t \in P_i$ and $T(i) = 1$, i.e. $\bar{\omega} \cap P_i = (y_1^i) = \{\bar{x}_{t(1)}\}$, then

$$\beta_i(\bar{x}_{t(1)+1}) = \varphi_i^{\bar{x}_0} \tag{10.10}$$

(c4) if $\bar{x}_t \in P_i$ and $T(i) = 2$, i.e. $\bar{\omega} \cap P_i = (y_1^i, y_2^i) = (\bar{x}_{t(1)}, \bar{x}_{t(2)})$, then

$$\beta_i(\bar{x}_{t(1)+1}) = \varphi_i^{\bar{x}_0} - \varphi_i^{\bar{x}_{t(1)+1}}; \quad \beta_i(\bar{x}_{t(2)+1}) = \varphi_i^{\bar{x}_{t(1)+1}} \tag{10.11}$$

(c5) if $\bar{x}_t \in P_i$ and $T(i) \geq 3$, i.e. $\bar{\omega} \cap P_i = (y_1^i, y_2^i, \ldots, y_{T(i)}^i) = (\bar{x}_{t(1)}, \bar{x}_{t(2)}, \ldots, \bar{x}_{t(T(i))})$, then

$$\beta_i(\bar{x}_{t(1)+1}) = \varphi_i^{\bar{x}_0} - \varphi^{\bar{x}_{t(1)+1}};$$

$$\beta_i(\bar{x}_{t(\lambda)+1}) = \varphi_i^{\bar{x}_{t(\lambda-1)+1}} - \varphi_i^{\bar{x}_{t(\lambda)+1}}, \ \lambda = 2, \ldots, T(i) - 1; \tag{10.12}$$

$$\beta_i(\bar{x}_{t(T(i))+1}) = \varphi_i^{\bar{x}_{t(T(i))-1)+1}}.$$

10.5 A-Incremental IDP May Imply Negative Current Payments

Let us use the game from Ex. 1 to demonstrate the A-incremental IDP implementation and properties and to reveal one possible disadvantage of this payment schedule. We will adopt the vector analogue of the Shapley value as an optimal cooperative solution when analysing Ex. 1.

Definition 10.6 ([22, 28]) The Shapley value of $\Gamma^{x_0}(N, V^{x_0})$ denoted by φ^{x_0} is defined for each player $i \in N$ as

$$\varphi_i^{x_0} = \sum_{S \subset N, i \in S} \frac{(n - |S|)!(|S| - 1)!}{n!} (V^{x_0}(S) - V^{x_0}(S \setminus \{i\})). \tag{10.13}$$

Example 10.1 (Continued) The values of the vector-valued ζ-characteristic function (10.4) for the game Γ^{x_0} are

S	{1}	{2}	{3}	{1, 2}	{1, 3}	{2, 3}	N
$V^{x_0}(S)$	0	0	0	18	0	0	126
	0	0	0	0	12	0	192

and the Shapley value for original game Γ^{x_0} is

$$\varphi^{x_0} = \begin{pmatrix} 45 & 45 & 36 \\ 66 & 60 & 66 \end{pmatrix}.$$

The vector-valued ζ-characteristic functions and the respective Shapley values for the subgames along the cooperative trajectory $\bar{\omega}$ can be constructed using the same approach.

The subgame $\Gamma^{x_1}(N, V^{x_1})$:

S	{1}	{2}	{3}	{1, 2}	{1, 3}	{2, 3}	N
$V^{x_1}(S)$	0	0	0	12	0	0	120
	0	0	0	0	0	12	180

$$\varphi^{x_1} = \begin{pmatrix} 42 & 42 & 36 \\ 56 & 62 & 62 \end{pmatrix}.$$

The subgame $\Gamma^{x_2}(N, V^{x_2})$:

S	{1}	{2}	{3}	{1, 2}	{1, 3}	{2, 3}	N
$V^{x_2}(S)$	0	0	0	6	6	0	114
	0	0	0	0	84	0	168

$$\varphi^{x_2} = \begin{pmatrix} 40 & 37 & 37 \\ 70 & 28 & 70 \end{pmatrix}.$$

The subgame $\Gamma^{x_3}(N, V^{x_3})$:

S	$\{1\}$	$\{2\}$	$\{3\}$	$\{1,2\}$	$\{1,3\}$	$\{2,3\}$	N
$V^{x_3}(S)$	0	0	0	0	6	0	108
	0	0	0	0	72	12	156

$$\varphi^{x_3} = \begin{pmatrix} 37 & 34 & 37 \\ 60 & 30 & 66 \end{pmatrix}.$$

The subgame $\Gamma^{x_4}(N, V^{x_4})$:

S	$\{1\}$	$\{2\}$	$\{3\}$	$\{1,2\}$	$\{1,3\}$	$\{2,3\}$	N
$V^{x_4}(S)$	0	0	0	72	0	0	102
	60	0	0	90	72	0	144

$$\varphi^{x_4} = \begin{pmatrix} 46 & 46 & 10 \\ 95 & 29 & 20 \end{pmatrix}.$$

The subgame $\Gamma^{x_5}(N, V^{x_5})$:

S	$\{1\}$	$\{2\}$	$\{3\}$	$\{1,2\}$	$\{1,3\}$	$\{2,3\}$	N
$V^{x_5}(S)$	0	0	0	66	0	0	96
	60	0	0	90	60	12	132

$$\varphi^{x_5} = \begin{pmatrix} 43 & 43 & 10 \\ 85 & 31 & 16 \end{pmatrix}.$$

The subgame $\Gamma^{x_6}(N, V^{x_5})$:

S	$\{1\}$	$\{2\}$	$\{3\}$	$\{1,2\}$	$\{1,3\}$	$\{2,3\}$	N
$V^{x_6}(S)$	30	0	0	60	60	0	90
	60	0	0	90	90	0	120

$$\varphi^{x_6} = \begin{pmatrix} 60 & 15 & 15 \\ 90 & 15 & 15 \end{pmatrix}.$$

Applying the A-incremental IDP (10.10), (10.12), (10.13) we obtain the following current payments along the optimal cooperative path $\bar{\omega} = (\bar{x}_0 = y_1^1, \ \bar{x}_1 = y_1^2, \ \bar{x}_2 = y_2^1, \ \bar{x}_3 = y_1^3, \ \bar{x}_4 = y_3^1, \ \bar{x}_5 = y_2^2, \ \bar{x}_6 = y_4^1, \ \bar{x}_7) : \beta_1(\bar{x}_1) = \varphi_1^{\bar{x}_0} - \varphi_1^{\bar{x}_1} = $

$$\begin{pmatrix} 3 \\ 10 \end{pmatrix}, \; \beta_1(\bar{x}_3) = \varphi_1^{\bar{x}_1} - \varphi_1^{\bar{x}_3} = \begin{pmatrix} 5 \\ -4 \end{pmatrix}, \; \beta_1(\bar{x}_5) = \varphi_1^{\bar{x}_3} - \varphi_1^{\bar{x}_5} = \begin{pmatrix} -6 \\ -25 \end{pmatrix}, \; \beta_1(\bar{x}_7) =$$

$$\varphi_1^{\bar{x}_5} = \begin{pmatrix} 43 \\ 85 \end{pmatrix}, \; \text{since } T(1) = 4;$$

$$\beta_j(\bar{x}_1) = \beta_j(x_3) = \beta_j(x_5) = \beta_j(x_7) = \begin{pmatrix} 0 \\ 0 \end{pmatrix}, \; j = 2, 3;$$

$$\beta_2(\bar{x}_2) = \varphi_2^{\bar{x}_0} - \varphi_2^{\bar{x}_2} = \begin{pmatrix} 8 \\ 32 \end{pmatrix}, \; \beta_2(\bar{x}_6) = \varphi_2^{\bar{x}_2} = \begin{pmatrix} 37 \\ 28 \end{pmatrix} \text{ since } T(2) = 2;$$

$$\beta_j(\bar{x}_2) = \beta_j(\bar{x}_6) = \begin{pmatrix} 0 \\ 0 \end{pmatrix}, \; j = 1, 3;$$

$$\beta_3(\bar{x}_4) = \varphi_3^{\bar{x}_0} = \begin{pmatrix} 36 \\ 66 \end{pmatrix} \text{ since } T(3) = 1; \; \beta_j(\bar{x}_4) = \begin{pmatrix} 0 \\ 0 \end{pmatrix}, \; j = 1, 2.$$

The efficiency condition for the player 1 and criterion 2 takes the form:

$$\sum_{t=1}^{7} \beta_{1/2}(\bar{x}_t) = \sum_{\lambda=1}^{4} \beta_{1/2}(\bar{x}_{t(\lambda)+1}) = 10 - 4 - 25 + 85 = 66 = \varphi_1^{\bar{x}_0}.$$

The time consistency equations for the player 1 and criterion 2 take the form:

$$\tau = 1 : \sum_{\lambda=1}^{1} \beta_{1/2}(\bar{x}_{t(\lambda)+1}) + \varphi_1^{\bar{x}_1} = 10 + 56 = 66 = \varphi_1^{\bar{x}_0};$$

$$\tau = 2 : \sum_{\lambda=1}^{2} \beta_{1/2}(\bar{x}_{t(\lambda)+1}) + \varphi_1^{\bar{x}_3} = 10 - 4 + 60 = 66 = \varphi_1^{\bar{x}_0};$$

$$\tau = 3 : \sum_{\lambda=1}^{3} \beta_{1/2}(\bar{x}_{t(\lambda)+1}) + \varphi_1^{\bar{x}_5} = 10 - 4 - 25 + 85 = 66 = \varphi_1^{\bar{x}_0}.$$

As it was mentioned in the Introduction, the A-incremental IDP, as well as the classical incremental IDP may imply negative current payments to some players at some nodes (see [4, 9, 20] for details). Thus, in Example 10.1 the A-incremental IDP implies negative payments to player 1 at $\bar{x}_3$ and $\bar{x}_5$.

10.6　Refined A-Incremental IDP and Its Implementation

Below we introduce a refinement of A-incremental payment schedule that is designed to satisfy assumption RIM, efficiency, time consistency inequality and non-negativity constraint.

We will use an auxiliary integer variable $a_{i/k}(\lambda)$ to denote the number of nodes $\bar{x}_{t(\tau)+1}$ on the optimal cooperative path $\bar{\omega}$ from $\bar{x}_\tau$ to $\bar{x}_{t(\lambda)+1}$ for which $\beta_{i/k}(\bar{x}_{t(\tau)+1}) = 0$ after the last positive current payment (one may call it the payment delay variable). We assume that $t^i(0) = -1$ for any i.

Definition 10.7 The *refined A-incremental imputation distribution procedure* $\hat{\beta} = \{\hat{\beta}_{i/k}(\bar{x}_t)\}, t = 0, \ldots, T; i \in N$ is formulated as follows

(c1) $\hat{\beta}_{i/k}(\bar{x}_0) = 0, i = 1, \ldots, n; k = 1, \ldots r$;

(c2)   if $\bar{x}_t \in P_i, t = 0, \ldots, T - 1$, then $\hat{\beta}_j(\bar{x}_{t+1}) = 0$ for all $j \in N\backslash\{i\}$;

(c3)   if $\bar{x}_t \in P_i$ and $T(i) = 1$, i.e. $\bar{\omega} \cap P_i = (y_1^i) = \{\bar{x}_{t(1)}\}$, then

$$\hat{\beta}_{i/k}(\bar{x}_{t(1)+1}) = \varphi_{i/k}^{\bar{x}_0} \tag{10.14}$$

(c4)   if $\bar{x}_t \in P_i$ and $T(i) = 2$, i.e. $\bar{\omega} \cap P_i = (y_1^i, y_2^i) = (\bar{x}_{t(1)}, \bar{x}_{t(2)})$, then

$$\hat{\beta}_{i/k}(\bar{x}_{t(1)+1}) = \max\{\varphi_{i/k}^{\bar{x}_0} - \varphi_{i/k}^{\bar{x}_{t(1)+1}}, 0\} \tag{10.15}$$

$$\hat{\beta}_{i/k}(\bar{x}_{t(2)+1}) = \varphi_{i/k}^{\bar{x}_0} - \hat{\beta}_{i/k}(\bar{x}_{t(1)+1}) \tag{10.16}$$

(c5)   if $\bar{x}_t \in P_i$ and $T(i) \geq 3$, i.e. $\bar{\omega} \cap P_i = (y_1^i, y_2^i, \ldots, y_{T(i)}^i) = (\bar{x}_{t(1)}, \bar{x}_{t(2)}, \ldots, \bar{x}_{t(T(i))})$, then

Step 1 ($\lambda = 1$):

$$\hat{\beta}_{i/k}(\bar{x}_{t(1)+1}) = \max\{\varphi_{i/k}^{\bar{x}_0} - \varphi^{\bar{x}_{t(1)+1}}, 0\}; \tag{10.17}$$

- if $\hat{\beta}_{i/k}(\bar{x}_{t(1)+1}) > 0$, then $a_{i/k}(1) = 0$ (no delay in payment compared to A-incremental IDP);
- if $\hat{\beta}_{i/k}(\bar{x}_{t(1)+1}) = 0$, then $a_{i/k}(1) = 1$ (the delay in payment for a one step).

Step 2 ($\lambda = 2$):

$$\hat{\beta}_{i/k}(\bar{x}_{t(2)+1}) = \max\{\varphi_{i/k}^{\bar{x}_{t(1-a_{i/k}(1))+1}} - \varphi_{i/k}^{\bar{x}_{t(2)+1}}, 0\}, \tag{10.18}$$

- if $\hat{\beta}_{i/k}(\bar{x}_{t(2)+1}) > 0$, then $a_{i/k}(2) = 0$;
- if $\hat{\beta}_{i/k}(\bar{x}_{t(2)+1}) = 0$, then $a_{i/k}(2) = a_{i/k}(1) + 1$.

Step λ ($\lambda = 2, \ldots, T(i) - 1$):

$$\hat{\beta}_{i/k}(\bar{x}_{t(\lambda)+1}) = \max\{\varphi_{i/k}^{\bar{x}_{t(\lambda-1-a_{i/k}(\lambda-1))+1}} - \varphi_{i/k}^{\bar{x}_{t(\lambda)+1}}, 0\}, \tag{10.19}$$

- if $\hat{\beta}_{i/k}(\bar{x}_{t(\lambda)+1}) > 0$, then $a_{i/k}(\lambda) = 0$ (no delay in payment);
- if $\hat{\beta}_{i/k}(\bar{x}_{t(\lambda)+1}) = 0$, then $a_{i/k}(\lambda) = a_{i/k}(\lambda - 1) + 1$ (the delay in payment at $\bar{x}_{t(\lambda)+1}$ for $a_{i/k}(\lambda)$ steps).

Step $\lambda = T(i)$:

$$\hat{\beta}_{i/k}(\bar{x}_{t(T(i))+1}) = \max\{\varphi_{i/k}^{\bar{x}_0} - \sum_{\lambda=1}^{T(i)-1} \hat{\beta}_{i/k}(\bar{x}_{t(\lambda)+1}), 0\} = \varphi_{i/k}^{\bar{x}_{t(T(i)-1-a_{i/k}(T(i)-1))+1}}$$

$$\tag{10.20}$$

By the construction of this refined payment schedule the following proposition holds.

Proposition 10.1 *Refined A-incremental IDP satisfies assumption RIM, efficiency condition (10.7), non-negativity constraint and time consistency inequality (10.8).*

Now we will apply the refined A-incremental algorithm (10.14)–(10.20) to the game from Example 10.1.

Example 10.1 (Continued) Note that if the A-incremental IDP implies non-negative current payments to the ith player w.r.t. criterion k at all nodes along the cooperative trajectory $\bar{\omega}$, then $\hat{\beta}_{i/k}(\bar{x}_t) = \beta_{i/k}(\bar{x}_t)$, $\bar{x}_t \in \bar{\omega}$. Hence, the current payments to the player 2 and 3 according to the refined A-incremental IDP $\hat{\beta}$ will not change compared to the A-incremental IDP β.

Let us now consider the payments to the player $i = 1$:

$$\hat{\beta}_{1/1}(\bar{x}_1) = \max\{\varphi_{1/1}^{\bar{x}_0} - \varphi_{1/1}^{\bar{x}_1}; 0\} = 3, a_{1/1}(1) = 0;$$
$$\hat{\beta}_{1/2}(\bar{x}_1) = \max\{\varphi_{1/2}^{\bar{x}_0} - \varphi_{1/2}^{\bar{x}_1}; 0\} = 10, a_{1/2}(1) = 0;$$
$$\hat{\beta}_{1/1}(\bar{x}_3) = \max\{\varphi_{1/1}^{\bar{x}_1} - \varphi_{1/1}^{\bar{x}_3}; 0\} = \max\{5; 0\} = 5, a_{1/1}(2) = 0;$$
$$\hat{\beta}_{1/2}(\bar{x}_3) = \max\{\varphi_{1/2}^{\bar{x}_1} - \varphi_{1/2}^{\bar{x}_3}; 0\} = max\{-4; 0\} = 0, a_{1/2}(2) = 1;$$
$$\hat{\beta}_{1/1}(\bar{x}_5) = \max\{\varphi_{1/1}^{\bar{x}_3} - \varphi_{1/1}^{\bar{x}_5}; 0\} = \max\{-6; 0\} = 0, a_{1/1}(3) = 1;$$
$$\hat{\beta}_{1/2}(\bar{x}_5) = \max\{\varphi_{1/2}^{\bar{x}_1} - \varphi_{1/2}^{\bar{x}_5}; 0\} = max\{-29; 0\} = 0, a_{1/2}(3) = 2;$$
$$\hat{\beta}_{1/1}(\bar{x}_7) = \varphi_{1/1}^{\bar{x}_3} = 37;$$
$$\hat{\beta}_{1/2}(\bar{x}_7) = \varphi_{1/2}^{\bar{x}_1} = 56.$$

All the payments to player $i = 1$ are non-negative now. Note that the current payments at $\bar{x}_7$ are less than the relevant payments according to the simple A-incremental IDP.

The efficiency condition for the player 1 and criterion 2 now takes the form:

$$\sum_{t=1}^{7} \hat{\beta}_{1/2}(\bar{x}_t) = \sum_{\lambda=1}^{4} \hat{\beta}_{1/2}(\bar{x}_{t(\lambda)+1}) = 10 + 0 + 0 + 56 = 66 = \varphi_1^{\bar{x}_0}.$$

The time consistency inequalities for the player 1 and criterion 2 take the form:

$$\tau = 1 : \sum_{\lambda=1}^{1} \hat{\beta}_{1/2}(\bar{x}_{t(\lambda)+1}) + \varphi_1^{\bar{x}_1} = 10 + 56 \geq 66 = \varphi_1^{\bar{x}_0};$$

$$\tau = 2 : \sum_{\lambda=1}^{2} \hat{\beta}_{1/2}(\bar{x}_{t(\lambda)+1}) + \varphi_1^{\bar{x}_3} = 10 + 0 + 60 \geq 66 = \varphi_1^{\bar{x}_0};$$

$$\tau = 3 : \sum_{\lambda=1}^{3} \hat{\beta}_{1/2}(\bar{x}_{t(\lambda)+1}) + \varphi_1^{\bar{x}_5} = 10 + 0 + 0 + 85 \geq 66 = \varphi_1^{\bar{x}_0}.$$

Note that the refined A-incremental IDP may not necessarily satisfy balance condition (10.9). Let us for instance consider the balance condition in Example 10.1 for $t = 4$ and $k = 2$:

$$\sum_{\tau=1}^{4} \sum_{i=1}^{3} \hat{\beta}_{i/2}(\bar{x}_\tau) = 108 > \sum_{\tau=1}^{4} \sum_{i=1}^{3} h_{i/2}(\bar{x}_\tau) = 48.$$

As it was firstly noted in [9], in general it is impossible to design a time consistent IDP which satisfies both the balance condition and non-negativity constraint.

10.7   Conclusion

When analyzing Example 10.1, we adopt the Shapley value as an optimal imputation and use the RL algorithm for choosing a unique Pareto optimal solution (to find optimal cooperative trajectory and to construct vector-valued characteristic function). It is worth noting that the provided algorithm to calculate the refined A-incremental IDP as well as Proposition 10.1 remains valid if the players employ another optimal imputation, other approach to calculate the characteristic function and other time consistent rule for choosing a particular Pareto optimal solution, for instance, the rule of minimal sum of relative deviations from the ideal payoffs vector [13].

Note that, since the set of active players in extensive form game changes while the game is evolving along the optimal path, multistage game could be considered as an example of the so-called "games with changing conditions". The RIM assumption and the proposed refined A-incremental payment schedule allows taking into account this specific feature of a n-person multistage game. It is worth noting that similar assumptions could be implied implicitly in some ancient texts—cf., for instance, the so-called "History of King David's ascent to power" in connection with David's activity at the beginning of his career (see, e.g.: [30]). The detailed interdisciplinary analysis of the relevant motivation for "optimal" behaviour could be an interesting issue for further research.

Acknowledgments We would like to thank the anonymous Reviewer for the valuable comments. The research of the first and the third author was funded by RFBR under the research project 18-00-00727 (18-00-00725). The research of the fourth author was funded by RFBR under the research project 18-00-00727 (18-00-00628).

References

1. Climaco, J., Romero, C., Ruiz, F.: Preface to the special issue on multiple criteria decision making: current challenges and future trends. Intl. Trans. Oper. Res. **25**, 759–761 (2018) http://dx.doi.org/10.1111/itor.12515
2. Finus, M.: Game Theory and International Environmental Cooperation. Edward Elgar Ed (2001).
3. Gromova, E. V., Petrosyan, L. A.: On an approach to constructing a characteristic function in cooperative differential games. Autom. Remote Control **78**(9), 1680–1692 (2017)
4. Gromova, E.V., Plekhanova, T.M.: On the regularization of a cooperative solution in a multistage game with random time horizon. Discrete Appl. Math. **255**, 40–55 (2019) http://dx.doi.org/10.1016/j.dam.2018.08.008
5. Haurie, A.: A note on nonzero-sum diferential games with bargaining solution. J. Optim. Theory Appl. **18**, 31–39 (1976)
6. Haurie, A., Krawczyk, J. B., Zaccour, G.: Games and Dynamic Games. Scientific World, Singapore (2012)
7. Kuhn, H.: Extensive games and the problem of information. Ann. Math. Stud. **28**, 193–216 (1953)
8. Kuzyutin, D.: On the consistency of weak equilibria in multicriteria extensive games. In: Petosyan, L.A., Zenkevich, N.A. (eds.) Contributions to Game Theory and Management, Vol. V, pp. 168–177 (2012)
9. Kuzyutin, D., Nikitina, M.: Time consistent cooperative solutions for multistage games with vector payoffs. Oper. Res. Lett. **45**(3), 269–274 (2017) http://dx.doi.org/10.1016/j.orl.2017.04.004
10. Kuzyutin, D., Nikitina, M.: An irrational behavior proof condition for multistage multicriteria games. In: Proceedings of the Consrtuctive Nonsmooth Analysis and Related Topics (Dedic. to the Memory of V.F. Demyanov), CNSA 2017, pp. 178–181. IEEE (2017)
11. Kuzyutin, D., Gromova, E., Pankratova, Ya.: Sustainable cooperation in multicriteria multistage games. Oper. Res. Lett. **46**(6), 557–562 (2018). http://dx.doi.org/10.1016/j.orl.2018.09.004
12. Kuzyutin, D., Pankratova, Y., Svetlov, R.: A-subgame concept and the solutions properties for multistage games with vector payoffs. In: Petrosyan, L., Mazalov, V., Zenkevich, N. (eds.) Frontiers of Dynamic Games. Static & Dynamic Game Theory: Foundations & Applications. Birkhäuser, Cham (2019) http://dx.doi.org/10.1007/978-3-030-23699-1_6

13. Kuzyutin, D., Smirnova, N., Gromova, E.: Long-term implementation of the cooperative solution in multistage multicriteria game. Oper. Res. Perspect. **6**, 100107 (2019) http://dx.doi.org/10.1016/j.orp.2019.100107
14. Mendoza, G.A., Martins, H.: Multi-criteria decision analysis in natural resource management: A critical review of methods and new modelling paradigms. For. Ecol. Manage. **230**, 1–22 (2006)
15. Parilina, E., Zaccour, G.: Node-consistent core for games played over event trees. Automatica **55**, 304–311 (2015)
16. Parilina, E., Zaccour, G.: Node-consistent Shapley value for games played over event trees with random terminal time. J. Optim. Theory Appl. **175**(1), 236–254 (2017)
17. Petrosyan, L.: Stable solutions of differential games with many participants. Vestn. Leningr. Univ. **19**, 46–52 (1977) (in Russian)
18. Petrosyan, L., Danilov, N.: Stability of the solutions in nonantagonistic differential games with transferable payoffs. Vestn. Leningr. Univ. **1**, 52–59 (1979) (in Russian)
19. Petrosyan, L.A., Kuzyutin, D.V.: On the stability of E-equilibrium in the class of mixed strategies. Vestn. St.Petersburg Univ. Math. **3**(15), 54–58 (1995) (in Russian)
20. Petrosyan, L., Kuzyutin, D.: Games in Extensive Form: Optimality and Stability. Saint Petersburg University Press (2000) (in Russian)
21. Petrosyan, L., Zaccour, G.: Time-consistent Shapley value allocation of pollution cost reduction. J. Econ. Dyn. Control **27**(3), 381–398 (2003)
22. Pieri, G., Pusillo, L.: Multicriteria partial cooperative games. Appl. Math. **6**, 2125–2131 (2015)
23. Podinovskii, V., Nogin, V.: Pareto-Optimal Solutions of Multicriteria Problems. Nauka (1982) (in Russian)
24. Puerto, J., Perea, F.: On minimax and Pareto optimal security payoffs in multicriteria games. J. Math. Anal. Appl. **457**(2), 1634–1648 (2018). http://dx.doi.org/10.1016/j.jmaa.2017.01.002
25. Reddy, P., Shevkoplyas, E., Zaccour, G.: Time-consistent Shapley value for games played over event trees. Automatica **49**(6), 1521–1527 (2013)
26. Rettieva, A.: Cooperation in dynamic multicriteria games with random horizons. J. Glob. Optim. (2018). http://dx.doi.org/10.1007/s10898-018-0658-6
27. Sedakov, A.: On the strong time consistency of the core. Autom. Remote Control **79**(4), 757–767 (2018)
28. Shapley, L.: A value for n-person games. In: Kuhn, H., Tucker, A.W. (eds.) Contributions to the Theory of Games, II, pp. 307–317. Princeton University Press, Princeton (1953)
29. Shapley, L.: Equilibrium points in games with vector payoffs. Nav. Res. Logist. Q. **6**, 57–61 (1959)
30. Tantlevskij, I.: King David and His Epoch in the Bible and History. 3rd edn. Revised and Enlarged. RChGA Publishing House (2020) (in Russian)
31. Voorneveld, M., Vermeulen, D., Borm, P.: Axiomatizations of Pareto equilibria in multicriteria games. Games Econom. Behav. **28**, 146–154 (1999)

Chapter 11
Complementarity of Goods and Cooperation Among Firms in a Dynamic Duopoly

Mario Alberto Garcia Meza and Cesar Gurrola Rios

Abstract We construct a simple model to show how complementarities between goods yield a possibility for cooperation between rival firms. To show this, we use a simple dynamic model of Cournot oligopoly under sticky prices. While cooperation in an oligopoly model with sticky prices is not feasible, there exists a feasible cooperation when good are perfect complements and not substitutes.

Keywords Differential games · Cooperative games · Sticky prices · Cournot duopoly

11.1 Introduction

While competition is often assumed to deliver the best social outcome, there are reasons to try to promote cooperation among firms. A classical example where less competition is desirable is when firms are parte of a supply chain and integration results in avoiding the problem of double marginalization (cf. [15]). An often overlooked market situation that requires the need of cooperation is when small firms do not have direct economies of scale and therefore cannot compete with big businesses unless they cooperate. In [6], the authors describe for example the situation many small businesses in poor countries encounter, with small, unprofitable businesses, facing the paradox of having a high marginal return and a small overall return.

Nevertheless, achieving cooperation requires that the marginal returns of being in a coalition are higher than those of individual efforts. Moreover, business contracts of cooperation may require a long time commitment, and therefore it should be clear

M. A. G. Meza (✉) · C. G. Rios
Universidad Juarez del Estado de Durango, Durango, Mexico
e-mail: mario.agm@ujed.mx

L. A. Petrosyan et al. (eds.), *Frontiers of Dynamic Games*,
Static & Dynamic Game Theory: Foundations & Applications,
https://doi.org/10.1007/978-3-030-51941-4_11

for all participants that such a commitment is profitable throughout the entire period of the coalition for it to be sustainable.

While substitute goods hardly find a feasible (profitable) solution for cooperation among firms, complementarity is one way to achieve it. In this article we give an example on how complementarity can achieve such results. A good A is complementary to another B when its popularity is linked with the demand of B. In other words

Definition 11.1 Let $q_j(p_i)$ with $j \neq i$ be the quantity supplied by firm j to the market. Firm's j's offered product is a complement of firm's i's, if

$$\frac{\partial p_i}{\partial q_j} > 0. \tag{11.1}$$

Where $p_i(t)$ represents the price of product i.

As a result, there might be incentives for the firms for cooperation. In this paper we use as an example an oligopoly with sticky prices. In Sect. 11.2 we set up the baseline model and compare its results of cooperation in Sect. 11.3. In Sect. 11.4, we make the case for complimentary goods as an instance when cooperation is possible by and example.

The critical variable to look for to determine the feasibility of cooperation is the profit obtained. In this case we observe the stable quantities and prices that firms use in their optimal strategies under the model.

11.2 Non-Cooperative Oligopoly

For a first approach, we set up the model of a sticky price duopoly without cooperation and with a demand function similar to the model in a Cournot oligopoly model [4]. While the initial models on sticky prices date back to [14], further instances can be found in [5] and [3]. Consider an oligopoly composed by $N = 1, 2, \ldots, n$ firms competing in the period of time $\mathscr{T} = [0, \infty)$. Each firm can choose the quantity $q_i(t)$ they offer to the market at the period of time $t \in \mathscr{T}$, considering that the market price $p(t)$ is determined by the demand equation

$$\hat{p}(t) = a - Q(t), \tag{11.2}$$

where $Q(t) = \sum_{i \in 1}^{n} b_i q_i$. For the price to be positive, the sum of the quantities offered to the market has to be less than a. Naturally, there are no incentives for the firms to offer more if that entails a negative price. The coefficient b_i is an indicator of the complementarity of goods. For the current section and Sect. 11.3 we will assume that $b_1 = b_2 = \cdots = b_n = 1$, which can be interpreted as a market in which all goods are perfect substitutes. Whenever $b_i \neq b_j$ there is an implicit complementarity between goods for $i \neq j$.

Proposition 11.1 *We say that two goods* i *and* j *are* perfect complements *whenever* $b_i + b_j = 0$.

Note that if a good i is a perfect complement of j in the terms described by Proposition 11.1 then its price increases with the quantity of j and Eq. (11.1) holds.

Nevertheless, this dynamic competition model considers the existence of *menu costs*. Although the concept of menu cost was introduced by [13] as a consideration to price adjustment in a macroecomic environment, [1] showed that price stickiness can be a result of bounded rationality when firms refuse to update their price unless the benefit is high enough [2]. The price dynamics thus is expressed by the differential equation

$$\dot{p}(t) = s[a - Q(t) - p(t)], \qquad p(0) = p_0, \tag{11.3}$$

where s is a speed-of-adjustment parameter and $p(t)$ is the current price. The interpretation of this equation is that the dynamics of the price are determined by the difference between the price determined by the market forces and the current price. A speed parameter that tends to infinity would then reveal the case of a market that updates the price immediately.

To know the instant profit of a firm that acts in a non-cooperative scenario, we need to state their profit function. Thus, each firm has as their objective to maximize

$$J_i = \int_0^\infty e^{-\rho t} \left[p(t) - c - \frac{q_i(t)}{2} \right] q_i(t) dt, \tag{11.4}$$

subject to (11.3). Here, the cost of production has the same structure for both firms and is equal to $C(t) = c + \frac{q_i(t)}{2}$ and ρ is the discount factor.

A concept of solution for this kind of problem is a Nash equilibrium. In particular, we can consider an open-loop Nash equilibrium or a Feedback solution [16]. An open loop Nash equilibrium would be conceptually equivalent to say that the firms observe her rivals and their own behavior and decide all the trajectory of their strategy *a priori* whereas a feedback solution requires a continuous response from the players. The feedback solution uses dynamic programming techniques to find optimal strategies and, while somehow more computational exhaustive it tends to achieve more precise answers (cf. [16]).

Here we will find feedback solutions for the problem. Thus, we state the problem above in terms of the Hamilton Jacobi Bellman (HJB) equation

$$\rho V_i = \max_{q_i \geq 0} \left\{ \left[p(t) - c - \frac{q_i(t)}{2} \right] q_i(t) + s \frac{\partial V_i}{\partial p} [a - Q(t) - p(t)] \right\}. \tag{11.5}$$

Maximizing the right part of the equation yields the feedback quantities for firms:

$$q_i^*(t) = \begin{cases} p(t) - c - s\frac{\partial V}{\partial p} & \text{if } \frac{\partial V_i}{\partial p} < (p(t) - c)/s, \\ 0 & \text{if } \frac{\partial V_i}{\partial p} \geq (p(t) - c)/s. \end{cases} \tag{11.6}$$

These values point to the optimal strategies that all players should execute in order to obtain the optimal possible profit under oligopoly conditions. Note that the optimal quantity is still a function of the market price $p(t)$ and the term $\frac{\partial V_i}{\partial p}$. This last one is usually interpreted as a *shadow price*, that is, a monetary value assigned to costs that are unknowable or difficult to directly calculate (see [11]).

The result obtained in (11.6) can thus be inserted into Eq. (11.5) and solve for the value function, which yields

$$V_i = \frac{1}{\rho}\left[s\frac{\partial V_i}{\partial p}\left(\sum_{j=1}^{n} \frac{\partial V_j}{\partial p} + a + nc - (n+1)p\right) - \frac{(\frac{\partial V_i}{\partial p} - c + p)(\frac{\partial V_i}{\partial p} + c - p)}{2}\right], \tag{11.7}$$

provided that $q_i(t)$ is positive, otherwise

$$V_i = \frac{s(a - p)}{\rho}\frac{\partial V_i}{\partial p}. \tag{11.8}$$

Unless stated otherwise, from this point, we shall focus on the positive solution. Since we are facing a nonhomogeneous Clairaut differential system, we can use as an ansatz for the value function a linear equation

$$V_i(p) = \alpha + \gamma p + \eta p^2. \tag{11.9}$$

To find out the values of coefficients α, γ and η. Consider that (11.9) implies

$$\frac{\partial V_i}{p(t)} = \gamma + \eta p. \tag{11.10}$$

Note that, since there is a unique market price for goods, be them substitutes or complements, the value function will have the same structure for all players. Since they are symmetrical players in all other senses, this result is acceptable, but might be useful to keep in mind for modeling situations in which the good is not homogeneous. In other words, the demand function described in Eq. (11.2) is the only source of information about the complementarities of the goods.

In order to solve the equation, we acknowledge the equivalence between Eq. (11.7) and (11.9) and substitute (11.10). The roots of the polynomial are obtained by solving the following system of equations:

$$\alpha = \frac{n}{2\rho} \left(2\gamma ns(a + cn + \gamma n^2 s) + c^2 - \gamma^2 n^2 s^2 \right),$$

$$\gamma = \frac{a\eta ns + c\eta n^2 s - c}{-2\eta n^3 s^2 + \eta n^2 s^2 + n^2 s + ns + \rho},$$

(11.11)

$$\eta = \frac{2n^2 s + 2ns + \rho \pm \sqrt{4n^4 s^2 + 4n^2 \rho s + 8n^2 s^2 + 4n\rho s + \rho^2}}{4n^2 s^2 (2n - 1)}.$$

Therefore, a solution of the system is obtained by getting the values of the parameters. This solution coincides with solutions in [3] and [8] and can be used to compute the real value of the feedback quantities. Recall that the optimal quantities can be stated by substituting (11.10) into (11.6).

$$q_i^*(t) = p(t) - c - s[\gamma + 2\eta p(t)].$$

(11.12)

Note that η has two possible values. It can be shown that only when the sign of the square root on the right side of the equation is negative yields a stable solution for the system. To find this stable solution, we should equate to zero the left hand side of Eq. (11.3), and solve for $p(t)$. This process yields

$$\bar{p} = \frac{a + cn + \gamma ns}{n + 1 - 2\eta ns},$$

(11.13)

which can be introduced in the value of $q_i^*(t)$ to obtain

$$\bar{q}_i = \frac{2a\eta s - a + c + \gamma s}{2\eta ns - n - 1}.$$

(11.14)

This is the general feedback solution before the introduction of the solution values of the parameters η and γ. For example, consider the solution of the stable quantity when $n = 3$:

$$q_{i \in \{1,2,3\}} = \frac{(a - c) \left(4\rho + 7s + \sqrt{(\rho + 8s)^2 - 20s^2} \right)}{16\rho + 33s + 4\sqrt{(\rho + 8s)^2 - 20s^2}}$$

(11.15)

which has real valued optimal quantities as long as

$$s > \frac{\pm \rho(\sqrt{5} - 4)}{22}.$$

(11.16)

Since the discount factor is usually a measure of the value of money across time, it only makes sense that ρ is a positive number, and since s only makes sense as well as a positive number,[1] then there is a solution for all possible values of s, regardless of the discount factor.

11.3 Cooperation in Business

While cooperation is possible by the firms, that does not mean that the incentives for it are set. In fact, in this section we will explore why there does not exist a cooperative solution in such a model and what can we do about it.

A cooperation between firms would mean that each firm has as a goal to maximize the joint profit of the grand coalition

$$J(q_1, \ldots, q_n) = \int_0^\infty e^{-\rho t} \sum_{i=1}^n \left[\left(p(t) - c - \frac{q_i(t)}{2} \right) q_i(t) \right] dt \qquad (11.17)$$

subject to the sticky price dynamics described in Eq. (11.3). Analogously to the problem of non-cooperative firms, this problem can be translated into the corresponding HJB equation

$$\rho V = \max_{q_1, \ldots, q_n} \left\{ \sum_{i=1}^n \left[p(t) - c - \frac{q_i(t)}{2} \right] q_i(t) + \sum_{i=1}^n s \frac{\partial V}{\partial p} \left(a - \sum_{i=1}^n q_i(t) - p(t) \right) \right\}.$$

$$(11.18)$$

Maximizing the right hand side of the equation thus yields

$$q_i(Q) = p(t) - c + s \frac{\partial V}{\partial p}. \qquad (11.19)$$

Note that while in the HJB equation (11.5) every firm is maximizing their own individual value function V_i, in a cooperative problem the maximization is done jointly by all firms as if it were a monopoly.

Inserting the optimal quantities in Eq. (11.19) of the coalition in equation into the HJB equation, we obtain the value function

$$\rho V^C = ns \frac{\partial V}{\partial p} \left(a - p(t) - \frac{\partial V}{\partial p} n^2 s \right) + n(c - p(t))^2 - \frac{n^2}{2} \left(ns \frac{\partial V}{\partial p} - c + p \right)^2.$$

---

[1] A negative number of s would be interpreted as a reversal in the process of convergence of prices. This would mean that the prices are drifting ever apart from the natural prices of the market.

The same as in the non-cooperative problem, we can observe that the value function as a function of the price can be stated as a linear quadratic function. Therefore, we introduce as an informed guess the Eq. (11.9) into the HJB equation. This allows us to state that a solution to the system can be obtained by solving for the parameters

$$\alpha = \frac{n}{2\rho}[2\gamma s(a + cn^2 - \gamma n^2 s) - c^2 n + 2c^2 - \gamma^2 n^3 s^2],$$

$$\gamma = \frac{n\left(2a\eta s + 2c\eta n^2 s + cn - 2c\right)}{2\eta n^4 s^2 + 4\eta n^3 s^2 + n^3 s + ns + \rho}, \tag{11.20}$$

$$\eta = -\frac{2n^3 s + 2ns + \rho \pm \sqrt{24n^4 s^2 + 4n^3 \rho s + (2ns + \rho)^2}}{4n^3 s^2 (n + 2)}.$$

Note that, again this is a set of two linear systems of equations, depending on the sign of the squared root in η. Let η^+ be the value of η whenever the sign before the squared root is plus and η^- the value when the sign is minus.

An important issue to consider in this case is whether there exists a stable solution for the control and the prices. In the case of the system with η^+ this condition does not hold, and therefore we will focus on the system where $\eta = \eta^-$.

Let $\bar{q}_n^C$ be the stable quantity that optimizes profit while on cooperation. For example, consider the case where $n = 3$, then

$$\bar{q}_3^C = \frac{(a - c)(4\rho + 45s + \sqrt{(\rho + 60)^2 - 1620s^2})}{5(3\rho + 33s - \sqrt{(\rho + 60)^2 - 1620s^2})}. \tag{11.21}$$

It is possible to verify that $\bar{q}_n^C > q_i$ described in Eq. (11.15) and that, in the limit, the profits of the firm with cooperation are *lower* than those without cooperation. This result is consistent regardless of the use of the parameters.

In a practical sense, we can infer that firms do not have any interest of engaging in cooperation under these circumstances. While there might be some exogenous reasons to cooperate (e.g. there is a war of attrition going on with some other agent), this is not captured by the current model.

Although this result is natural, since the nature of the basic market is non-cooperative, it is important to note that there exists an important interest in the results of cooperation among firms for business reasons. This cooperation retains a for-profit motivation and can be used for internationalization of small firms [7], development of strategic sectors in an economy (see [10] and [17]) or to implement cooperative marketing strategies for a complementary sector or members of a franchise [9].

11.4 Complementary Goods

The possibility of cooperation depends on the comparison of profits obtained by firms under cooperation and without it. While the Cournot model found in Sects. 11.2 sand 11.3 does not yield any profitable solution for cooperation, we explore the possibility that such cooperation can be achieved when the goods are complimentary.

Proposition 11.2 *Cooperation among firms is feasible when profits satisfies the property of* superadditivity, *i.e.:*

$$\pi^C \geq \pi_i + \pi_j \tag{11.22}$$

for different firms i and j.

A simple way to show this feasibility comes from the implementation of complementary goods into the market with sticky prices. This can be achieved by the use of Proposition 11.1. For simplicity, consider a simple modification to the Cournot model where $n = 2$, $b_1 = 1$ and $b_2 = -1$. This implies that there are only two firms and one is a perfect complement of the other. This implies slight modifications in the statement of the problem that are omitted from this article, but can be easily computed from the equations previously displayed.[2]

Considering a similar approach as in previous sections, the problems to solve are stated in the terms of a dynamic programming problem with the same price dynamics as a restriction. In a similar way, we find that the HJB equations can be stated and solved by stating an informed guess about their shape. As before, a suitable ansatz is the linear function displayed in Eq. (11.9). The use of the stated values for b_1 and b_2 implies that the system to solve is reduced to

$$\alpha = \frac{a\gamma s + \frac{c^2}{2} + \frac{3\gamma^2 s^2}{2}}{\rho},$$

$$\gamma = \frac{2a\eta s - c}{-6\eta s^2 + \rho + s}, \tag{11.23}$$

$$\eta = \frac{\rho + 2s \pm \sqrt{\rho^2 + 4\rho s - 8s^2}}{12s^2}.$$

As before, we focus on the version of the problem with real solutions. Naturally, the same version with η^- arises for solution of the system. The result of this process yields

$$\gamma = \frac{a\rho + 2as - a\sqrt{(\rho + 2s)^2 - 10s^2} - 6cs}{3s(\rho + \sqrt{(\rho + 2s)^2 - 10s^2})},$$

---

[2]A complete detail of the calculations made for this paper can be found in the repository in https://github.com/MariusAgm/Oligopolios/tree/master/Jupyter. This includes Jupyter notebooks with the results obtained in the paper and some comprobations of the claims of the paper.

and

$$\eta = \frac{\rho + 2s - \sqrt{(\rho + 2s)^2 - 10s^2}}{12s^2}.$$

The value of alpha is omitted due to reasons of space. To find the value of the stable price and quantity, we introduce the value functions and the previous values of γ and η, then we find the stable price by stating that $\dot{p}(t) = 0$ in Eq. (11.3) and solving for $p(t)$. The resulting stable price is then plugged into the optimal quantities to obtain:

$$\bar{q} = \frac{a\rho - 10as + 5a\sqrt{(\rho + 2s)^2 - 10s^2} - c\rho + 10cs - 5c\sqrt{(\rho + 2s)^2 - 10s^2}}{9\left(\rho + \sqrt{(\rho + 2s)^2 - 10s^2}\right)}.$$

(11.24)

Which has real solutions as long as $s < \frac{\rho(1 - \sqrt{3})}{4}$.

It is important to note that the existence of real solutions is dependent on the size of s. Recall that a small size of the speed parameter implies a slower recuperation toward the market price or higher menu costs from the firms. Moreover, it is possible to find instances where the profits of cooperation are higher than competition under complementary goods, when the speed parameter is sufficiently low in comparison to the discount rate ρ.

It only makes sense that the rate of adjustment of the market prices s is related with the discount rate ρ. If the speed of adjustment is low enough for the firms to collect benefits, there is a window of opportunity for firms to profit from a lower discount rate.

As a result, we must only plug the stable prices and quantities to obtain the stable profit for the firms with and without cooperation and with complementary and substitute goods. The analysis of such profits yields as a result that, while cooperation does not show superadditivity in the case of perfect substitutes, we can find an instance of the property when the goods are complementary.

This paper only shows the result of duopoly with complementary goods. In further research, it is important to verify if the same result holds in a set of n good. This would require a more complex system of complementarities in which the set of coefficients in the demand function would be arranged in a matrix. Moreover, it would require that the price of the goods would not be unique but a complete vector

$$\begin{bmatrix} p_1 \\ p_2 \\ \vdots \\ p_n \end{bmatrix} = A - \begin{bmatrix} b_{11} & \cdots & b_{nn} \\ \vdots & \ddots & \vdots \\ b_{n1} & \cdots & b_{nn} \end{bmatrix} \begin{bmatrix} q_1 \\ q_2 \\ \vdots \\ q_n \end{bmatrix}.$$

(11.25)

This results in a totally different kind of interaction between firms and a different market structure that can be further studied.

11.5   Discussion

While competition remains one of the more important ways to allocate resources in an efficient way in the context of a market economy, cooperation among firms remain a key issue for economists and policymakers (see, for example the work of Roemer in this particular field [12]).

The correct identification of the mechanisms that yield cooperation is thus an important area for policy and lawmaking. This is particularly true nowadays while we are in the look for novel ways of structuring markets that remain profitable for businesses while at the same time yield a higher social return.

We envision thus, future work on the field of identification of market structures that allow cooperation. In the context of the present work, it is important to identify if cooperation remains feasible in a general sense under complementarity and the conditions that allow it. This in turn leads the way for the identification of the conditions of market structure that yield cooperative markets.

References

1. Akerlof, G.A., Yellen, J.L.: A near-rational model of the business cycle, with wage and price inertia. Q. J. Econ. **100**, 823–838 (1985)
2. Akerlof, G.A., Yellen, J.L.: Can small deviations from rationality make significant differences to economic equilibria? Am. Econ. Rev. **75**(4), 708–720 (1985)
3. Cellini, R., Lambertini, L.: Dynamic oligopoly with sticky prices: closed-loop, feedback, and open-loop solutions. J. Dyn. Control Syst. **10**(3), 303–314 (2004)
4. Cournot, A.A.: Recherches sur les principes mathématiques de la théorie des richesses (1838)
5. Dockner, E.: On the relation between dynamic oligopolistic competition and long-run competitive equilibrium. Eur. J. Polit. Econ. **4**(1), 47–64 (1988)
6. Duflo, E., Banerjee, A.: Poor Economics. PublicAffairs (2011)
7. Felzensztein, C., Deans, K.R., Dana, L.P.: Small firms in regional clusters: local networks and internationalization in the Southern Hemisphere. J. Small Bus. Manag. **57**, 496–516 (2019)
8. Fershtman, C., Kamien, M.I.: Dynamic duopolistic competition with sticky prices. Econometrica J. Econometric Soc. **55**, 1151–1164 (1987)
9. Garcia-Meza, M.A., Gromova, E.V., López-Barrientos, J.D.: Stable marketing cooperation in a differential game for an oligopoly. Int. Game Theory Rev. **20** 1750028 (2018)
10. Gazel, M., Schwienbacher, A.: Entrepreneurial Fintech Clusters. Available at SSRN: https://ssrn.com/abstract=3309067 or http://dx.doi.org/10.2139/ssrn.3309067 (2019)
11. Mas-Colell, A., Whinston, M.D., Green, J.R.: Microeconomic Theory. Oxford University Press (1995)
12. Roemer, J.E.: What Is Socialism Today? Conceptions of a Cooperative Economy. Cowles Foundation Discussion Paper No. 2220 (2020). Available at SSRN: https://ssrn.com/abstract=3524617 or http://dx.doi.org/10.2139/ssrn.3524617

13. Sheshinski, E., Weiss, Y.: Inflation and costs of price adjustment. Rev. Econ. Stud. **44**(2), 287–303 (1977)
14. Simaan, M., Takayama, T.: Game theory applied to dynamic duopoly problems with production constraints. Automatica **14**(2), 161–166 (1978)
15. Spengler, J.J.: Vertical integration and antitrust policy. J. Polit. Econ. **58**(4), 347–352 (1950)
16. Yeung, D.W., Petrosjan, L.A.: Cooperative Stochastic Differential Games. Springer Science & Business Media (2006)
17. Zhu, X., Liu, Y., He, M., Luo, D., Wu, Y.: Entrepreneurship and industrial clusters: evidence from China industrial census. Small Bus. Econ. **52**, 595–616 (2019)

Chapter 12
Cooperative Solutions for the Eurasian Gas Network

Ekaterina Orlova

Abstract We relate three solutions for cooperative games, the Shapley value, the nucleolus and the core. We use an empirical case study, provided in Hubert and Orlova (2018) to analyze the liberalization of network access in the European gas market. For these games the Shapley value is not in the core. To obtain a differentiated picture of the (in)stability of an allocation, we propose the $n\epsilon$-core which is a generalization of the strong ϵ-core, and define three stability measures. We find that the liberalization of network access increases the degree of instability of the Shapley value for all three metrics. The nucleolus is a unique point in the core, hence often used to characterize stable imputations. We show that liberalization compresses the core, but not always the nucleolus corresponds well to the shifts in the minimal and maximal values which players might receive in the core.

Keywords Network access · Natural gas · Shapley value · Nucleolus · Core

12.1 Introduction

In a recent series of papers, cooperative game theory has been used to assess the power structure in the network for natural gas and how it is affected by investments in new pipelines or changes in access regulation.[1] In this applied

---

This paper is part of larger collaborative research project on the Eurasian gas network which was developed and supervised by Prof. Dr. Franz Hubert and to which Onur Cobanli made essential contributions. The draft of this paper was written in the 2014 at Humboldt University Berlin.

[1][11] consider the strategic relevance of various options to expand the network. [10] investigate three pipeline projects in detail: Nord Stream, South Stream, and Nabucco. [1] considers pipeline projects for the Central Asian region. [12] and [13] look at the liberalization of pipeline access

---

E. Orlova (✉)
RANEPA, Moscow, Russia

L. A. Petrosyan et al. (eds.), *Frontiers of Dynamic Games*,
Static & Dynamic Game Theory: Foundations & Applications,
https://doi.org/10.1007/978-3-030-51941-4_12

literature dis-aggregated network models are being used which are calibrated with real data. While the results are typically robust with respect to the assumptions on parameters, the findings turn out to be very sensitive depending on which solution from cooperative game theory, the Shapley value or the nucleolus, is used to obtain the power index. Moreover, the Shapley value appears to fit economic intuition and the empirical evidence better than the nucleolus [10, 11, 13]. At the cost of some simplification, the basic story of these papers is one of 'cutting out the middlemen'. Either a new pipeline can bypass a transit country or access to an existing pipeline is liberalized. Intuition suggests that in both cases the owner of the bottleneck facility is weakened, while customers and gas producers would be strengthened.

[10] find for the Shapley value that new pipelines weaken those transit countries which they allow to circumvent, while producers and importers gain. For the nucleolus, in contrast, these pipelines appear to be essentially irrelevant.

[13] study the effect of granting third party access to pipelines within EU on the balance of power between 'local champions' within EU, acting as middlemen, EU customers, and external natural gas suppliers.[2] In [13] we distinguish between the liberalization of access to transmission networks and liberalization of access to distribution systems. As a result, three market structures are considered. Before the onset of reforms, in the fragmented market, regional champions control local production, LNG imports, access to both transmission and distribution systems. The first step of reform is opening of access to trunk pipes that provides *free transit* of gas within EU and creates the integrated market. In the integrated market regional champions lose control over transmission, but keep control over access to local customers. The second step of reform opens access to distribution networks that provide *access to customers* in a region and leads to the fully liberalized market. In the liberalized market champions retain control only over local production and LNG imports, but they are not needed for access to local customers.

One of the results obtained in [13] refers to the influence of *full liberalization* on the players outside the EU. Under the Shapley value the customers gain less than the champions lose and one third of champions' losses leaks away to the group of external gas suppliers. For the nucleolus, in contrast, [13] obtain *pure redistribution* from champions to customers while outside producers gain nothing. Therefore, results in the Shapley value case support concerns of the critics of liberalization policy that such a policy might strengthen outside producers.[3] While the limited

---

within the European Union, with the first paper emphasizing regional effects and cartels, while the second paper's focus is on customers versus local champions.

[2] The development of liberalization process was mainly determined by several consecutive Directives of the EU Commission: Directive 98/30/EC, also known as the First Gas Directive [3], Directive 2003/55/EC, known as the Second Gas Directive [4], Directive 2009/73/EC [5] which refers to the Third Energy Package.

[3] Incumbents pointed out the strong import dependency of the EU in the gas industry and argued that

there is a need for a limited number of strong market players in order to deal with the high level of concentration of gas producers outside the European Union [2, Second Phase, p. 207].

empirical evidence supports the Shapley value as the more appropriate solution for this network [10, 11], at least two major questions remain.

First, the Shapley value, if not in the core, may be an unlikely outcome because it lacks stability. We find that for none of the variants considered in [13] the core is ever empty, but the Shapley value never belongs to the respective core. Thus, we find that the Shapley value is unstable for each market structure. This result, however, does not tell us anything about the degree of instability of the Shapley value and how it depends on the market structure. Stability is black or white. In an applied analysis we need measures of stability that can be used, once an allocation is not in the core. In this paper we introduce stability measures and analyze the degree of instability of the Shapley value depending on the market structure.

Second, the nucleolus is just one element of the core and may be misleading when used to measure the impact of some change on the set of stable allocations. In this paper we study whether the effect of liberalization on the nucleolus is a good indicator of the impact of reform on the core.

Therefore, using the model of the Eurasian gas network developed in [13], in this paper we focus on different questions. In [13] we analyze power redistribution from the reform, applying the Shapley value and the nucleolus. Here we focus on the stability issue in the Shapley value case and discuss how the change of the nucleolus is related to the change of the core.

In an applied analysis it might be useful to have a metric which allows for different degrees of stability. We take into account that it might be more difficult to set up larger coalitions, but do it in a different manner as compared to the weak ϵ-core [20]. We approximate the core by relaxing the strong ϵ-core concept with respect to the size of deviating coalitions. For a group of players the decision to deviate from the proposed allocation should involve not only the computation of the respective value function, but also the agreement about the rent sharing. We suggest to consider whether the allocation is stable with respect to the set of coalitions, the size of which can be limited from above. The more we restrict the size of coalitions, the larger becomes the relaxed core, as the payoff allocation has to satisfy the smaller number of conditions. We call the relaxed core the $n\epsilon$-core.

The $n\epsilon$-core enables to introduce three stability measures related to the coalition size and the costs. In general, the first metric is based on the minimal costs of establishing a coalition for a given upper bound on the size of coalitions. The second metric refers to the minimal number of players, which are necessary for setting up a coalition to veto the payoff for a given costs of establishing a coalition. The third metric is a probabilistic one. It is based on the probability of picking up a deviating coalition for the given costs and for the given upper bound on the coalition size. To analyze how the instability of the Shapley value changes with liberalization, we apply the three measures to our real life model. We find that liberalization increases the degree of instability. Opening of access to pipelines increases the minimal costs of setting up a coalition that provide the stability of the Shapley value, decreases the minimal number of players in a deviating coalition and raises the probability of selecting such a coalition if we select coalitions at random.

To study whether the impact of reform on the nucleolus is a good indicator of the influence on the core, we proceed in two steps. As the first step, we analyze the effect of liberalization on the core. To deal with the numerous inequalities, characterizing the core, we partially describe it by computing the minimal and the maximal gains of players in the core. Then, we discuss whether the change of the nucleolus is a good indicator of the change of the core.

We find that liberalization compresses the core. The core in case of the fully liberalized market is contained in the core of the integrated market, which, in its turn, is contained in the core of the fragmented market. According to this compression, the full liberalization shrinks the range of values between the minimal and the maximal payoffs of all players in the core. The impact of full liberalization is dominated by the second step of reform for the EU champions and the customers, but by the first step for the countries outside EU and for the EU regions without champions and customers. For the champion and the customers in a region the compression of the range is of the same magnitude, but is determined by different factors. For all champions the compression is a result of the decrease of maximal gains in the core. For all customers the range decreases because the minimal gains increase.

We are interested in how the change of the nucleolus is related to the change of the core.[4] We find that in the fragmented market the nucleolus of a player tends to be centrally located, i.e. in the middle between the minimal and the maximal payoffs in the core. For each step of reform for a number of players the nucleolus and the respective midpoint shift into the same direction. However, for each step we find examples of movement into the opposite direction and cases when the nucleolus changes, but the range is not affected. Overall, it is difficult to infer the pattern of the impact on the core from the change of the nucleolus. At the same time, as the core compresses, the nucleolus becomes a more precise estimate of a point in the core in the liberalized market.

To the best of our knowledge, there are no applied studies devoted to ϵ-cores and stability issues which are calibrated with real data. In addition, the paper contributes to the quantitative studies using the cooperative approach. Application of the cooperative game theory to the real world problems is mainly limited to the *voting games* [19] and the *cost allocation problems* [21]. While the Shapley value is the most widely used measure of voting power from the cooperative game theory [8, 18], in the literature on the cost allocation problems various solutions are applied, including the nucleolus, the Shapley value and the core.[5] For example, in the series

---

[4]Two findings about the compression of the range: (i) that the total effect of reform is dominated by the second step only for the EU champions and the customers and (ii) that the losses of the champion and the customers in a region are of the same magnitude, correspond to the results in case of nucleolus. In [13] we find that in case of nucleolus the second step of reform dominates the effect of full liberalization only for the EU champions and the customers, and that full liberalization leads to pure redistribution of power between the champion and the customers in a region.

[5]See [9] for a detailed review of studies, applying cooperative game theory to the cost allocation problems.

of papers the landing fees for Birmingham airport were computed using the Shapley value and the nucleolus. [15] find that the fees in the Shapley value case are larger than the actual charges for the smallest and the largest aircrafts. [14] receives the similar results for the nucleolus. [16] find that the structure of movement fees based on the Shapley value approximates the actual structure of charges better than the set of fees given the nucleolus. Comparison of the cost allocations resulted from the different solutions with the real tariffs was also conducted by [6, 7] and [22]. Our paper refers to the empirical studies comparing the Shapley value, the nucleolus and the core.

The paper is organized as follows. In Sect. 12.2 we describe the $n\epsilon$-core and introduce stability measures, in Sect. 12.3 we compare the Shapley value, the nucleolus and the core according to the amount of power allocated to players. We study the core of the games and report the influence of liberalization on the minimal and the maximal values achievable in the core in Sect. 12.4. In Sect. 12.5 we relate the nucleolus and the core. In Sect. 12.6 we relate the Shapley value and the core by studying the degree of instability of the Shapley value.

12.2   The $n\epsilon$-Core and Stability Measures

The inter-dependencies among the players in the Eurasian gas network can be represented by a game in value function form $\Gamma = (N, v)$, where N is the set of players and the value function $v : 2^{|N|} \to R_+$ gives the maximal payoff, which a coalition $S \subseteq N$ can achieve [13]. Let x be a vector of payoffs. We denote the set of payoff vectors which are efficient $\Sigma_{i \in N} x_i = v(N)$ as $X^*(\Gamma)$. Let $x(S) = \Sigma_{i \in S} x_i$, $S \subseteq N$ be the corresponding payment to a group of players S. If the excess $e(x, S) = v(S) - x(S)$ is positive for a coalition, these players could block or veto x.

[20] proposed a useful generalization of the core, the so called strong ϵ-core, which requires that the gains from blocking x must not be larger than a threshold ϵ, formally $c(\epsilon) = \{x \in X^*(\Gamma) : e(x, S) \leq \epsilon, \forall S \subset N\}$. The strong ϵ-core is the set of payoffs that cannot be vetoed by any coalition if establishing a coalition entails a fixed cost of ϵ (a negative ϵ indicates a bonus). The authors defined also the weak ϵ-core by making the costs of setting up a coalition proportional to the size of coalition. Formally, $c_w(\epsilon) = \{x \in X^*(\Gamma) : e(x, S) \leq \epsilon|S|, \forall S \subset N\}$.

Here we propose to relax the strong ϵ-core with respect to the *coalition size n* in a different way than the weak ϵ-core. We allow for fixed costs of setting up a coalition and control for the number of players in a deviating coalition. We introduce an upper bound on the size of coalitions which provides stability of an allocation.

Our approach can be motivated by the following thought experiment. Consider a game (N, v). The players have to agree on a proposed payoff x. As a first step, every single player checks whether the offer is individually acceptable. In total this requires the computation of $|N|$ values $v(\{i\})$. Next, pairs of players consider whether to object x. To do so another $|N|(|N| - 1)$ values have to be computed and, upon finding that the excess is large enough, a pair would have to agree on how to share before seriously blocking the proposal.[6] Then we move on to groups of three players, then four and so on. As we reach ever larger coalitions, not only the number of necessary computations might grow, also the complexity of organizing the group will increase. Instead of relating coalition size to these cost in a particular way, we propose to account for the group size directly. Let $\mathscr{S}(n)$, $1 \leq n < |N|$ denote the set of coalitions which can be formed by permutations of at most n players: $\mathscr{S}(n) = \{S \subset N : |S| \leq n, \ S \neq \emptyset, N\}$. We define the $n\epsilon$-core as $c(n, \epsilon) = \{x \in X^*(\Gamma) : e(x, S) \leq \epsilon, \forall S \in \mathscr{S}(n)\}$. Besides the fixed cost of setting up a coalition to veto x, the $n\epsilon$-core can also account for the fact that it might be more costly to set up larger coalitions. The larger we select ϵ and the smaller we select n, the larger will be the $n\epsilon$-core. Obviously, $c(1, 0)$ is equivalent to individual rationality: $x_i \geq v(\{i\})$, $i \in N$. The strong ϵ-core is $c(|N| - 1, \epsilon)$ and $c(|N| - 1, 0)$ yields the core.

With the $n\epsilon$-core we have two dimensions to measure the stability of a given payoff x. For a given n we can look for the minimal $\epsilon^*(x, n)$ so that $x \in c(n, \epsilon^*)$ or we can ask for the minimal $n^*(x, \epsilon)$ so that $x \in c(n^* - 1, \epsilon)$. In other words, $n^*(x, \epsilon)$ denotes the minimal number of players which are necessary to veto a payoff x.

Finally, we take $c(n, \epsilon)$ as given. For a payoff vector x not in $c(n, \epsilon)$, we assess the 'degree' of instability by comparing the number of coalitions which could gain from vetoing x to the total number of coalitions formed by permutations of at most n players. Let $\hat{\mathscr{S}} = \{S : S \in \mathscr{S}(n) \text{ and } e(x, S) > \epsilon\}$. The larger the fraction $f(x, n, \epsilon) = |\hat{\mathscr{S}}|/|\mathscr{S}(n)|$ is, the more likely it is that we pick a coalition rejecting x if we select coalitions at random.

12.3   Concepts: Power Allocation

In this section we study how the Shapley value, the nucleolus and the core are related to each other with respect to the power in our real world game. We use the same model of the Eurasian gas network as in [13]. We refer the interested reader to [13] for the definition of players (we have 20 players), for details of the model calibration and the value function calculation, for the description of games determined by the three access regimes. As both the Shapley value and the nucleolus were computed in [13], here we have to characterize the core.[7]

According to the definition of the core, its characterization involves $2^{|N|} - 2$ (in our case over a million) inequalities, and hence, it is of limited practical use as such. To deal with such large set of inequalities we introduce the partial description of the

---

[6]For the game with $|N|$ players there are $|N|(|N| - 1)/2$ pairs of players, but each player in a pair has to implement the calculation so that in total $|N|(|N| - 1)$ values will be computed. For the groups of k players the total number of computations is $kc_k^{|N|}$.

[7]The analyzed games are not convex, but the core is never empty for the analyzed games.

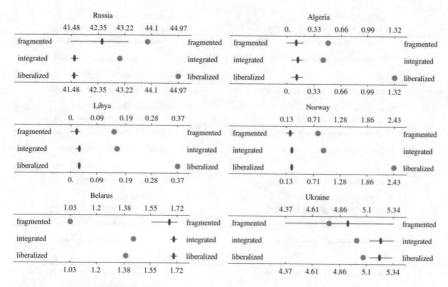

Fig. 12.1 Different solutions for external producers and transit countries. The grey bar presents the min-max range of a player in the core. Blue elliptical disks and red circles present the nucleolus and the Shapley values, respectively. All figures are in percentage of the total surplus

core. The partial description involves finding the minimal and the maximal gains of a player in the core. In the following we refer to the interval of values between the minimum and the maximum as to the 'min-max range' of a player in the core. We compute the min-max range for all players. It is important to note that one has to be careful with the interpretation of any vector with coordinates taken from the min-max ranges. Not all such vectors will belong to the core.

We report the results for the short-sighted scenario, when investment options are not available for a coalition, and a high value of demand intercept.[8] The results are robust to changes of parameters (see Appendix).

All results are presented in the graphs (see Figs. 12.1, 12.2, 12.3). For each player and for all three market structures we depict the min-max range as the grey bar. All figures are given as percentage of the total surplus. We also present the nucleolus and the Shapley value as the blue elliptical disks and the red circles, respectively. Trivially, as the nucleolus belongs to the core, it lies in the min-max range.

The Shapley value assigns more power to Russia, Norway, Algeria and Libya than the nucleolus (Fig. 12.1). Moreover, all external producers get larger shares than the corresponding maximal values achievable in the core. Both Belarus and Ukraine have less power under the Shapley value as compared to the nucleolus. In addition, Belarus is assigned less power than the respective minimal values in the core. The same pattern holds for Ukraine in the integrated and liberalized markets.

---

[8]These are the basic parameter settings in [13].

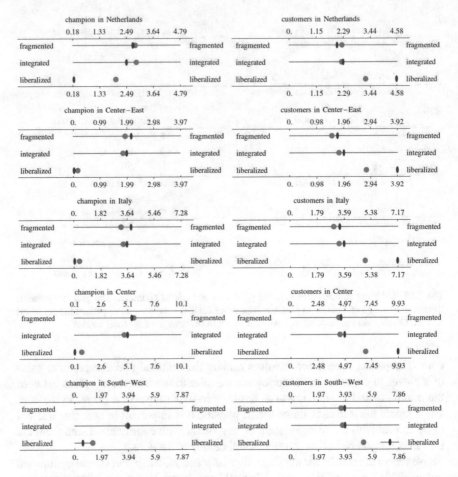

Fig. 12.2 Different solutions for EU champions and customers

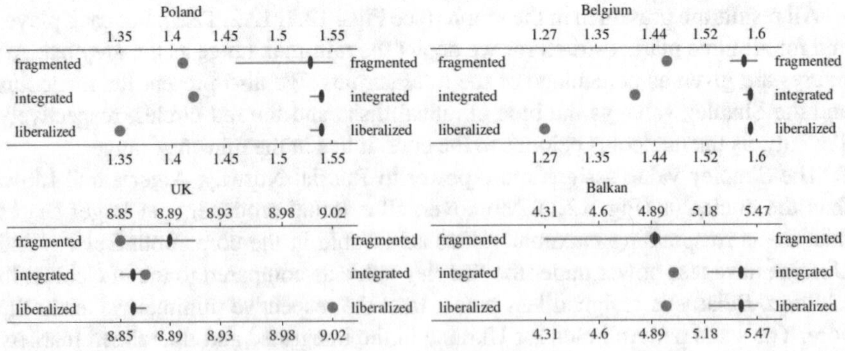

Fig. 12.3 Different solutions for EU regions

In the fragmented market the share of Ukraine under the Shapley value falls into the min-max range.

For the EU champions and the customers the results depend on the market structure (Fig. 12.2). In the fragmented and integrated markets the Shapley values belong to the respective min-max ranges. The relation of the power under the Shapley value and the nucleolus depends on the player and the access regime. For example, the Shapley value assigns less power to the champions in Italy, South-West and Center-East regions, but allocates more power to Netherlands' champion than the nucleolus. The latter holds for the champion in Center region only in the fragmented market. In the liberalized market the Shapley value assigns more power to all EU champions than the nucleolus. Moreover, their shares are larger than the corresponding maximal values in the core. The opposite pattern emerges for the EU customers. They appear less powerfull under the Shapley value than under the nucleolus and receive less than their minimal values in the core.

We present the results for the EU regions without champions and customers in Fig. 12.3. With few exceptions, we observe that the Shapley value allocates less power to these regions as compared to the nucleolus and that it does not belong to the min-max range.

12.4   Liberalization: Compression of the Core

As it can be seen from Figs. 12.1, 12.2, 12.3, liberalization compresses the min-max range for all players. If the compression results in a small scope between the minimal and the maximal values, then we will not find the allocation in the core which is very different from the nucleolus.

To understand why we observe the compression of the core, consider two games $\Gamma^0 = (N, v^0)$ and $\Gamma^1 = (N, v^1)$ with non-empty cores. Let games Γ^0 and Γ^1 have the same set of players and the same values of grand coalition: $v^0(N) = v^1(N)$ (as well as the same values of the empty set: $v^0(\emptyset) = v^1(\emptyset) = 0$). For other coalitions the value function of Γ^1 either increases or does not change in comparison to the value function of Γ^0: $v^1(S) \geq v^0(S) \ \forall S \neq N$. We denote the core for the game Γ^0 as c^0 and for the game Γ^1 as c^1. Then, by definition of the core, $c^1 \subseteq c^0$.

Consecutive liberalization of access to the transmission and distribution systems does not change the total surplus due to calibration [13], but either increases the value function or does not change the value function for any other coalition: $v^2(S) \geq v^1(S) \geq v^o(S) \ \forall S \neq N$. In other words, the value function never decreases for any coalition. We know, that the core is not empty for each market structure. Then, obviously, the core compresses with each step of liberalization. The core in the fragmented market case contains the core in the integrated market case, and the latter contains the core in the fully liberalized market case. Compression of the core is reflected in the change of minimal and maximal values that players get in the core.

Typically, it is difficult to explain why the minimal and the maximal values change. But occasionally the effect is simple. The maximal value of a player in the core might be given by his contribution to the grand coalition. The player cannot require a higher payment, as then the rest of the players will 'kick' him out and form the coalition on their own. If for either two market structures the maximum of a player is equal to the respective contribution, it is enough to consider the effect of reform on the contribution. The minimal value of a player in the core, in the simplest case, is given by his stand alone value. In other words, by the amount that the player can assure on his own. Liberalization does not influence the stand alone values of players. If for either two market structures the minimum of a player is equal to his stand alone value, then trivially there is no effect on the minimum. If the minimum of a player is determined by the binding individual rationality constraint only for the initial market structure, we proceed as following. For the new access regime we find the coalitions corresponding to the binding constraints. Then we consider the effect of reform on the values of such coalitions.

We study the effect of each step of reform and the overall impact of full liberalization on the minimal value, the maximal value and the difference between the two. We refer to the difference between the maximal and the minimal values as to the range or the span. Results are presented in Table 12.1. All figures are expressed as percentage of the redistributed amount resulted from the full liberalization.[9] In the columns 2–4 we report the influence of liberalization of access to the high-pressure pipelines on all three values. In the columns 5–7 we report the incremental effect of liberalization of access to the distribution systems. In the columns 8–10 the impact of full liberalization is shown. The range can be reduced either because of the increase of minimal value or the decrease of maximal value, or because of both changes. We report the impact on the minimal and the maximal values and point out the change, which is the most important for the compression.

We start analysis with the first step of liberalization. For all EU champions, except the champion in South-West, we observe a modest compression. The decrease of maximal values tends to be more significant than the increase of minimal values.[10] In the fragmented market for all champions the maximal payoffs are determined by the respective contributions to the grand coalition. The pattern is similar in the integrated market.[11] Consequently, we may consider the effect of the first step on the contributions. Opening of access to trunk pipes decreases

[9]The redistributed amount from the full liberalization is equal to the sum of benefits of those players who gain from two steps of reform. For the estimates of redistribution given nucleolus see [13].

[10]The only exception is the champion in Netherlands, for the champion in Center-East the two effects are shown as equal due to rounding.

[11]In the integrated market only for the champion in Center the calculated maximal value is less than the respective contribution. But the difference between the respective contribution and the maximum is minor.

Table 12.1 Impact of liberalization on the minimal/maximal values in the core

	Change of minimal/maximal values in the core [% of redistribution]								
	Step 1: transmission			Step 2: distribution			Two steps together		
	Δmin / Δmax / Δspan			Δmin / Δmax / Δspan			Δmin / Δmax / Δspan		
Outside countries									
Russia	0.1	−8.9	−9.0	0.0	−0.1	−0.1	0.1	−9.0	−9.1
Belarus	0.6	0.0	−0.6	0.0	0.0	0.0	0.6	0.0	−0.6
Ukraine	4.2	0.0	−4.2	0.0	0.0	0.0	4.2	−0.1	−4.2
Algeria	0.3	0.0	−0.3	0.0	0.0	0.0	0.3	0.0	−0.3
Libya	0.1	0.0	−0.1	0.0	0.0	0.0	0.1	0.0	−0.1
Norway	0.5	0.0	−0.6	0.0	0.0	0.0	0.5	−0.1	−0.6
Netherlands									
Champion	0.1	0.0	−0.1	0.0	−25.2	−25.2	0.2	−25.2	−25.4
Customers	0.0	0.0	0.0	25.2	0.0	−25.2	25.2	0.0	−25.2
Center-East[a]									
Champion	0.1	−0.1	−0.2	0.0	−21.7	−21.7	0.1	−21.8	−21.9
Customers	0.0	0.0	0.0	21.6	0.0	−21.6	21.6	0.0	−21.6
Italy									
Champion	0.2	−0.4	−0.6	0.0	−39.6	−39.6	0.2	−40.0	−40.2
Customers	0.0	0.0	0.0	39.4	0.0	−39.4	39.4	0.0	−39.4
Center[b]									
Champion	0.1	−0.4	−0.4	0.0	−54.8	−54.8	0.1	−55.2	−55.3
Customers	0.0	−0.1	−0.1	54.6	0.0	−54.6	54.6	−0.1	−54.7
South-West[c]									
Champion	0.0	0.0	0.0	0.0	−37.0	−37.0	0.0	−37.0	−37.0
Customers	0.0	0.0	0.0	36.3	0.0	−36.3	36.3	0.0	−36.3
Poland	0.4	0.0	−0.4	0.0	0.0	0.0	0.4	0.0	−0.4
Belgium	0.1	0.0	−0.2	0.0	0.0	0.0	0.2	0.0	−0.2
United Kingdom	0.0	−0.2	−0.3	0.0	0.0	0.0	0.0	−0.2	−0.3
Turkey & Balkan[d]	4.9	0.0	−4.9	0.0	0.0	0.0	5.0	0.0	−5.0

[a] Austria, Czech Republic, Slovakia, Hungary, Serbia and Slovenia
[b] Germany, Switzerland, Denmark and Luxembourg
[c] France, Spain and Portugal
[d] Romania, Bulgaria and Greece

contributions of champions to the grand coalition as in the integrated market the gas can be shipped freely within EU.[12]

For the EU customers, except the customers in Center, there is no compression of the range. In the fragmented and integrated markets for all customers the minimal values are determined by the respective stand alone values. As a result, we do not observe any impact on the minimal values. The maximal values of all customers,

---

[12]The contribution of the champion in Netherlands is not affected by the first step of reform.

except the customers in Center, are determined by the corresponding contributions to the grand coalition.[13] Neither step of reform has impact on the contribution of a customer to the grand coalition. As a result, we observe the minor decrease of the range only for the customers in Center.

The span compresses for the producers and the transit countries outside EU. For all countries, except Russia, the range decreases due to the increase of minimal values. Only for North-African countries the increase of minimal values corresponds to the simple case. In the fragmented market the minimal values of Algeria and Libya are determined by the respective stand alone values. In the integrated market producers can ship gas freely within EU and the minimal values are determined by the coalitions with regions which could not be accessed in the fragmented market. Opening of access to trunk pipes increases the values of such coalitions making the individual rationality constraints non-binding. For Russia the compression is the strongest, but the decrease of maximal value cannot be explained by the simple case. For other producers and transit countries the maximal values are either determined by the respective contributions or are slightly less than the contributions. Neither step of reform has impact on the contribution of a supplier or a transit country. As a result, if the maximal values are affected, they decrease only slightly.

The incremental impact of liberalization of access to the distribution systems on *the range* varies for the different groups of players. For all EU champions and customers the span decreases. The compression resulted from the second step of reform is much larger than from the first step. Therefore, the total impact on the range is clearly dominated by the second step for the champions and the customers. The pattern is different for all other players. The incremental compression is either zero or close to zero, and hence, tends to be substantially less than from the opening of access to high-pressure pipelines. Therefore, the total effect is dominated by the first step of reform for these players. In the following we find the main factors of compression for the EU champions and the customers.

For all champions the impact of the second step of reform on the minimal values is essentially zero. The maximal values drop substantially. Recall, that in the integrated market the maximal values tend to be determined by the respective contributions to the grand coalition. In the liberalized market this pattern holds for all champions. In contrast to the integrated market, in the liberalized market customers can be reached without a champion. Hence, the share of the total surplus which the champion can require decreases substantially. As a result, for the EU champions the compression from the full liberalization is determined by the decrease of maximal values from the opening of access to distribution networks.

The pattern is reversed for the EU customers. The impact on the maximal values is essentially zero, but the minimal values increase a lot. In the integrated market the minimal values are determined by the respective stand alone values. In the liberalized market the individual rationality constraints become non-binding. The

---

[13] Only in the integrated market for the customers in Center the calculated maximal value is slightly less than the respective contribution.

minimal values increase because opening of access to distribution networks raises the values of coalitions with customers which could not be accessed by producers in the integrated market. As a result, for the EU customers the compression from the full liberalization is determined by the increase of minimal values from the opening of access to low-pressure pipelines.

We report the results for the last column of Table 12.1. The range compresses significantly for the EU champions and the customers. Though we observe the different factors of compression, the magnitude of the loss in the span is approximately the same for the champion and the customers in a region. For example, for the champion in Netherlands the range decreases by 25.4 percentage points. The decrease for the customers is equal to 25.2 percentage points. In comparison to the champions and the customers, the magnitude of losses in the span is low for all other players. The largest compression is observed for Russia and is equal to 9.1 percentage points. It is more than two times less than the lowest shrinkage within the group of EU champions and customers.[14]

12.5   Liberalization: The Nucleolus and the Core

In [13] we find that under the nucleolus the total effect of reform on power is dominated by the second step only for the EU champions and the customers. For all other players the total effect is dominated by the first step of reform. We receive the similar result for the compression of the core. In addition, given the nucleolus, there is pure redistribution of power between the champion and the customers in a region. We find that the min-max range compresses for both the champion and the customers in a region and the losses in the range are of the same magnitude. At a first glance, these findings allow us to assume that the change of the nucleolus is a good indicator of the impact of liberalization on the core.

To check this hypothesis, for each player we consider the direction of the movement of the nucleolus as compared to the shift of the midpoint of the min-max range. The idea behind such measurement is the following. On the one hand, in the fragmented market for most of the players the nucleolus is centrally located in the min-max range (see Figs. 12.1, 12.2, 12.3).[15] On the other hand, the movement of the midpoint depicts the pattern of compression of the min-max range. If the increase of minimal value is larger than the decrease of maximal value, then the center shifts to the right. If the opposite holds, then the center shifts to the left. In the previous section we discussed the dominant effects of compression. Thus, we consider the impact of reform on the nucleolus to be a good indicator of the effect

---

[14]The lowest compression of the span within the group of EU champions and customers is equal to 21.6 percentage points and corresponds to the customers in Center-East region.

[15]The difference between the center and the nucleolus is larger than 10% only for Belarus, Belgium, Poland and UK. For these players the nucleolus is shifted to the right endpoint of the min-max range.

on the core, if the nucleolus is shifted into the same direction as the midpoint of the respective min-max range.

We start analysis with the first step of reform. The nucleolus and the respective center move into the same direction for two-third of the players. This pattern holds for all players in the group of outside producers and transit countries and for all players in the group of EU regions without champions and customers (see Figs. 12.1 and 12.3). Within the group of EU champions and customers the values shift into the same direction only for the champions in Center, Center-East and Italy. In some cases the compression 'forces' the nucleolus to move into the same direction. Consider, for example, Russia. The decrease of maximum is larger than the increase of minimum so that the midpoint shifts to the left. The nucleolus moves into the same direction. As the maximal value in the integrated market is less than the nucleolus in the fragmented market, the movement into the opposite direction is not possible. For the rest of the players we observe two cases. First, the nucleolus shifts into the opposite direction as compared to the respective midpoint. Second, the midpoint is not affected, but the nucleolus changes. The first case holds for the champions in Netherlands and South-West and for the customers in Center. The second case holds for all other customers.

These two cases are also fulfilled for the second step of reform. The nucleolus moves into the opposite direction for Algeria. For Belarus the min-max range is not affected, but the nuclelous decreases. For all other players the values shift into the same direction. The champions and the customers are exposed to the largest shift of both the nucleolus and the center. In contrast to the integrated market, in the liberalized market for the champions and the customers the nucleolus is not always centrally located in the respective min-max range. Nevertheless, the magnitude of the shift of the nucleolus and the midpoint is approximately the same due to the substantial compression of the min-max range. In other words, the nucleolus becomes a more precise estimate of a point in the core in the liberalized market.

Overall, in the fully liberalized market, as compared to the fragmented market, for all players the nucleolus and the midpoint move into the same direction. But for each step of reform we find exceptions. We find players, for which the values shift into the opposite direction as well as cases when the midpoint is not affected, but the nucleolus changes. Taken together, not always the impact of reform on the nucleolus corresponds well to the effect on the core.

12.6 Liberalization: Degree of Instability of the Shapley Value

For each access regime there are a number of coalitions which find it profitable to deviate from the Shapley value. This means that for neither market structure the Shapley value is in the core and, hence for neither market structure the Shapley value is stable. In this section we study the degree of instability of the Shapley value

Table 12.2 Impact of liberalization on stability measures

	Stability measures		
	Fragmented	Integrated	Liberalized
$\epsilon^*(\phi, |N| - 1)/ \sum \phi_{EU}^0$	1.7	2.2	8.0
$n^*(\phi, 0) - 1$	8	1	1
$f(\phi, |N| - 1, 0)$	0.0002	0.0019	0.0907

depending on the access regime. To implement this analysis we relate the Shapley values to the measures of stability introduced in Sect. 12.2. The results for the three metrics are presented in Table 12.2. The first metric to consider is the minimal costs of setting up a coalition of any size $\epsilon^*(\phi, |N| - 1)$, such that it is not profitable to deviate from the Shapley value. We report $\epsilon^*(\phi, |N| - 1)$ as percentage of the joint share of EU players in the fragmented market in the first row of Table 12.2. In the second row, for zero costs of establishing a coalition, we present the second metric. The second measure is the maximal number of players $n^*(\phi, 0) - 1$, such that all coalitions, formed by permutations of at most this number, cannot block the Shapley value. In the third row we find $f(\phi, |N| - 1, 0)$, the fraction of coalitions which could gain from vetoing the Shapley value. Computation of the third metric involves assumption that setting up a coalition of any size would not cost anything.

We start analysis with the first metric. In the fragmented market the costs of establishing a coalition have to constitute at least 1.7% of the joint rent of the EU players. Then rejection of the Shapley value becomes unprofitable for all coalitions. Opening of access to trunk pipes raises the value of threshold up to 2.2%. In the liberalized market the costs increase up to 8%, which is several times larger than the corresponding values in both the fragmented and integrated markets. With liberalization a coalition gets access to resources that were unavailable in the fragmented market and, hence, can gain more than before the reform. As a result, the costs of establishing a coalition have to increase in order to make the deviation from the respective Shapley value unprofitable.

Now we turn to the second metric, related to the size of deviating coalitions. In the following calculations we set ϵ equal to zero. For each market structure we search for $n^*(\phi, 0)$, the minimal number of players necessary for setting up a coalition to veto the corresponding Shapley value. In the fragmented market $n^*(\phi, 0) = 9$, so that the coalitions, formed by permutations of at most 8 players, cannot improve by acting on their own. In other words, in the fragmented market almost half of the players has to be in a coalition to be able to veto the Shapley value. We find two coalitions with 9 players for which the excess is positive. The two coalitions include different types of players: EU champions and customers, outside producers, transit countries for Russian gas and Turkey & Balkan region.[16] In the integrated and liberalized markets the size of deviating coalitions diminishes. In

---

[16]The coalitions are: {Algeria, Turkey & Balkan, Belarus, customers in Center-East, champion in Center-East, customers in Italy, champion in Italy, Russia, Ukraine} and {Turkey & Balkan,

both cases the minimal number of players in a deviating coalition is equal to 2. In the integrated market Belgium and Libya find it more profitable to cooperate on their own rather than to accept their Shapley values. In the liberalized market the customers in Center region can veto the Shapley value together with either the champion in Netherlands or with Norway. Thus, while in the fragmented market the deviation requires bargaining between relatively large sets of players, in the integrated/liberalized market the deviation from the Shapley value can be profitable even when there are only two players in a coalition.

The third metric is the fraction of coalitions, which could block the Shapley value. In the following calculations we set ϵ equal to zero and consider the set of coalitions which can be formed by permutations of at most 19 players ($n = |N| - 1 = 19$). We find the ratio of the number of coalitions which can veto the Shapley value and the total number of coalitions. The fraction of deviating coalitions increases when we move from the fragmented to the integrated and liberalized markets. In the fragmented market, the fraction is the lowest and is close to zero. In the integrated market, the share increases, but it is still less than 1% of the total number of relevant coalitions. The fraction increases further in the liberalized market, so that with probability 9% we can pick a coalition rejecting the Shapley value if we select coalitions at random.

Therefore, as liberalization provides access to new resources for a coalition, the attractiveness to act on its own increases in the integrated/liberalized market in comparison to the fragmented market. As a result, the instability of the Shapley value raises with liberalization with respect to all three measures. For the two measures, the costs of establishing a coalition and the fraction of deviating coalitions, the second step of reform dominates the first step with respect to the increase of the degree of instability. The opposite holds for the second criteria.

12.7 Conclusion

When applying cooperative game theory to the real world problems, we have to decide on how to solve the game. Though in the simple models different solutions may provide the same results, it can be completely misleading for the more complicated cases. In [13] we consider the Eurasian natural gas supply system and study the impact of opening access to transmission and distribution networks on the balance of power between regional champions, customers and external producers. In this case the Shapley value and the nucleolus yielded different results with respect to the power redistribution.

In general, the choice of the concept is complicated by the fact that the Shapley value and the nucleolus have different merits and shortcomings. The nucleolus

---

Belarus, customers in Center-East, champion in Center-East, customers in Italy, champion in Italy, Libya, Russia, Ukraine}.

presents the stable imputation for the game with the non-empty core, which is not necessarily true for the Shapley value. While the Shapley concept features not only the aggregate, but also the strong monotonicity, the nucleolus does not satisfy even the property of the aggregate monotonicity [17, 23].

In this paper we use the model of the Eurasian natural gas supply system to relate the Shapley value and the nucleolus to the core. We examine the degree of instability of the Shapley value and how it depends on the market structure. We study whether the effect of liberalization on the nucleolus is a good indicator of the influence on the core.

To evaluate the degree of instability of a payoff allocation which is not in the core, we propose several stability measures. We relax the strong ϵ-core concept by taking into account the size of deviating coalitions. Using the $n\epsilon$-core one can study whether the payoff allocation is stable with respect to the set of coalitions, the size of which is bounded from above. We introduce three stability measures related to the coalition size and the costs of setting up a coalition. We find that liberalization increases the instability of the Shapley value for all criteria.

To study whether the change of the nucleolus might be considered as an indicator of the change of the core in our model, we first analyze the impact of liberalization on the core. We find that liberalization consecutively compresses the core. The compression is depicted in the decrease of the range of values between the minimum and the maximum of a player in the core. As the nucleolus tends to be centrally located in the min-max range in the fragmented market, we compare the direction of the shift of the nucleolus with the movement of the respective midpoint. For each step of reform we find players characterized by the movement of values into the opposite direction. In addition, we also find examples when the min-max range is not affected, but the nucleolus changes. Hence, it is difficult to judge about the change of the core on the basis of the impact of liberalization on the nucleolus.

Taken together, the Shapley value suits better for the application to the Eurasian natural gas system. As it is pointed out in [13], the results under the Shapley value correspond to the intuition derived from the middleman story. Though the instability of the Shapley value increases for all criteria, the degree of increase differs between the metrics. The first measure, the minimal costs, and the third measure, the fraction of deviating coalitions, provide less sharp results than the second measure. Liberalization consecutively increases the minimal costs of establishing a coalition, but even in the fully liberalized market this amount does not exceed 10% of the joint rent of EU players. The fraction of deviating coalitions in the liberalized market never exceeds 20%. Simultaneously, according to the second metric, only two players are enough to reject the Shapley value in the fully liberalized market. Looking at all three metrics together, we find it easier to apply the Shapley value in the fragmented market. So that taking into account the second criteria, application of the Shapley value in the integrated and liberalized markets requires more caution.

Appendix

As in [13], we assess the robustness of our results by considering three more variants: a high value of demand intercept and the far-sighted scenario, a low value of intercept and the short-sighted scenario, the low value of intercept and the far-sighted scenario. We will discuss the robustness of our results in the same order as they are reported in the main text.

Power Allocation

We start analysis with the comparison of concepts according to the power allocation (see Figs. 12.4, 12.5, 12.6, 12.7, 12.8, 12.9, 12.10, 12.11, 12.12). With minor modifications, all previous statements from the main text could be repeated for each of three scenarios. For example, for all variants of parameters it holds that the Shapley value assigns more power to all outside producers than the nucleolus and the core. It also holds that Belarus and Ukraine have less power in the Shapley value case as compared to the nucleolus. Moreover, for Belarus the shares are less than the respective minimal values in the core. Only for Ukraine it depends on the scenario and the access regime whether the Shapley value falls into the min-max range.

For all scenarios, in the fragmented and integrated markets, the Shapley values of champions and customers belong to the respective min-max ranges. All results for the liberalized market can be repeated. For all scenarios the Shapley value assigns

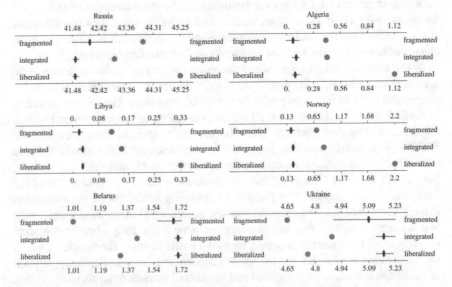

Fig. 12.4 Different solutions for external producers and transit countries (far-sighted scenario, high intercept). The grey bar presents the min-max range of a player in the core. Blue elliptical disks and red circles present the nucleolus and the Shapley values, respectively. All figures are in percentage of the total surplus

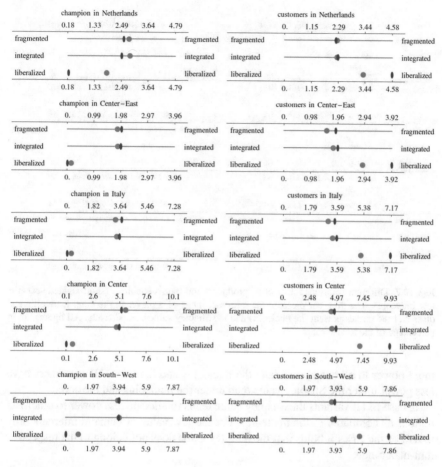

Fig. 12.5 Different solutions for EU champions and customers (far-sighted scenario, high intercept)

Fig. 12.6 Different solutions for EU regions (far-sighted scenario, high intercept)

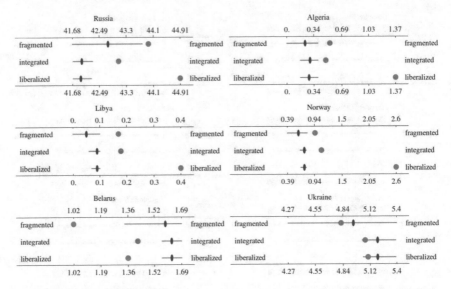

Fig. 12.7 Different solutions for external producers and transit countries (short-sighted scenario, low intercept). The grey bar presents the min-max range of a player in the core. Blue elliptical disks and red circles present the nucleolus and the Shapley values, respectively. All figures are in percentage of the total surplus

more power to all champions than the nucleolus and the core. All customers have less power under the Shapley value than under the nucleolus and the core.

In case of all variants the Shapley value tends to allocate less power to other EU regions as compared to the nucleolus. For the low value of demand intercept only in half of the cases it holds that the Shapley value does not belong to the respective min-max range.

Liberalization: Compression of the Core

The impact of liberalization on the minimal values, the maximal values and the range is presented in Tables 12.3, 12.4, 12.5. For all scenarios it holds that the total effect on the range is dominated by the second step of reform only for the EU champions and the customers, but by the first step of reform for all other players. For the champions the compression of the range is determined by the decrease of maximal values resulted from the second step of reform. For the customers the compression is determined by the increase of minimal values from the second step. The statements concerning the influence of liberalization on the minimum and the maximum hold for all scenarios.[17] For example, in the fragmented and

---

[17]Minor modifications in the statements might refer to the maximal values. In the basic scenario for a number of maximal values it holds that the value is equal to the respective contribution. With the change of parameters some of these maximal values become slightly less than the respective contributions.

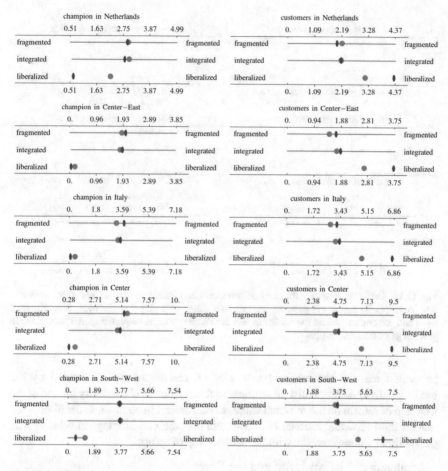

Fig. 12.8 Different solutions for EU champions and customers (short-sighted scenario, low intercept)

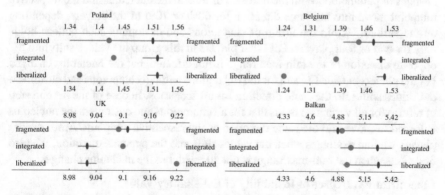

Fig. 12.9 Different solutions for EU regions (short-sighted scenario, low intercept)

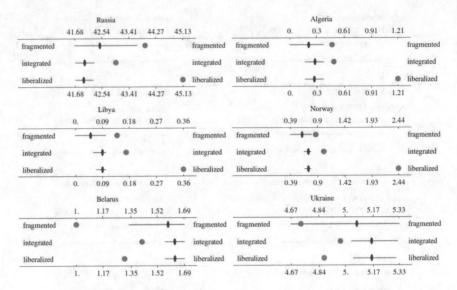

Fig. 12.10 Different solutions for external producers and transit countries (far-sighted scenario, low intercept). The grey bar presents the min-max range of a player in the core. Blue elliptical disks and red circles present the nucleolus and the Shapley values, respectively. All figures are in percentage of the total surplus

integrated markets the minimal values of all customers are determined by the binding individual rationality constraints so that we do not observe any impact of the first step of reform on the minimal values of customers. In contrast, in the liberalized market the individual rationality constraints become non-binding. In other words, the minimal values increase with the second step of reform.

Liberalization: The Nucleolus and the Core

The main results concerning the relation of the nucleolus and the core are robust to changes of parameters. With the first step of reform the nucleolus and the respective midpoint move into the same direction for 60% or 70% of the players, depending on the variant. Within the group of champions and customers such pattern holds only for two or four players. For other players in this group the values shift into the opposite direction or the min-max range is not affected, but the nucleolus changes. For the players outside EU and for the EU regions without champions and customers the values shift into the same direction for all scenarios. In case of the second step of reform for all variants it holds that for all champions and customers the nucleolus is forced to move into the same direction as the respective midpoint. Among other players we find examples when the values shift into the opposite direction. We also find cases when the min-max range is not affected, but the nucleolus changes.

Liberalization: Degree of Instability of the Shapley Value

Results concerning the degree of instability of the Shapley value are robust to changes of parameters (Tables 12.6, 12.7, 12.8). For all scenarios it holds that

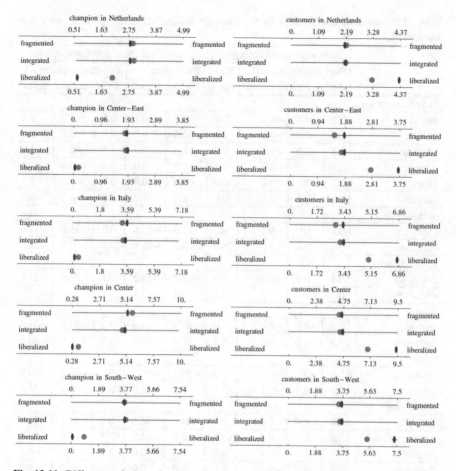

Fig. 12.11 Different solutions for EU champions and customers (far-sighted scenario, low intercept)

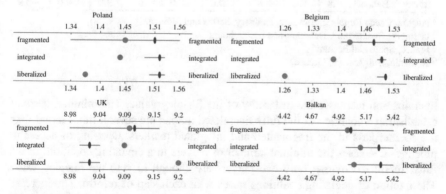

Fig. 12.12 Different solutions for EU regions (far-sighted scenario, low intercept)

Table 12.3 Impact of liberalization on the minimal/maximal values in the core (far-sighted scenario, high intercept)

	Change of minimal/maximal values in the core [% of redistribution]								
	Step 1: transmission			Step 2: distribution			Two steps together		
	Δmin / Δmax / Δspan			Δmin / Δmax / Δspan			Δmin / Δmax / Δspan		
Outside countries									
Russia	0.1	−6.7	−6.7	0.0	−0.1	−0.1	0.1	−6.7	−6.8
Belarus	0.6	0.0	−0.6	0.0	0.0	0.0	0.6	0.0	−0.6
Ukraine	1.3	0.0	−1.3	0.0	0.0	0.0	1.3	−0.1	−1.4
Algeria	0.3	0.0	−0.3	0.0	0.0	0.0	0.3	0.0	−0.3
Libya	0.1	0.0	−0.1	0.0	0.0	0.0	0.1	0.0	−0.1
Norway	0.5	0.0	−0.6	0.0	0.0	0.0	0.5	−0.1	−0.6
Netherlands									
champion	0.1	0.0	−0.1	0.0	−25.8	−25.8	0.2	−25.8	−25.9
customers	0.0	0.0	0.0	25.8	0.0	−25.8	25.8	0.0	−25.8
Center-East[a]									
champion	0.1	−0.1	−0.2	0.0	−22.2	−22.2	0.1	−22.3	−22.4
customers	0.0	0.0	0.0	22.1	0.0	−22.1	22.1	0.0	−22.1
Italy									
champion	0.2	−0.4	−0.6	0.0	−40.5	−40.6	0.2	−41.0	−41.2
customers	0.0	0.0	0.0	40.3	0.0	−40.3	40.3	0.0	−40.3
Center[b]									
champion	0.1	−0.4	−0.5	0.0	−56.1	−56.1	0.1	−56.5	−56.5
customers	0.0	−0.1	−0.1	55.9	0.0	−55.9	55.9	−0.1	−56.0
South-West[c]									
champion	0.0	0.0	0.0	0.0	−44.1	−44.1	0.0	−44.1	−44.1
customers	0.0	0.0	0.0	43.7	0.0	−43.8	43.7	0.0	−43.8
Poland	0.4	0.0	−0.4	0.0	0.0	0.0	0.4	0.0	−0.4
Belgium	0.1	0.0	−0.2	0.0	0.0	0.0	0.2	0.0	−0.2
United Kingdom	0.0	−0.2	−0.3	0.0	0.0	0.0	0.0	−0.2	−0.3
Turkey & Balkan[d]	3.8	0.0	−3.8	0.0	0.0	0.0	3.9	0.0	−3.9

[a] Austria, Czech Republic, Slovakia, Hungary, Serbia and Slovenia
[b] Germany, Switzerland, Denmark and Luxembourg
[c] France, Spain and Portugal
[d] Romania, Bulgaria and Greece

liberalization increases the instability of the Shapley value. The minimal costs of establishing a coalition in the fully liberalized market are several times larger than the counterparts in the fragmented and integrated markets. Opening of access to pipelines decreases the minimal number of players in a deviating coalition. In the fully liberalized market only two players are enough to veto the Shapley value. The fraction of deviating coalitions raises with each step of reform. The increase realized from the second step of liberalization is larger than from the first step for all scenarios.

Table 12.4 Impact of liberalization on the minimal/maximal values in the core (short-sighted scenario, low intercept)

| | Change of minimal/maximal values in the core [% of redistribution] | | | | | | | | |
| | Step 1: transmission | | | Step 2: distribution | | | Two steps together | | |
	Δmin / Δmax / Δspan			Δmin / Δmax / Δspan			Δmin / Δmax / Δspan		
Outside countries									
Russia	0.1	−8.8	−9.0	0.0	−0.2	−0.2	0.1	−9.0	−9.1
Belarus	1.3	0.0	−1.3	0.0	0.0	0.0	1.3	0.0	−1.3
Ukraine	4.5	−0.1	−4.6	0.1	0.0	−0.1	4.6	−0.1	−4.7
Algeria	1.0	0.0	−1.0	0.0	0.0	0.0	1.0	0.0	−1.0
Libya	0.3	0.0	−0.3	0.1	0.0	−0.1	0.4	0.0	−0.4
Norway	1.6	−0.1	−1.7	0.0	0.0	0.0	1.6	−0.1	−1.7
Netherlands									
champion	0.5	0.0	−0.5	0.0	−25.4	−25.5	0.5	−25.4	−25.9
customers	0.0	0.0	0.0	25.3	0.0	−25.3	25.3	0.0	−25.3
Center-East[a]									
champion	0.3	−0.2	−0.5	0.0	−22.1	−22.1	0.3	−22.3	−22.6
customers	0.0	0.0	0.0	21.7	0.0	−21.7	21.7	0.0	−21.7
Italy									
champion	0.5	−1.3	−1.8	0.1	−40.3	−40.4	0.6	−41.6	−42.2
customers	0.0	0.0	0.0	39.6	0.0	−39.6	39.6	0.0	−39.6
Center[b]									
champion	0.2	−1.2	−1.4	0.0	−55.7	−55.7	0.2	−56.9	−57.1
customers	0.0	−0.3	−0.3	55.1	0.0	−55.2	55.1	−0.3	−55.4
South-West[c]									
champion	0.0	−0.1	−0.1	0.0	−37.8	−37.9	0.0	−38.0	−38.0
customers	0.0	0.0	0.0	36.7	0.0	−36.8	36.7	−0.1	−36.8
Poland	0.9	−0.2	−1.1	0.1	0.0	−0.1	1.0	−0.2	−1.1
Belgium	0.5	−0.1	−0.6	0.0	0.0	0.0	0.5	−0.1	−0.6
United Kingdom	0.1	−0.7	−0.8	0.0	0.0	0.0	0.1	−0.7	−0.8
Turkey & Balkan[d]	3.4	0.0	−3.4	0.2	0.0	−0.2	3.6	0.0	−3.6

[a] Austria, Czech Republic, Slovakia, Hungary, Serbia and Slovenia
[b] Germany, Switzerland, Denmark and Luxembourg
[c] France, Spain and Portugal
[d] Romania, Bulgaria and Greece

Table 12.5 Impact of liberalization on the minimal/maximal values in the core (far-sighted scenario, low intercept)

	Change of minimal/maximal values in the core [% of redistribution]								
	Step 1: transmission			Step 2: distribution			Two steps together		
	Δmin / Δmax / Δspan			Δmin / Δmax / Δspan			Δmin / Δmax / Δspan		
Outside countries									
Russia	0.1	−8.4	−8.5	0.0	−0.2	−0.2	0.1	−8.5	−8.7
Belarus	1.4	0.0	−1.4	0.0	0.0	0.0	1.4	0.0	−1.4
Ukraine	2.2	−0.1	−2.3	0.1	0.0	−0.1	2.3	−0.1	−2.4
Algeria	1.0	0.0	−1.0	0.0	0.0	0.0	1.0	0.0	−1.0
Libya	0.3	0.0	−0.3	0.1	0.0	−0.1	0.5	0.0	−0.5
Norway	1.6	−0.1	−1.7	0.0	0.0	0.0	1.6	−0.1	−1.7
Netherlands									
champion	0.5	0.0	−0.5	0.0	−26.0	−26.0	0.5	−26.0	−26.5
customers	0.0	0.0	0.0	25.9	0.0	−25.9	25.9	0.0	−25.9
Center-East[a]									
champion	0.3	−0.2	−0.5	0.0	−22.6	−22.6	0.3	−22.8	−23.1
customers	0.0	0.0	0.0	22.1	0.0	−22.1	22.1	0.0	−22.1
Italy									
champion	0.5	−1.3	−1.9	0.1	−41.2	−41.3	0.6	−42.5	−43.2
customers	0.0	0.0	0.0	40.4	0.0	−40.5	40.4	0.0	−40.5
Center[b]									
champion	0.2	−1.3	−1.5	0.0	−56.9	−56.9	0.2	−58.2	−58.4
customers	0.0	−0.3	−0.3	56.3	0.0	−56.4	56.3	−0.3	−56.6
South-West[c]									
champion	0.0	−0.1	−0.1	0.0	−44.3	−44.3	0.0	−44.4	−44.4
customers	0.0	0.0	0.0	43.3	0.0	−43.3	43.3	−0.1	−43.3
Poland	0.9	−0.2	−1.1	0.1	0.0	−0.1	1.0	−0.2	−1.1
Belgium	0.5	−0.1	−0.6	0.0	0.0	0.0	0.5	−0.1	−0.6
United Kingdom	0.1	−0.7	−0.8	0.0	0.0	0.0	0.1	−0.7	−0.8
Turkey & Balkan[d]	3.4	0.0	−3.4	0.2	0.0	−0.2	3.6	0.0	−3.6

[a] Austria, Czech Republic, Slovakia, Hungary, Serbia and Slovenia
[b] Germany, Switzerland, Denmark and Luxembourg
[c] France, Spain and Portugal
[d] Romania, Bulgaria and Greece

Table 12.6 Impact of liberalization on stability measures (far-sighted scenario, high intercept)

	Stability measures		
	Fragmented	Integrated	Liberalized
$\epsilon^*(\phi, |N| - 1)/\sum \phi_{EU}^0$	1.5	2.1	7.6
$n^*(\phi, 0) - 1$	6	1	1
$f(\phi, |N| - 1, 0)$	0.0003	0.0027	0.1775

Table 12.7 Impact of liberalization on stability measures (short-sighted scenario, low intercept)

	Stability measures		
	Fragmented	Integrated	Liberalized
$\epsilon^*(\phi, |N| - 1)/\sum \phi_{EU}^0$	1.0	1.4	7.1
$n^*(\phi, 0) - 1$	8	2	1
$f(\phi, |N| - 1, 0)$	0.0001	0.0011	0.0820

Table 12.8 Impact of liberalization on stability measures (far-sighted scenario, low intercept)

	Stability measures		
	Fragmented	Integrated	Liberalized
$\epsilon^*(\phi, |N| - 1)/\sum \phi_{EU}^0$	1.0	1.2	6.6
$n^*(\phi, 0) - 1$	7	2	1
$f(\phi, |N| - 1, 0)$	0.0001	0.0014	0.1400

Acknowledgments I am very grateful to my supervisor Prof. Dr. Franz Hubert, without whom this work would be impossible. We thank Johannes H. Reijnierse for providing us with MATLAB code for calculating the nucleolus.

References

1. Cobanli, O.: Central Asian gas in Eurasian power game. Energy Policy **68**, 348–370 (2014)
2. DG Competition Report on Energy Sector Inquiry. SEC(2006) 1724, Brussels
3. Directive 98/30/EC of the European Parliament and of the Council of 22 June 1998 concerning common rules for the internal market in natural gas. Official Journal of the European Communities, L204
4. Directive 2003/55/EC of the European Parliament and of the Council of 26 June 2003 concerning common rules for the internal market in natural gas and repealing Directive 98/30/EC. Official Journal of the European Union, L176
5. Directive 2009/73/EC of the European Parliament and of the Council of 13 July 2009 concerning common rules for the internal market in natural gas and repealing Directive 2003/55/EC. Official Journal of the European Union, L211(94)
6. Engevall, S., Göthe-Lundgren, M., Värbrand, P.: The traveling salesman game: An application of cost allocation in a gas and oil company. Ann. Oper. Res. **82**, 203–218 (1998)
7. Engevall, S., Göthe-Lundgren, M., Värbrand, P.: The heterogeneous vehicle-routing game. Transp. Sci. **38**(1), 71–85 (2004)
8. Felsenthal, D.S., Machover, M.: A priori voting power: What is it all about? Polit. Stud. Rev. **2**, 1–23 (2004)
9. Fiestras-Janeiro, M.G., García-Jurado, I., Mosquera, M.A.: Cooperative games and cost allocation problems. Top **19**(1), 1–22 (2011)
10. Hubert F., Cobanli, O.: Pipeline power: A case study of strategic network investments. Rev. Netw. Econ. **14**(2), 75–110 (2016)
11. Hubert, F., Ikonnikova, S.: Investment options and bargaining power in the eurasian supply chain for natural gas. J. Ind. Econ. **59**(1), 85–116 (2011)
12. Hubert, F., Orlova, E.: Competition or countervailing power for the European gas market. Discussion Paper (2014)
13. Hubert, F., Orlova, E.: Network access and market power. Energy Econ. **76**, 170–185 (2018)

14. Littlechild, S.: A simple expression for the nucleolus in a special case. Int. J. Game Theory **3**(1), 21–29 (1974)
15. Littlechild, S.C., Owen, G.: A simple expression for the Shapley value in a special case. Manag. Sci. **20**(3), 370–372 (1973)
16. Littlechild, S.C., Thompson, G.F.: Aircraft landing fees: A game theory approach. Bell J. Econ. **8**(1), 186–204 (1977)
17. Megiddo, N.: On the nonmonotonicity of the bargaining set, the kernel and the nucleolus of a game. SIAM J. Appl. Math. **27**(2), 355–358 (1974)
18. Montero, M.: On the nucleolus as a power index. Homo Oeconomicus **22**(4), 551–567 (2005)
19. Shapley, L.S., Shubik, M.: A method for evaluating the distribution of power in a committee system. Am. Polit. Sci. Rev. **48**(03), 787–792 (1954)
20. Shapley, L.S., Shubik, M.: Quasi-cores in a monetary economy with nonconvex preferences. Econometrica **34**(4), 805–827 (1966)
21. Shubik, M.: Incentives, decentralized control, the assignment of joint costs and internal pricing. Manag. Sci. **8**(3), 325–343 (1962)
22. Thomas, L.C.: Dividing credit-card costs fairly. IMA J. Manag. Math. **4**(1), 19–33 (1992)
23. Young, H.P.: Monotonic solutions of cooperative games. Int. J. Game Theory **14**(2), 65–72 (1985)

Chapter 13
Optimal Incentive Strategy in a Continuous Time Inverse Stackelberg Game

Dmitry B. Rokhlin and Gennady A. Ougolnitsky

Abstract We consider a continuous time dynamic incentive problem in the case of one leader and one follower. Follower's ε-optimal strategy is determined via an auxiliary control problem. The main result is similar to that obtained by the authors for a stochastic discrete time model. We give an illustrative example concerning a non-renewable resource extraction problem.

Keywords Incentive strategy · Inverse Stackelberg game · Continuous time · Resource extraction

13.1 Introduction

The theory of incentives is an important research stream for several decades [10–13]. The adequate mathematical model of an incentive mechanism is provided by inverse Stackelberg games [16, 17]. In these games the leader (she) reports to the follower (he) her strategy as a function of his control actions, and maximizes her payoff on the set of the best responses of the follower. The most comprehensive approach to the solution of this difficult mathematical problem was proposed by

D. B. Rokhlin (✉)
I.I. Vorovich Institute of Mathematics, Mechanics and Computer Sciences of Southern Federal University and Regional Scientific and Educational Mathematical Center of Southern Federal University, Rostov-on-Don, Russia
e-mail: dbrohlin@sfedu.ru

G. A. Ougolnitsky
I.I. Vorovich Institute of Mathematics, Mechanics and Computer Sciences of Southern Federal University, Rostov-on-Don, Russia
e-mail: gaugolnickiy@sfedu.ru

L. A. Petrosyan et al. (eds.), *Frontiers of Dynamic Games*, Static & Dynamic Game Theory: Foundations & Applications, https://doi.org/10.1007/978-3-030-51941-4_13

201

Germeyer in the static case [3] and developed by Kononenko in the dynamic case [6–8]. Following this approach, the leader rewards the follower if he cooperates, and punishes him, otherwise. Kononenko [6–8] has proved that such control mechanism forms an ε-optimal strategy of the leader.

Novikov and Shokhina [15] and Novikov [14] used this idea in his theory of control in organizational systems for the model with specific payoff functions. Namely, the leader maximizes the difference between her revenue and incentive payments, meanwhile the follower maximizes the difference between these payments and his labor cost. In the static case it was shown that in the optimal control mechanism the leader compensates the follower his cost (with an incentive surplus), if the latter cooperates and accepts leader's optimal plan, and refuses to pay, otherwise. The leader optimal plan is determined as a solution of an auxiliary optimization problem of maximization of the difference between leader's revenue and follower's cost.

Rokhlin and Ougolnitsky [19] generalized this result for an incentive model with Markov dynamics and discounted optimality criteria in the case of complete information, discrete time and infinite planning horizon. In this model, the leader influences the follower by selecting an incentive function that depends on the system state and the actions of the follower, who employs closed-loop control strategies. System dynamics, revenues of the leader and costs of the follower depend on the system state and follower's actions. It was shown that finding an approximate solution of the inverse Stackelberg game reduces to the solution of an auxiliary optimal control problem with the objective function equal to the difference between the revenue of the leader and the cost of the follower. An ε-optimal strategy of the leader is an economic incentive for the follower to implement the strategy, which is optimal in this auxiliary problem.

In this paper we consider a continuous time setup of the incentive problem characterized above. Section 13.2 contains the problem formulation. Section 13.3 exposes the main result. Section 13.4 describes an illustrative example. The leader payoff function in this example includes an additional term, related to the intention to reach a desirable state (a form of the homeostasis condition). Section 13.5 concludes.

13.2   Problem Formulation

Consider a controlled dynamical system

$$\dot{x}_t = b(t, x_t, \alpha_t), \quad x_0 = y, \quad \alpha_t \in A, \quad t \in [0, T].$$

where $b = (b_1, \ldots, b_d)$, $x_t \in \mathbb{R}^d$ and $A \subset \mathbb{R}^m$ is a compact set. There are two players: the leader and the follower. It is assumed that their payoffs are given by

$$J_1(x, c, \alpha) = \int_0^T (f(t, x_t, \alpha_t) - c(t, x_t, \alpha_t)) \, dt + \varphi(x_T),$$

$$J_2(x, c, \alpha) = \int_0^T (c(t, x_t, \alpha_t) - g(t, x_t, \alpha_t)) \, dt.$$

respectively. Here α is the control of the follower, and c is the incentive function selected by the leader.

Let us say that $h = h(t, x, a)$ is a *Caratheodory function* on $[0, T] \times \mathbb{R}^d \times A$, if h is Borel measurable in t and continuous in (x, a) [1, Definition 4.50]. Denote by $Cr = Cr([0, T] \times \mathbb{R}^d \times A)$ the class of Caratheodory functions. Clearly, Cr is a linear space. We impose the following conditions:

- $b_i, f, g, c \in Cr([0, T] \times \mathbb{R}^d \times A)$;
- φ is continuous;
- f, g, c are uniformly bounded;
- b is Lipschitz in x uniformly in (t, a):

$$\|b(t, x, a) - b(t, y, a)\| \leq L\|x - y\|, \tag{13.1}$$

where $\| \cdot \|$ is the Euclidean norm in $\mathbb{R}^d$;
- $c \geq 0$, $g(t, x, 0) = 0$.

The set of incentive functions c, satisfying the above conditions is denoted by $\mathscr{C}$.

A *strict control* is a Borel measurable mapping from $[0, T]$ to A. The set of strict controls is denoted by $\mathscr{A}$. In general, to get an optimal solution this set should be enlarged. A *relaxed control* is a positive Borel measure μ on $[0, T] \times A$ such that $\mu(\cdot, A)$ is the Lebesgue measure λ on $[0, T]$. The set of relaxed controls is denoted by $\mathscr{A}^r$. Any relaxed control μ admits a representation $\mu(dt, da) = dt q_t(da)$, where q is a probability kernel from $[0, T]$ to A. A strict control $\alpha \in \mathscr{A}$ is identified with the relaxed control $dt \delta_{\alpha_t}(da)$, where δ_a is the Dirac measure, concentrated at a.

The set $\mathscr{A}^r$ is endowed with the weak topology, induced by the neighborhoods of zero

$$\left| \int_{[0,T] \times A} \eta(t, a) \mu(dt, da) \right| < \varepsilon, \tag{13.2}$$

where η is a continuous function and $\varepsilon > 0$. This topology coincides with the narrow topology, generated by the neighborhoods (13.2), where η is a Caratheodory function: that is, it is Borel measurable in t and continuous in a [22, Theorem 3]. In these topologies the set $\mathscr{A}^r$ is compact and metrizable [9, Lemma 3.3], and the set $\mathscr{A}$ is dense in $\mathscr{A}^r$ [22, Proposition 8].

The Lipschitz condition (13.1) ensures that the equation

$$x_t = y + \int_{[0,t] \times A} b(s, x_s, a) \mu(ds, da), \quad t \in [0, T] \tag{13.3}$$

has a unique continuous solution x for any $\mu \in \mathscr{A}^r$. Denote by C^d the set of continuous $\mathbb{R}^d$-valued functions with the uniform norm $\|x\|_C = \max_{t \in [0,T]} \|x_t\|$.

Now let us give a formal description of the game:

- The leader selects an incentive function $c \in \mathscr{C}$.
- The follower finds an optimal solution (x^*, μ^*) of the problem

$$J_2(x, c, \mu) = \int_{[0,T] \times A} (c(t, x_t, a) - g(t, x_t, a)) \mu(dt, da) \to \max_{(x,\mu) \in C^d \times \mathscr{A}^r}, \tag{13.4}$$

where x satisfies (13.3), and implements the related the optimal control μ^*.

- The leader gets the reward

$$J_1(x^*, c, \mu^*) = \int_{[0,T] \times A} (f(t, x_t^*, a) - c(t, x_t^*, a)) \mu^*(dt, da) + \varphi(x_T^*).$$

Denote by $R(c)$ the set of optimal solutions of (13.4). Leader's aim is to maximize

$$G(c) = \inf_{(x^*, \mu^*) \in R(c)} J_1(x^*, c, \mu^*)$$

over $c \in \mathscr{C}$.

Since the leader considers the worst-case scenario, this is a weak Stackelberg game: [2]. Furthermore, it may be classified as an inverse Stackelberg game (see [16, 17]), as long as leader's strategies depend on the strategies of the follower.

Let us call $V_L = \sup_{c \in \mathscr{C}} G(c)$ the value of the leader. An element $c^\varepsilon \in \mathscr{C}$ is called an ε-*Stackelberg solution* if $G(c^\varepsilon) \geq V_L - \varepsilon$.

Note, that we consider the terminal term $\varphi(x_T)$ only for the leader, keeping in mind that usually only she is interested in controlling the state of the system: see an example in Sect. 13.4. In this example x_t is the amount of available resource, α_t is the intensity of resource extraction, f, g are the instantaneous leader's gain and follower's cost respectively. The terminal term φ measures the deviation of the resource amount from the desired level.

Consider an auxiliary problem:

$$J(x, \mu) = \int_{[0,T] \times A} (f(t, x_t, a) - g(t, x_t, a)) \mu(dt, da) + \varphi(x_T) \to \max_{(x,\mu) \in C^d \times \mathscr{A}^r}, \tag{13.5}$$

where x satisfies (13.3), and let $V = \sup_{(x,\mu)\in C^d\times\mathscr{A}^r} J(x, \mu)$ be its value. The objective function (13.5) represents the hypothetical leader's gain, which she could receive by extracting resource by herself and facing the related costs. It appears that V coincides with the value of the leader V_L. Moreover, an $(\epsilon + \delta)$-Stackelberg solution $c^{\varepsilon,\delta}$ of the leader consists in covering follower's cost with an incentive premium, which is proportional to δ and stimulates an ϵ-optimal solution of (13.5). All these assertions are the content of Theorem 13.2, which is the main result of the paper.

13.3 The Main Result

The next theorem contains technical and essentially known results, concerning the solvability of a general deterministic continuous-time the optimal control problem.

Theorem 13.1 *Assume that $h \in Cr([0, T] \times \mathbb{R}^d \times A)$ is a bounded function, and $\chi : \mathbb{R}^d \mapsto \mathbb{R}$ is a continuous function. Then*

(i) the problem

$$J(x, \mu) = \int_{[0,T]\times A} h(t, x_t, a)\mu(dt, da) + \chi(x_T) \to \max_{(x,\mu)\in C^d\times\mathscr{A}^r}, \quad (13.6)$$

where x satisfies (13.3), is solvable;

(ii) for any $\varepsilon > 0$ there exists an ε-optimal strict control $\alpha^\varepsilon \in \mathscr{A}$:

$$J(x^\varepsilon, \alpha^\varepsilon) \geq J^* - \varepsilon,$$

where x^ε is the solution of (13.3), corresponding to α^ε, and J^* is the optimal value of (13.6).

(iii) if for all $(t, x) \in [0, T] \times \mathbb{R}^d$ the set

$$\{(b(t, x, a), z) : a \in A, \ h(t, x, a) \geq z\} \subset \mathbb{R}^d \times \mathbb{R} \quad (13.7)$$

is convex, then there exists an optimal strict control $\alpha^* \in \mathscr{A}$.

Proof

(i) Let $(x^n, \mu^n) \in C^d \times \mathscr{A}^r$ be a maximizing sequence:

$$J(x^n, \mu^n) \to J^*, \quad x_t^n = y + \int_{[0,t]\times A} b(s, x_s^n, a)\mu^n(ds, da). \quad (13.8)$$

We may assume that $\mu^n \to \mu^* \in \mathscr{A}^r$ in the narrow topology by the compactness of $\mathscr{A}^r$. The sequence x^n satisfies the inequality

$$\|x_t^n - x_s^n\| = \left\| \int_{[s,t] \times A} b(u, x_u^n, a) \mu^n(du, da) \right\| \leq K(t - s),$$

with some constant $K \geq |b|$. It follows that the sequence x^n is uniformly bounded and equicontinuous. By the Arzela-Ascoli theorem there exists a subsequence, converging to $x^* \in C^d$. Without loss of generality we can assume that $x^n \to x^*$ in C^d and $\|x^n\|_C \leq \gamma$.

Assume that $|h| \leq d_\gamma$ on $[0, T] \times B_\gamma \times A$, where $B_\gamma = \{y \in \mathbb{R}^d : \|y\| \leq \gamma\}$. We claim that

$$\int_{[0,T] \times A} h(t, x_t^n, a) \mu^n(dt, da) \to \int_{[0,T] \times A} h(t, x_t^*, a) \mu^*(dt, da). \quad (13.9)$$

The function $(t, a) \mapsto h(t, x_t^*, a)$ is measurable in t and continuous in a. Moreover, $|h| \leq d_\gamma$. By the definition of the narrow topology,

$$\int_{[0,T] \times A} h(t, x_t^*, a) \mu^n(dt, da) \to \int_{[0,T] \times A} h(t, x_t^*, a) \mu^*(dt, da). \quad (13.10)$$

Furthermore, by the Scorza-Dragoni theorem [18, Theorem 2.5.19] for any $\varepsilon > 0$ there exists a compact set $K_\varepsilon \subset [0, T]$ with $\lambda(K_\varepsilon^c) \leq \varepsilon$, $K_\varepsilon^c = [0, T] \backslash K_\varepsilon$ such that $h|_{K_\varepsilon \times \mathbb{R}^d \times A}$ is continuous. Since the function h is uniformly continuous on $K_\varepsilon \times B_\gamma \times A$, for any $\varepsilon > 0$ there exists $\delta > 0$ such that

$$|h(t, x_t^n, a) - h(t, x_t^*, a)| \leq \varepsilon \quad \text{if } \|x^n - x^*\|_C \leq \delta.$$

Thus,

$$\int_{K_\varepsilon \times A} |h(t, x_t^n, a) - h(t, x_t^*, a)| \mu^n(dt, da) \leq T\varepsilon \quad \text{for } \|x^n - x^*\|_C \leq \delta,$$

$$(13.11)$$

$$\int_{K_\varepsilon^c \times A} |h(t, x_t^n, a) - h(t, x_t^*, a)| \mu^n(dt, da) \leq 2d_\gamma \int_{K_\varepsilon^c \times A} \mu^n(dt, da) = 2\varepsilon d_\gamma.$$

$$(13.12)$$

Using the relations (13.10), (13.11), and (13.12), from the inequality

$$\left| \int_{[0,T] \times A} h(t, x_t^n, a) \mu^n(dt, da) - \int_{[0,T] \times A} h(t, x_t^*, a) \mu^*(dt, da) \right|$$

$$\leq \int_{[0,T] \times A} |h(t, x_t^n, a) - h(t, x_t^*, a)| \mu^n(dt, da)$$

$$+ \left| \int_{[0,T] \times A} h(t, x_t^*, a) \mu^n(dt, da) - \int_{[0,T] \times A} h(t, x_t^*, a) \mu^*(dt, da) \right|$$

we get

$$\limsup_{n \to \infty} \left| \int_{[0,T] \times A} h(t, x_t^n, a) \mu^n(dt, da) \right.$$
$$\left. - \int_{[0,T] \times A} h(t, x_t^*, a) \mu^*(dt, da) \right| \leq (T + 2d_\gamma)\varepsilon,$$

thus proving the claim (13.9).

The same argumentation shows that

$$\int_{[0,t] \times A} b(s, x_s^n, a) \mu^n(ds, da) \to \int_{[0,t] \times A} b(s, x_s^*, a) \mu^*(ds, da)$$

We have justified the passage to the limit in (13.8), which implies that (x^*, μ^*) is a solution of (13.6):

$$x_t^* = y + \int_{[0,t] \times A} b(s, x_s^*, a) \mu^*(ds, da),$$

$$J(x^*, \mu^*) = \int_{[0,T] \times A} h(t, x_t^*, a) \mu(dt, da) + \chi(x_T^*) = J^*.$$

(ii) Take a sequence $\alpha^n \in \mathscr{A}$, converging to an optimal $\mu^* \in \mathscr{A}$ in the narrow topology. As in the proof of (i), considering α^n instead of μ^n, we can argue that the optimal value J^* can be approximated with arbitrary accuracy by $J(x^n, \alpha^n)$.

(iii) This is the well-known result and (13.7) is known as the Roxin condition: see [4] (Theorem 3.6 and condition (3.4)) for a rather general result or [9, Theorem 3.6] for an exposition. □

Recall that $V = \sup_{(x,\mu) \in C^d \times \mathscr{A}^r} J(x, \mu)$ is the value of the auxiliary problem (13.5). Let $\rho : A \times A \mapsto \mathbb{R}_+$ be any metric on A.

Theorem 13.2 *Under the adopted assumptions the following holds true.*

(i) $V_L = V$.

(ii) *Let $\alpha^\varepsilon \in \mathscr{A}$ be a ε-optimal solution of the auxiliary problem (13.5). Then*

$$c^{\varepsilon,\delta}(t,x,a) = g(t,x,a) + \frac{\delta}{T}(1 - \rho(a,\alpha_t^\varepsilon))^+, \tag{13.13}$$

where $z^+ := \max\{z, 0\}$ is an $(\varepsilon + \delta)$-Stackelberg solution, and

$$R(c^{\varepsilon,\delta}) = \{dt\delta_{\alpha_t^\varepsilon}(da)\}.$$

Proof By Theorem 13.1, $R(c) \neq \emptyset, c \in Cr$. For $(x^*, \mu^*) \in R(c)$ we have

$$G(c) \leq J_1(x^*,c,\mu^*) \leq J_1(x^*,c,\mu^*) + J_2(x^*,c,0)$$
$$\leq J_1(x^*,c,\mu^*) + J_2(x^*,c,\mu^*) = J(x^*,\mu^*) \leq V.$$

Hence,

$$V_L \leq V. \tag{13.14}$$

Furthermore, consider the functional of the follower for the incentive function (13.13):

$$J_2(x, c^{\varepsilon,\delta}, \mu) = \frac{\delta}{T}\int_{[0,T]\times A}(1 - \rho(a,\alpha_t^\varepsilon))^+ \mu(dt, da)$$
$$= \frac{\delta}{T}\int_{[0,T]\times A}(1 - \rho(a,\alpha_t^\varepsilon))^+ q_t(da)dt.$$

Note that an ε-optimal solution $\alpha^\varepsilon \in \mathscr{A}$ of (13.5) exists by Therorem 13.1.

Clearly, if $(x, dtq_t(da)) \in R(c^{\varepsilon,\delta})$, then $q_t(da) = \delta_{\alpha_t^\varepsilon}(da)$ λ-a.e., since this is the unique control, providing zero value to the last expression. In other words, α^ε is the optimal control of the follower. Denote by x^ε the related trajectory. Since

$$c^{\varepsilon,\delta}(t, x_t^\varepsilon, \alpha_t^\varepsilon) = g(t, x_t^\varepsilon, \alpha_t^\varepsilon) + \frac{\delta}{T},$$

we get

$$V_L \geq G(c^\varepsilon) = J_1(x^\varepsilon, c^{\varepsilon,\delta}, \alpha^\varepsilon)$$
$$= \int_{[0,T]\times A}\left(f(t, x_t^\varepsilon, \alpha_t^\varepsilon) - g(t, x_t^\varepsilon, \alpha_t^\varepsilon) - \frac{\delta}{T}\right)dt + \varphi(x_T^\varepsilon)$$
$$= J(\alpha^\varepsilon) - \delta \geq V - \varepsilon - \delta.$$

Together with (13.14) this inequality implies both assertions of the theorem. $\square$

The form of the second term (the incentive premium) in (13.13) is not so essential, but in accordance with the economic meaning, it should be non-negative. The closer the actions of the follower are to the desired control α^ε, communicated by the leader, the larger will be incentive premium, which is proportional to δ. Note also that the leader always communicates to the follower a strict control strategy. We believe that in practice it is impossible to communicate a relaxed strategy.

13.4   Example

Consider a non-renewable resource extraction problem. Let x_t be the amount of remaining resource, and denote by α_t the intensity of its extraction:

$$\dot{x}_t = -\alpha_t, \quad \alpha_t \in [0, \bar{\alpha}], \quad t \in [0, T]; \quad x_0 = y. \tag{13.15}$$

Assume that the leader obtains instantaneous profit by selling the resource at the market price $P(t)$. Besides the profit, the leader is interested in setting the resource level as close as possible to y_1 at some predefined time moment T:

$$J_1(x, c, \alpha) = \int_0^T (P(t)\alpha_t - c(t, x_t, \alpha_t)) \, dt - \frac{B}{2}(x_T - y_1)^2 \to \max. \tag{13.16}$$

A problem similar to (13.15), (13.16), but with unspecified horizon and the hard boundary condition $x_T = y_1$, was considered in [5, Chapter 10].

Follower's gain is simply the difference between the incentive function and the extraction cost:

$$J_2(x, c, \alpha) = \int_0^T (c(t, x_t, \alpha_t) - C(t)\alpha_t) \, dt \to \max.$$

In this example it is enough to use only strict controls.

Consider the auxiliary problem (13.5):

$$-J(x, \alpha) = \int_0^T (C(t) - P(t))\alpha_t \, dt + \frac{B}{2}(x_T - y_1)^2 \to \min. \tag{13.17}$$

To solve the problem (13.15), (13.17) let us apply the Pontryagin maximum principle. Consider the Hamiltonian

$$H = -\lambda a - (C(t) - P(t))a.$$

The adjoint equation

$$\dot{\lambda}_t = -H_x = 0$$

shows that λ is constant. An optimal control α^* satisfies the maximum principle:

$$\alpha_t^* \in \arg \max_{a \in [0,\bar{\alpha}]} H(t, a, \lambda) = \begin{cases} \bar{\alpha}, & P(t) - C(t) > \lambda, \\ 0, & P(t) - C(t) < \lambda. \end{cases}$$

Let us assume that the equation $t \mapsto P(t) - C(t) = \lambda$ has only finite number of solutions for any constant λ. Then $\alpha_t^* \in \{0, \bar{\alpha}\}$ except may be finite number of points $\tau_i \in [0, T]$:

$$\tau_0 = 0 < \tau_1 < \cdots < \tau_k = T, \quad k \geq 1,$$

and the values $\alpha_t^* = 0$ and $\alpha_t^* = 1$ alternate on the neighbor intervals (τ_{i-1}, τ_i), (τ_i, τ_{i+1}) if $k \geq 2$. Thus,

$$\alpha_t^* = \bar{\alpha} \cdot I_S(t), \quad S = \bigcup_{i \in J} (\tau_i, \tau_{i+1}), \quad I_S(t) = \begin{cases} 1, & t \in S, \\ 0, & 0 \notin S \end{cases} \tag{13.18}$$

for some set of indexes $J \subset \{0, \ldots, k-1\}$.

According to Theorem 13.2 the leader should select an incentive function

$$c(t, x, a) = C(t)a + \frac{\delta}{T} \left(1 - |a - \alpha_t^*|/\bar{\alpha} \right)^+$$

(we put $\rho(a, b) = |a - b|/\bar{\alpha}$). Then follower's optimal strategy coincides with α^*.

To make the problem more interesting assume that the true instantaneous cost $\tilde{C}$ of the follower differs from the estimate C of the leader. Then follower's payoff will be

$$\tilde{J}_2(x, c, \alpha^*) = \int_0^T (c(t, x_t, \alpha_t) - \tilde{C}(t)\alpha_t) \, dt$$

$$= \int_0^T \left(C(t)\alpha_t - \tilde{C}(t)\alpha_t + \frac{\delta}{T}(1 - |a - \alpha_t^*|/\bar{\alpha})^+ \right) dt \to \max. \tag{13.19}$$

We interested under what conditions the follower will implement the strategy α^*, which the leader tried to communicate. Writing down the relations of the maximum principle for the problem (13.15), (13.19), we see that the adjoint variable equals to zero and the follower optimal strategy $\hat{\alpha}_t$ satisfies the condition

$$\hat{\alpha}_t \in \arg \max_{a \in [0,\bar{\alpha}]} \left((C(t) - \tilde{C}(t))a + \frac{\delta}{T}(1 - |a - \alpha_t^*|/\bar{\alpha})^+ \right).$$

Substituting α^* from (13.18), we get

$$\widehat{\alpha}_t \in \arg \max_{a \in [0,\overline{\alpha}]} \left((C(t) - \widetilde{C}(t))a + \frac{\delta}{T}(1 - a/\overline{\alpha}) \right), \quad t \in (\tau_{i-1}, \tau_i), \ i \notin J,$$

$$\widehat{\alpha}_t \in \arg \max_{a \in [0,\overline{\alpha}]} \left((C(t) - \widetilde{C}(t))a + \frac{\delta}{T}a/\overline{\alpha} \right), \quad t \in (\tau_{i-1}, \tau_i), \ i \in J.$$

The desired condition $\widehat{\alpha} = \alpha^*$ is equivalent to the inequalities

$$C(t) - \widetilde{C}(t) - \frac{\delta}{T\overline{\alpha}} < 0, \quad t \in (\tau_{i-1}, \tau_i), \ i \notin J,$$

$$C(t) - \widetilde{C}(t) + \frac{\delta}{T\overline{\alpha}} > 0, \quad t \in (\tau_{i-1}, \tau_i), \ i \in J.$$

Thus, the condition $\widehat{\alpha} = \alpha^*$ will be satisfied if

$$\delta > T\overline{\alpha} \sup_{t \in [0,T]} |C(t) - \widetilde{C}(t)|. \tag{13.20}$$

If this condition fails, the leader may encounter an undesirable behavior of the follower.

Under the condition (13.20) the gain of the leader equals to

$$J_1(x^*, c, \alpha^*) = \int_0^T (P(t)\alpha_t^* - c(t, x_t^*, \alpha_t^*)) \, dt - \frac{B}{2}(x_T^* - y_1)^2$$

$$= \int_0^T \left(P(t)\alpha_t^* - C(t)\alpha_t^* \right) dt - \frac{B}{2}(x_T^* - y_1)^2 - \delta = V(C) - \delta.$$

We write $V(C)$ for the optimal value of the auxiliary problem (13.15) and (13.17), where follower's cost is estimated as $C(t)a$ by the leader. For the leader, the magnitude of δ is related to the tradeoff between the controlled decrease of the payoff and the protection against an undesirable behavior of the follower.

Note, that in fact there is no need to stimulate zero extraction strategy of the follower. That is, the leader can consider the incentive function

$$\widehat{c}(t, x, a) = c(x, t, a)I_S(t) = \left(C(t)a + \frac{\delta}{T}\frac{a}{\overline{\alpha}} \right) I_S(t).$$

Indeed, in this case the follower with any strictly positive cost $\widetilde{C}$ will select $\widehat{\alpha}_t = 0$, $t \in [0, T] \backslash S$. Under the condition

$$\delta > T\overline{\alpha} \sup_{t \in S} |C(t) - \widetilde{C}(t)|,$$

weaker than (13.20), this allows to guarantee somewhat larger lower estimate of leader's payoff:

$$J_1(x^*, \widehat{c}, \alpha^*) = \int_0^T \left(P(t)\alpha_t^* - \left(C(t)\overline{\alpha} + \frac{\delta}{T} \right) I_S(t) \right) dt - \frac{B}{2}(x_T^* - y_1)^2$$

$$= V(C) - \frac{\lambda(S)}{T}\delta.$$

13.5 Conclusion

We considered a dynamic incentive problem in continuous time for the case of one leader and one follower. It is shown that to implement an ε-optimal incentive mechanism, the leader should solve an optimal control problem to maximize the difference between her revenue and the cost of the follower, report a fixed optimal plan to the follower, and cover the costs of the follower with an incentive premium ε. This result is similar to that obtained in our paper [19] for a discrete-time stochastic model. From the other side, it is another generalization of a static case result obtained in [14, 15].

The dynamic case of several agents in discrete time was studied by Rokhlin and Ougolnitsky in [20, 21]. We plan a similar investigation for the continuous time model with several agents and different information structures. Also, the condition of homeostasis, which is now partially taken into account by the second term in the leader payoff function (13.16), seems to be important in applications and should be analyzed in details.

Acknowledgement The research is supported by the Russian Science Foundation, project 17-19-01038.

References

1. Aliprantis, C.D., Border, K.C.: Infinite Dimensional Analysis: A Hitchhiker's Guide. Springer, Berlin (2006)
2. Breton, M., Alj, A., Haurie, A.: Sequential Stackelberg equilibria in two-person games. J. Optim. Theory Appl. **59**(1), 71–97 (1988)
3. Germeier, Yu.B.: Non-Antagonistic Games. Reidel Publishing Co., Dordrecht (1986)
4. Haussmann, U.G., Lepeltier, J.P.: On the existence of optimal controls. SIAM J. Control. Optim. **28**(4), 851–902 (1990)
5. Hritonenko N., Yatsenko Y.: Mathematical Modeling in Economics, Ecology and the Environment. Springer, New York (2013)
6. Kononenko, A.F.: Game-theory analysis of a two-level hierarchical control system. USSR Comput. Math. Math. Phys. **14**(5), 72–81 (1974)
7. Kononenko, A.F.: On multi-step conflicts with information exchange. USSR Comput. Math. Math. Phys. **17**(4), 104–113 (1977)

8. Kononenko, A.F.: The structure of the optimal strategy in controlled dynamic systems. USSR Comput. Math. Math. Phys. **20**(5), 13–24 (1980)
9. Lacker D.: Probabilistic compactification methods for stochastic optimal control and mean field games (2018) (Unpublished)
10. Laffont, J.-J., Martimort, D.: The Theory of Incentives: The Principal-Agent Model. Princeton University Press, Princeton (2002)
11. Myerson, R.: Incentive compatibility and the bargaining problem. Econometrica **47**, 61–73 (1979)
12. Myerson, R.: Optimal coordination mechanisms in generalized principal-agent models. J. Math. Econ. **10**, 67–81 (1982)
13. Myerson, R.: Mechanism design by an informed principal. Econometrica **51**, 1767–1798 (1983)
14. Novikov, D.: Theory of Control in Organizations. Nova Science Publishers, New York (2013)
15. Novikov, D.A., Shokhina, T.E.: Incentive mechanisms in dynamic active systems. Autom. Remote. Control. **64**, 1912–1921 (2003)
16. Olsder, G.J.: Phenomena in inverse Stackelberg games. Part 1: static problems. J. Optim. Theory Appl. **143**, 589–600 (2009)
17. Olsder, G.J.: Phenomena in inverse Stackelberg games. Part 2: dynamic problems. J. Optim. Theory Appl. **143**, 601–618 (2009)
18. Papageorgiou, N.S., Winkert P.: Applied Nonlinear Functional Analysis: An Introduction. de Gruyter, Berlin (2018)
19. Rokhlin, D.B., Ougolnitsky, G.A.: Stackelberg equilibrium in a dynamic stimulation model with complete information. Autom. Remote Control **79**, 701–712 (2018)
20. Rokhlin, D.B., Ougolnitsky, G.A.: Optimal incentive strategy in a discounted stochastic Stackelberg game. Contrib. Game Theory Manag. **12**, 273–281 (2019)
21. Rokhlin, D.B., Ougolnitsky, G.A.: Optimal incentive strategy in a Markov game with multiple followers. In: Petrosyan, L., Mazalov, V., Zenkevich, N. (eds.) Frontiers of Dynamic Games. Static & Dynamic Game Theory: Foundations & Applications, pp. 231–243. Birkhäuser, Basel (2019)
22. Valadier, M.: A course on young measures. Rend. Istit. Mat. Univ. Trieste **26**(suppl), 349–394 (1994)

Chapter 14
The Looking Forward Approach in a Differential Game Model of the Oil Market with Non-transferable Utility

Ovanes Petrosian, Maria Nastych, and Yin Li

Abstract The paper applies the Looking Forward Approach to analyze the world oil market within the framework of a differential game model of a quantity competition oligopoly. More precisely, the Looking Forward Approach is used to take into account dynamically updating information. Under information we understand the forecast of oil demand dynamics. We use a non-cooperative game modeling for the period from December 2015 to November 2016, because over this period the countries did not cooperate officially in what concerns the amounts of oil to be produced. For the period from December 2016 to May 2017, a non-transferable utility cooperative game modeling is adapted due to the agreement to reduce oil extraction signed by the largest oil exporters at the end of November 2016. We use both solutions, which correspond to the historical cooperative solution and the sub-game consistent solution proposed developed in the field of dynamic games. In order to define the parameters of the model, open source data is used, with the results of numerical simulations and comparison between the historical data and model data for both periods also presented.

Keywords Non-transferable utility game · Looking Forward Approach · Oil market · Differential game

O. Petrosian (✉)
Saint Petersburg State University, St. Petersburg, Russia
e-mail: petrosian.ovanes@yandex.ru

M. Nastych
National Research University Higher School of Economics, St. Petersburg, Russia
e-mail: manastych@hse.ru

Y. Li
Saint Petersburg State University, St. Petersburg, Russia

© The Editor(s) (if applicable) and The Author(s), under exclusive licence
to Springer Nature Switzerland AG 2020
L. A. Petrosyan et al. (eds.), *Frontiers of Dynamic Games*,
Static & Dynamic Game Theory: Foundations & Applications,
https://doi.org/10.1007/978-3-030-51941-4_14

215

14.1   Introduction

The paper is devoted to constructing a game theoretical model for the world oil market using the Looking Forward Approach. Game models with the Looking Forward Approach allow taking into account the variability of market demands, an adaptation of participants actions to the changing environment and actual planning horizons for demand. Oil market has a highly volatile prices, therefore the Looking Forward Approach is applied to the oil market and the resulting model is studied. We suppose that countries do not have or cannot use long-term forecasts for the parameters of the oil market. Therefore, their behavior can be modeled using the approach proposed.

The object of this paper is to simulate the oil market dynamics during two particular periods, from December 2015 to November 2016 and from December 2016 to May 2017. The OPEC countries and eleven non-OPEC countries reached a 6-month-long agreement about the reduction of oil production for the latter period at the summit in Vienna on November 30, 2016. The agreement was aimed to result in the growth of oil prices. These countries, taken together, account for over 60% of oil production in the world. To show the effect of the agreement, we consider two game theoretical models corresponding to the two periods mentioned. We suppose that the countries did not cooperate officially on the amounts of oil to be produced before November 30, 2016. Therefore, their behavior can be simulated using a non-cooperative game model. Hence, firstly, we build a non-cooperative game for the time interval from December 2015 to November 2016 and adapt it to the real oil price data. As an optimality principle, the feedback Nash equilibrium is used. Making an agreement on the quantity of oil production seems to be a case of purely cooperative behavior. Therefore, we construct a coalitional game model for the time interval from December 2016 to May 2017. The cooperation is modeled using a non-transferable utility game model. Subsequently, we construct a classical cooperative solution for this type of games, i.e. a sub-game consistent solution proposed in [38]. Furthermore, we model the cooperative agreement which was actually used by the countries and call it the historical solution. According to the historical agreement, the solution is based on the oil production quantities in November 2016. It prescribes players 1 and 2 to linearly lower the quantities of oil production to the level previously agreed upon. Obviously, the second solution is not sub-game consistent, though it allows us to model the latest real-life market agreement.

We present the results of the statistical simulation and comparison between the historical data and model data over both periods. In particular, we give a comparison of the theoretical trajectory of oil price between December 2015 and November 2016 with the statistical data. Over the period between December 2016 and May 2017, we compare the trajectory of oil price corresponding to the statistical data, the subgame consistent cooperative solution defined in [38] and the cooperative solution corresponding to the agreement signed by the largest oil exporters at the end of 2016. In order to define the parameters of the model, open source data on the world oil market is used. We used the International Energy Agency for monthly data on crude oil supply from January 2015 to May 2017, the Finam agency for monthly data on Brent and light oil prices from January 2015 to May 2017, Rystad Energy Ucube and oil market news for the cost of producing a barrel of oil in 2016.

The largest oil exporters reached the agreement to reduce oil production aiming at raising prices after November 30, 2016, and in all our models we consider all of them in groups of five to one players. This is not the first time when the OPEC has restricted oil extraction hoping to control the prices on oil. These actions of the members seem rational in a sense, as oil prices started to fall steadily in 2014. Thus, in our game models, all the members of the OPEC together are called player one. The other eleven countries which signed the agreement and which clearly have common interests, although with more freedom in decision-making as compared to the OPEC countries, are called player two. At the same time, the steady fall in oil prices was largely due to the US shale oil and gas revolution, which started in 2012. Therefore, our models include US shale and non-shale oil producing companies as the main market rivals, and we call them players three and four, respectively. All the other oil exporting countries constitute player five. We combine the countries into players from one to five according to the described rule, since in our models we consider only the price change process. Furthermore, such a combination greatly simplifies the computational process without exerting any serious effect on the plausibility of the results.

The global oil market is one of the most significant markets and a crucial component in growth rate and budget scenarios for some resource-dependent countries. In all fairness, this has been given a detailed coverage in literature. Most of the existing models consider the oil market as a market with imperfect competition. Market power models for the oil market explain the presence of the OPEC by cartel behavior, by a dominant firm, or by target behavior in most cases. Thus, Dahl and Yucel in [6] describe OPEC behavior as a loose coordination or duopoly. In the paper [31] the pattern of extraction in the oil market is shown to be inconsistent with either the patterns predicted by the competitive theory or the dominant firm-competitive fringe theory. Danielsen and Kim [7] provide significant evidence of cooperation among the OPEC countries. Smith in [34] asserts that the OPEC is much more than simply a non-cooperative oligopoly but less than a frictionless cartel. Youhanna in [39] affirms that a partial market sharing cartel model dominates over all other models. Gulen in [10] provides evidence of output coordination and suggests that the OPEC acted like a cartel in the 1980s (1982–1993). The author of [17] maintains that the OPEC behavior is consistent with the cartel theory. Bockem in [4] states that the crude oil market is best described as a price leader model, with the OPEC appearing to be the leader and all non-OPEC countries being regarded as price takers. The paper [30] claims that the OPEC behaves more like an oligopoly, with Saudi Arabia as a price leader and the largest producer. In the paper [1] the author asserts that the Nash–Cournot non-cooperative model can potentially explain the oil market better than the competitive one. Moran in [18] and Krasner in [13] analyze the main features of oil oligopoly. The authors in [16] give a selective survey of oligopoly models for energy production. In the paper [33] the author examines cartel formation in the world oil market under the Cournot setting. The authors in [3] also use the quantity competitive environment to model collisions and proportionate adjustment of production levels. Following this well-established tradition, we use the oligopoly quantity setting to model the oil market.

Existing differential games often rely on the assumption of time-invariant game structures for the derivation of equilibrium solutions. However, many events in the considerably distant future are intrinsically obscure and unknown. In this paper, information about the players' future payoffs will be revealed as the game proceeds. Making use of the newly obtained information, the players revise their strategies accordingly, and the process continues indefinitely. The Looking Forward Approach for differential games provides a more realistic and practical alternative to the study of classical differential games. It enables us to construct game theoretical models in which the game structure can change or update over time (time-dependent formulation) and the players do not possess full information about the change of the game structure while obtaining full information about the game structure on the truncated time interval. The length of the truncated time interval will be called the information horizon as it refers to the duration of the period when the information is available to the players. By the information about the game structure we understand information about the motion equation and payoff functions. The duration of the period for which this information is relevant is known in advance. The information about the game structure is updated at certain points. The Looking Forward Approach was mainly developed for cooperative games with transferable utility [9, 21–23], but there are also papers on non-cooperative differential games [36], dynamic games [37] and a game with non-transferable utility [25]. The key element of the Looking Forward Approach is the notion of a truncated subgame, which helps to model the behavior of the players between the time points of information updating. We define all optimal strategies and a corresponding trajectory for each of the truncated subgames. When the information updates, we define a corresponding solution for the next truncated subgame. The resulting strategies and the corresponding trajectory are defined by the combination of optimal strategies and trajectories in each truncated subgame. The concept of the Looking Forward Approach is new in game theory, especially in differential games, and it provides avenues for further study of differential games with dynamic updating. So far there have been no known attempts of constructing approaches to model conflict-controlled processes where information about the process updates dynamically in time.

The Looking Forward Approach shares a common ground with the Model Predictive Control theory developed within the framework of numerical optimal control. We analyze [8, 15, 24, 26, 32, 35] to obtain recent results in this area. The main problem that the Model Predictive Control solves mathematically is the provision of movement along the target trajectory under the conditions of random perturbations and unknown dynamical system. At each time step, the optimal control problem is solved by defining the controls which will lead the system to the target trajectory. The Looking Forward Approach, on the other hand, solves the problem of modeling the behavior of the players when the information about the process updates dynamically. It means that the Looking Forward Approach does not use a target trajectory, but resolves the issue of composing a trajectory which will be used by the players and of allocating the cooperative payoff along the trajectory. Another interesting class of games is connected to the class of differential games with continuous updating was considered in the papers [14, 27], here it is supposed that the updating process evolves continuously in time. In the paper [27], the system of Hamilton–Jacobi–Bellman equations are derived for the Nash equilibrium in a game with continuous updating. In the paper [14] the class of linear-quadratic differential games with continuous updating is considered and the explicit form of the Nash equilibrium is derived.

In the first part of the paper, the Looking Forward Approach is applied to the non-cooperative oligopoly differential model [5] of the oil market with the largest oil exporters and other oil-producing countries. Non-cooperative game theory deals with strategic interactions among multiple decision makers with the objective functions depending on the choices of all the players and suggests solution concepts for a case when players do not cooperate or make any arrangements about their actions. A player cannot simply optimize her own objective function independent from the choices of the other players. In 1950 and 1951 in [19, 20] by John Nash, such a solution concept was introduced, which is now called the Nash equilibrium. In the second part of the paper, we consider a partially cooperative differential game with non-transferable utility, which uses the non-cooperative game described above as the initial model. We assume that players one and two cooperate and the rest use Nash equilibrium strategies against them. The cooperative differential game theory offers socially convenient and group efficient solutions to different decision problems involving strategic actions. One of the fundamental questions in the theory of cooperative differential games with the non-transferable utility is the formulation of optimal behavior for players or economic agents, the design of Pareto optimal trajectories, the computation of the corresponding solution, and the analysis of its sub-game consistency. The well-known solution in the games with non-transferable utility is the Nash bargaining solution. Haurie analyzed the problem of dynamic instability of Nash bargaining solutions in differential games [11]. The notion of time consistency of differential games solutions was formalized mathematically by Petrosyan [28]. In the paper [38] the authors derive sub-game consistent solutions for a class of cooperative stochastic differential games with non-transferable utility. Another technique for the construction of a sub-game consistent solution is presented in [29].

The paper is divided into two sections. In Sect. 14.1, a non-cooperative game model with the Looking Forward Approach is constructed. We describe the initial game model, define the notion of truncated sub-game, define the feedback Nash equilibrium for each truncated sub-game, describe the process of information updating, introduce the notion of conditionally cooperative trajectory and the corresponding non-cooperative outcome, describe the results of the numerical simulation of oil price on the market for the time interval from December 2015 to November 2016. In Sect. 14.2 we construct a partial cooperative game model corresponding to the model described in Sect. 14.1. We define a set of Pareto optimal outcomes, the sub-game consistent solutions from [38] for each truncated sub-game, introduce a solution concept for the whole game and the conditionally Pareto optimal trajectory, study the sub-game consistency of the solution concept, construct the historical solution and, finally, compare the solution based on the sub-game consistent solution [38] and the historical solution. Further on, we model oil price dynamic for the time interval from December 2016 to May 2017 in the numerical simulation part.

14.2 Non-cooperative Game Model

14.2.1 Initial Game Model

Let us consider a differential game model of Cournot oligopoly [5] on the oil market. An oligopolistic market of n asymmetrical countries (players) belonging to the set $N = \{1, \ldots, n\}$, producing oil, and competing for the quantity produced q_i under price stickiness is given by the differential game $\Gamma(p_0, T - t_0)$ with prescribed duration $T - t_0$ and initial state $p(t_0) = p_0 \in P \subset R$.

In compliance with the model, the market price p_i evolves according to the differential equation:

$$\dot{p}(t) = s(\hat{p}(t) - p(t)), \ p(t_0) = p_0, \tag{14.1}$$

where $\hat{p}(t) \in P \subset R$ is the notional level of the price at the time t, $p(t)$ is its current level, and the parameter $s : 0 < s < 1$ is the speed of adjustment. Thus, prices adjust to the difference between its notional and current levels.

Further, we assume that the notional prices at any time t are defined by the linear inverse demand function

$$\hat{p}(t) = a - d \sum_{i \in N} q_i(t). \tag{14.2}$$

Players $i \in N$ choose quantity $q_i(t) \in U_i \subset R$ produced in order to maximize their profits:

$$K_i(p_0, T - t_0; q_1, \ldots, q_n) = \int_{t_0}^{T} e^{-\rho(t-t_0)} \left[q_i(t)(p(t) - c_i - g_i q_i(t)) \right] dt, \tag{14.3}$$

here, $0 \leq \rho \leq 1$ represents the positive discount rate, which is taken to be the same for all the periods and all the players in order to simplify the model and to equalize the players as symmetrical participants in the global capital market. $C_i(t) = c_i q_i(t) + g_i q_i^2(t)$ is the total cost function for each player i.

14.2.2 Truncated Subgame

Let us suppose that the information for the players is updated at fixed time instants $t = t_0 + j\Delta t$, $j = 0, \ldots, l$, where $l = \frac{T}{\Delta t} - 1$. During the time interval $[t_0 + j\Delta t, t_0 + (j+1)\Delta t]$, the players have certain information about the dynamics of the game (14.1) and the payoff function (14.5) within the time interval $[t_0 + j\Delta t, t_0 + j\Delta t + \overline{T}]$, where $\Delta t \leq \overline{T} \leq T$. At the instant $t = t_0 + (j+1)\Delta t$ the information about the game is updated and the same procedure repeats for the time interval with number $j + 1$.

To model this kind of behavior we introduce the following definition, where vectors $p_{j,0} = p(t_0 + j\Delta t)$, $p_{j,1} = p(t_0 + (j+1)\Delta t)$.

Definition 14.1 Let $j = 0, \ldots, l$. A truncated subgame $\bar{\Gamma}_j(p_{j,0}, t_0 + j\Delta t, t_0 + j\Delta t + \overline{T})$ is defined on the time interval $[t_0 + j\Delta t, t_0 + j\Delta t + \overline{T}]$. The motion equation and the initial condition of the truncated subgame $\bar{\Gamma}_j(p_{j,0}, t_0 + j\Delta t, t_0 + j\Delta t + \overline{T})$ have the following form:

$$\dot{p}(t) = s\left(a_j - d_j \sum_{i \in N} q_i^j(t) - p(t)\right), \quad p(t_0 + j\Delta t) = p_{j,0}. \tag{14.4}$$

The payoff function of player i in truncated subgame j is equal to

$$K_i^j(p_{j,0}, t_0 + j\Delta t, t_0 + j\Delta t + \overline{T}; q_1^j, \ldots, q_n^j) =$$

$$= \int_{t_0 + j\Delta t}^{t_0 + j\Delta t + \overline{T}} e^{-\rho(t - t_0)} \left[q_i^j(t)(p(t) - c_i - g_i q_i^j(t))\right] dt. \tag{14.5}$$

The motion equation and the payoff function on the time interval $[t_0 + j\Delta t, t_0 + j\Delta t + \overline{T}]$ coincide with that of the game $\Gamma(p_0, T - t_0)$ on the same time interval (Fig. 14.1).

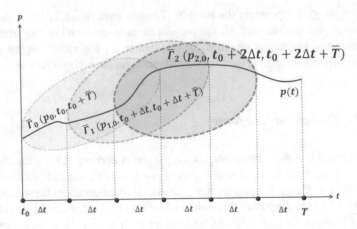

Fig. 14.1 Behavior of the players in the game with truncated information can be modeled using the truncated subgames $\bar{\Gamma}_j(p_{j,0}, t_0 + j\Delta t)$, $j = 0, \ldots, l$

14.2.3 Non-cooperative Outcome in Truncated Subgame

According to [2, 12], a non-cooperative Nash equilibrium solution of the game $\bar{\Gamma}_j(p_{j,0}, t_0 + j\Delta t, t_0 + j\Delta t + \bar{T})$ can be defined by the Hamilton–Jacobi–Bellman partial differential equations. Let us consider a family of subgames $\bar{\Gamma}_j(p(t), t, t_0 + j\Delta t + \bar{T})$ with the payoff structure (14.3) and the dynamics (14.15), starting at the time $t \in [t_0 + j\Delta t, t_0 + j\Delta t + \bar{T}]$ with the initial state $p(t)$. Let $q_j^{NE}(t, p) = (q_1^{jNE}(t, p), \ldots, q_n^{jNE}(t, p))$ for $t \in [t_0 + j\Delta t, t_0 + j\Delta t + \bar{T}]$ denote a set of feedback strategies that constitutes a Nash equilibrium solution to the game $\bar{\Gamma}_j(p(t), t, t_0 + j\Delta t + \bar{T})$ and $V_i^j(\tau, p) : [t, T] \times R^n \rightarrow R$ denote the value function of player $i \in N$ that satisfies the corresponding Hamilton–Jacobi–Bellman equations [2, 12].

Theorem 14.1 *Assume there exists a continuously differential function $V_i^j(t, p) :$ $[t_0 + j\Delta t, t_0 + j\Delta t + \bar{T}] \times R \rightarrow R$ satisfying the partial differential equation*

$$- V_t^{j,i}(t, p) = \max_{q_i^j} \left\{ e^{-\rho(t-t_0)} \left[q_i^j(p - c_i - g_i q_i^j) \right] + \right.$$

$$\left. + V_p^{j,i}(t, p) s\left(a - d\left[q_i^j(t) + \sum_{k \neq i} q_k^{jNE}(t) \right] - p(t) \right) \right\}, \quad i = 1, \ldots, n. \quad (14.6)$$

where $V_i^j(t_0 + j\Delta t + \overline{T}, p) = 0$. Let $q_j^{NE}(t)(t, p)$ denote the controls which maximize the right hand side of (14.6). Then $q_j^{NE}(t)(t, p)$ provides a feedback Nash equilibrium in the truncated subgame $\bar{\Gamma}_j(p_{j,0}, t_0 + j\Delta t, t_0 + j\Delta t + \overline{T})$.

Since the considered differential game is an LQ differential game, then the feedback Nash equilibrium is unique [2].

In this game model the Bellman function $V_i^j(t, p)$ can be obtained in the form:

$$V_i^j(t, p) = e^{-\rho(t-t_0)}\left[A_i^j(t)p^2 + B_i^j(t)p + C_i^j(t)\right], \quad i = \overline{1, n}. \tag{14.7}$$

Substituting (14.7) in (14.6) we can determine Nash equilibrium strategies in the following form:

$$q_i^{jNE}(t, p) = -\frac{(c_i - p) + sd\left[B_i^j(t) + 2A_i^j(t)p\right]}{2g_i}, \quad i = \overline{1, n}, \tag{14.8}$$

where functions $A_i^j(t)$, $B_i^j(t)$, $C_i^j(t)$, $t \in [t_0 + j\Delta t, t_0 + j\Delta t + \overline{T}]$ are defined by the system of differential equations:

$$\dot{A}_i^j(t) = A_i^j(t)[\rho + 2s] + \frac{(2A_i^j(t)d_js - 1)^2}{4g_i} - \sum_{k \neq i}\frac{A_i^j(t)d_js - 2A_i^j(t)A_k^j(t)d^2s}{g_k}$$

$$\dot{B}_i^j(t) = B_i^j(t)[\rho + s] - \frac{c_i}{2g_i} - 2A_i^j(t)a_js - \sum_{k \neq i}\frac{A_i^j(t)B_k^j(t)d_j^2s^2}{g_k} -$$

$$- \sum_{k \in N}\frac{B_i^j(t)d_js - A_i^j(t)c_kd_js - A_k^j(t)B_i^j(t)d_j^2s^2}{g_k}$$

$$\dot{C}_i^j(t) = C_i^j(t)\rho - B_i^j(t)a_js + \frac{c_i^2 + (B_i^j(t)d_js)^2}{4g_i} + \sum_{k \neq i}\frac{B_i^j(t)B_k^j(t)d_j^2s^2}{2g_k} +$$

$$+ \sum_{k \in N}\frac{B_i^j(t)c_kd_js}{2g_k}$$

with the boundary conditions $A_i^j(t_0 + j\Delta t + \overline{T}) = 0$, $B_i^j(t_0 + j\Delta t + \overline{T}) = 0$ and $C_i^j(t_0 + j\Delta t + \overline{T}) = 0$.

Substituting $q_j^{NE}(t, p)$ (14.8) into (14.15) yields the dynamics of a Nash equilibrium trajectory:

$$\dot{p}(t) = s\left(a - d\sum_{i \in N} q_i^{jNE}(t, p) - p(t)\right), \quad p(t_0 + j\Delta t) = p_{j,0}. \tag{14.9}$$

Let $p_j^{NE}(t)$ denote the solution of system (14.9).

14.2.4 Conditionally Non-cooperative Trajectory

Suppose that each truncated subgame $\bar{\Gamma}_j(p_{j,0}, t_0 + j\Delta t, t_0 + j\Delta t + \overline{T})$ develops along $p_j^{NE}(t)$ then the whole non-cooperative game with the Looking Forward Approach develops along:

Definition 14.2 Conditionally non-cooperative trajectory $\{\hat{p}_{NE}(t)\}_{t=t_0}^T$ is a combination of $p_j^{NE}(t)$ for each truncated subgame $\bar{\Gamma}_j(p_{j,0}^{NE}, t_0 + j\Delta t, t_0 + j\Delta t + \overline{T})$:

$$\{\hat{p}_{NE}^*(t)\}_{t=t_0}^T = \begin{cases} p_0^{NE}(t), \ t \in [t_0, t_0 + \Delta t), \\ \dots, \\ p_j^{NE}(t), \ t \in [t_0 + j\Delta t, t_0 + (j+1)\Delta t), \\ \dots, \\ p_l^{NE}(t), \ t \in [t_0 + l\Delta t, t_0 + (l+1)\Delta t]. \end{cases} \tag{14.10}$$

Along the conditionally non-cooperative trajectory the players receive payoff according to the following formula:

Definition 14.3 The resulting non-cooperative outcome for player $i = 1, \dots, n$ in the subgame of the game $\Gamma(p_0, T - t_0)$ with the Looking Forward Approach starting at $t \in [t_0 + j\Delta t, t_0 + j\Delta t + \overline{T}]$ has the following form:

$$\hat{V}_i(t, \hat{p}_{NE}(t)) = \sum_{m=j+1}^{l} \left[V_i^m(t_0 + m\Delta t, p_{m,0}^{NE}) - V_i^m(t_0 + (m+1)\Delta t, p_{m,1}^{NE}) \right] +$$

$$+ \left[V_i^j(t, p_j^{NE}(t)) - V_i^j(t_0 + (j+1)\Delta t, p_{j,1}^{NE}) \right], \ i \in N. \tag{14.11}$$

14.2.5   Numerical Simulation

The first game starts in December 2015 and lasts till the summit in Vienna in November 2016. We consider oil as a homogeneous product and appraise the demand function with the parameters of an average world oil price and the total world oil supply. We calculate average oil prices for each period based only on two major trading classifications, namely that of Brent crude and of light crude, which are accessible on the Finam agency data source. As the initial price, we take the average price in December 2015 which is equal to $p_0 = 34.51$.

Date	Total world supply MMBD	Brent, $ for barrel	Light $ for barrel	Average price, $ for barrel
12.2015	96.411	35.910	33.110	34.510
01.2016	95.875	36.640	33.990	35.315
02.2016	95.420	40.140	37.820	38.980
03.2016	95.294	47.320	45.990	46.655
04.2016	95.400	49.520	48.750	49.135
05.2016	95.187	49.740	48.640	49.190
06.2016	95.954	43.270	41.760	42.515
07.2016	96.891	46.970	45.000	45.985
08.2016	95.894	49.990	48.050	49.020
09.2016	96.001	48.510	46.970	47.740
10.2016	97.362	44.520	43.120	43.820
11.2016	97.241	47.034	45.908	46.471

To ensure that the slope of demand is negative, we decided not to regress the historical oil prices for the quantity produced and took into account the average price and the total world supply (in the table above) for only one previous period and a fixed choke price. Thus, we assume the parameter of demand a_j to be equal to 300 at each period, and, thus, the parameter d_j can be obtained from the (14.2) as

$$d_j = (a_j - \hat{p}(t-1)) \sum_{i \in N} q_i(t-1). \qquad (14.12)$$

The length of each Δt-time interval is 1 month. The players use the appraised demand with parameters a_j and d_j as the forecast for the next $\overline{T} = 3$ periods. We set values upon the parameters of cost function by using the total cost of producing a barrel and the average volumes of oil production for our players in 2016 and by fixing the parameter g_i at the level 0.7 for each player and each period. Both the parameters of c_i and g_i remain unchanged during the game. We assume that the speed of adjustment $s = 0.2$ and the discount factor $r = 10\%$.

Date	a	d
12.2015	300	2.717
01.2016	300	2.754
02.2016	300	2.761
03.2016	300	2.735
04.2016	300	2.659
05.2016	300	2.630
06.2016	300	2.635
07.2016	300	2.683
08.2016	300	2.622
09.2016	300	2.617
10.2016	300	2.628
11.2016	300	2.631

i	Producer	c	g
1	OPEC	3.169	0.7
2	Non-OPEC	17.333	0.7
3	US shale	20.238	0.7
4	US non-shale	18.182	0.7
5	Others	20.867	0.7

Figure 14.2 shows the comparison of the conditionally non-cooperative trajectory $\{\hat{p}_{NE}(t)\}_{t=t_0}^T$ and the historical average oil price dynamics. In Fig. 14.3 Nash feedback strategies (14.8) corresponding to the $\{\hat{p}_{NE}(t)\}_{t=t_0}^T$ and the historical quantities of production of oil are presented for each group of countries.

According to Figs. 14.2 and 14.3, we can suggest that the parameters c_i and g_i from the table below can be used for modeling the cooperative agreement in the time interval from December 2016 to May 2017.

We can see the payoff of players $\hat{V}_i(t, \hat{p}_{NE}(t)), i \in N$ corresponding to the Nash equilibrium strategies along the trajectory $\hat{p}_{NE}(t)$ on Fig. 14.4.

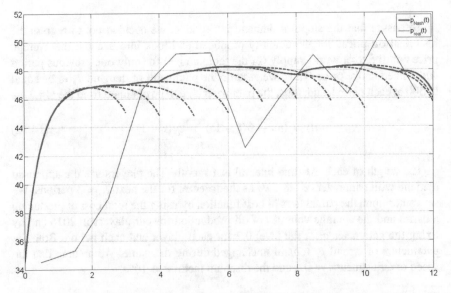

Fig. 14.2 Conditionally non-cooperative trajectory of the oil price $\hat{p}_{NE}(t)$ (thick solid line) with the Looking Forward Approach, historical oil price trajectory (thick dotted line)

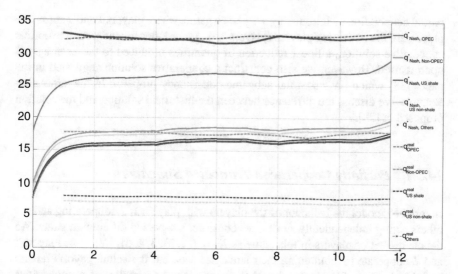

Fig. 14.3 Nash feedback strategies defined with the Looking Forward Approach (solid lines), and the corresponding historical quantities of oil production (dashed lines)

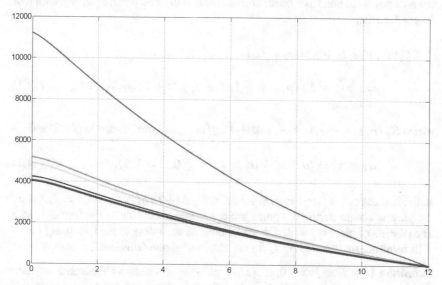

Fig. 14.4 Payoffs of players corresponding to the Nash equilibrium $\hat{V}_i(t, \hat{p}_{NE}(t))$

14.3 Partially Cooperative Game Model

As it was mentioned in the introduction, at the end of November 2016 two groups of countries, the OPEC and the non-OPEC, i.e. players 1 and 2 signed the cooperative agreement. By using the model constructed in Sect. 14.1, we build a model of such partial cooperation by introducing a 4-person game, where players 1 and 2 act as one player and the remaining players (players 3, 4, 5) act independently. As a historical cooperative solution, a linear reduction of quantities produced to the level agreed upon is used. However, we also construct a cooperative solution suggested in the paper [38] which possesses the subgame consistency property. In the simulation example, we discuss the difference between the historical solution and the solution proposed in [38].

14.3.1 Partially Cooperative Truncated Subgame

In order to model the behavior of the players with players 1, 2 cooperating and the others acting independently, at first we construct a 4-person differential game. We denote the set of players in this game as $\tilde{N} = \{(1, 2), 3, 4, 5\}$, $|\tilde{N}| = 4$. Players 1 and 2 cooperate by combining their strategies sets, but the actual payoffs remain individual. Thus, for players 1 and 2, we construct an analog of a cooperative differential game with non-transferable payoffs. Players 1, 2 orient themselves towards payoff defined as a linear combination of their respective payoffs with fixed α_j coefficients:

$$\tilde{K}_{1,2}^j(\alpha_j, p_{j,0}, t_0 + j\Delta t, t_0 + j\Delta t + \overline{T}; q_1^j, \ldots, q_5^j) =$$

$$= \sum_{i \in \{1,2\}} \alpha_i^j K_i^j(p_{j,0}, t_0 + j\Delta t, t_0 + j\Delta t + \overline{T}; q_1^j, \ldots, q_5^j), \qquad (14.13)$$

where $K_i^j(p_{j,0}, t_0 + j\Delta t, t_0 + j\Delta t + \overline{T}; q_1^j, \ldots, q_5^j)$ are defined in (14.3) and

$$\alpha_j \in \Lambda_j = \{\alpha : \alpha_1^j + \alpha_2^j = 1, \ \alpha_i^j \geq 0, \ i = 1, 2\}. \qquad (14.14)$$

Individual payoffs $\tilde{K}_i^j(\alpha_j, p_{j,0}, t_0 + j\Delta t, t_0 + j\Delta t + \overline{T}; q_1^j, \ldots, q_5^j), i = 1, 2$ which the players obtain during the game are calculated according to the formula (14.3). The payoffs of players $i = 3, 4, 5$ are calculated according to the formula (14.3).

It turns out that for any $\alpha_j \in \Lambda_j$ we obtain a unique differential game:

Definition 14.4 Let $j = 0, \ldots, l$. A partially cooperative truncated subgame $\tilde{\Gamma}_j^{\alpha_j}(p_{j,0}, t_0 + j\Delta t, t_0 + j\Delta t + \overline{T})$ is defined on the time interval $[t_0 + j\Delta t, t_0 + j\Delta t + \overline{T}]$. The motion equation and the initial condition of the truncated subgame $\tilde{\Gamma}_j^{\alpha_j}(p_{j,0}, t_0 + j\Delta t, t_0 + j\Delta t + \overline{T})$ have the following form:

$$\dot{p}(t) = s\left(a_j - d_j \sum_{i \in N} q_i^j(t) - p(t)\right), \ p(t_0 + j\Delta t) = p_{j,0}. \qquad (14.15)$$

The payoff function $\tilde{K}_{1,2}^j(\alpha_j, p_{j,0}, t_0 + j\Delta t, t_0 + j\Delta t + \overline{T}; q_1^j, \ldots, q_5^j)$ of players $\{1, 2\}$ in truncated subgame j is defined as in (14.13), the payoff function $\tilde{K}_i^j(\alpha_j, p_{j,0}, t_0 + j\Delta t, t_0 + j\Delta t + \overline{T}; q_1^j, \ldots, q_5^j)$ of players $i = 3, 4, 5$ is defined as in (14.3).

14.3.2   Pareto Optimal Outcomes

Suppose that players 1 and 2 can agree on the form of a joint payoff function, i.e. about the coefficients $\alpha_j \in \Lambda_j$. The set of outcomes for all possible $\alpha_j \in \Lambda_j$ for players 1 and 2 defines the set of Pareto optimal outcomes for players 1 and 2. Let us consider the feedback Nash equilibrium $q_{\alpha_j}^*(t, p)$ in the game $\tilde{\Gamma}_j^{\alpha_j}(p_{j,0}, t_0 + j\Delta t, t_0 + j\Delta t + \overline{T})$ of 4 persons. Then the payoffs in the Nash equilibrium for all possible values of α_j of players 1 and 2 form a set of Pareto-optimal outcomes $\tilde{K}_1^j(\alpha_j, p_{j,0}, t_0 + j\Delta t, t_0 + j\Delta t + \overline{T}; q_1^{*\alpha_j}, \ldots, q_5^{*\alpha_j})$ and $\tilde{K}_2^j(\alpha_j, p_{j,0}, t_0 + j\Delta t, t_0 + j\Delta t + \overline{T}; q_1^{*\alpha_j}, \ldots, q_5^{*\alpha_j})$. We define the feedback Nash equilibrium strategies of players as in [2].

We denote the value of players $\{1, 2\}$ in the feedback Nash equilibrium with a determined α_j by the function $W_{1,2}^{\alpha_j}(t, p)$, where t, p are the time and the initial state of the partially cooperative truncated subgame $\tilde{\Gamma}_j^{\alpha_j}(p_{j,0}, t_0 + j\Delta t, t_0 + j\Delta t + \overline{T})$ correspondingly. By the function $W_i^{\alpha_j}(t, p)$, $i = 3, 4, 5$ we denote the payoff of players $i = 3, \ldots, 5$ using the Nash equilibrium strategies $q_{\alpha_j}^*(t, p)$. Sufficient conditions for the solution and the feedback Nash equilibrium are given by the following assertion.

Theorem 14.2 *Assume there exist continuously differential functions* $W_{1,2}^{\alpha_j}(t, p)$, $W_3^j(t, p)$, $W_4^j(t, p)$, $W_5^j(t, p)$: $[t_0 + j\Delta t, t_0 + j\Delta t + \overline{T}] \times R \to R$ *satisfying the following partial differential equation*

$$-W_t^{1,2,\,\alpha_j}(t, p) = \max_{q_1^j, q_2^j} \left\{ \sum_{i=1}^{2} e^{-\rho(t-t_0)} \alpha_i^j \left[q_i^j (p - c_i - g_i q_i^j) \right] + \right.$$

$$\left. + W_p^{1,2,\,\alpha_j}(t, p) s \left(a - d \left[\sum_{i=1}^{2} q_i^j(t) + \sum_{k=3}^{5} q_k^{*\alpha_j}(t) \right] - p(t) \right) \right\},$$

$$-W_t^{i,\,\alpha_j}(t, p) = \max_{q_i^j} \left\{ e^{-\rho(t-t_0)} \left[q_i^j (p - c_i - g_i q_i^j) \right] + \right.$$

$$\left. + W_p^{i,\,\alpha_j}(t, p) s \left(a - d \left[q_i^j(t) + \sum_{k \neq i}^{5} q_k^{*\alpha_j}(t) \right] - p(t) \right) \right\}, \ i = 3, \ldots, 5, \ (14.16)$$

where $W_{1,2}^{\alpha_j}(t_0 + j\Delta t + \overline{T}, p) = 0$, $W_i^{\alpha_j}(t_0 + j\Delta t + \overline{T}, p) = 0$, $i=3,\ldots,5$. Then $q_{\alpha_j}^*(t, p)$ provides a feedback Nash equilibrium in the partial cooperative truncated subgame $\tilde{\Gamma}_j(p_{j,0}, t_0 + j\Delta t, t_0 + j\Delta t + \overline{T})$.

In this game model the Bellman function $W_{1,2}^{\alpha_j}(t, p)$, $W_i^{\alpha_j}(t, p)$, $i = 3,\ldots,5$ can be obtained in the form:

$$W_{1,2}^{\alpha_j}(t, p) = e^{-\rho(t-t_0)}\left[A_j^{1,2}(t)p^2 + B_j^{1,2}(t)p + C_j^{1,2}(t)\right],$$

$$W_i^{\alpha_j}(t, p) = e^{-\rho(t-t_0)}\left[A_j^i(t)p^2 + B_j^i(t)p + C_j^i(t)\right], \ i =, 3, 4, 5.$$

$$(14.17)$$

Feedback Nash equilibrium strategies for the partial cooperative truncated subgame:

$$q_i^{*\alpha_j}(t, p) = -\frac{(c_i - p) + ds\left[B_j^{1,2}(t) + 2A_j^{1,2}(t)p\right]}{2\alpha_i g_i}, \ i = 1, 2,$$

$$q_i^{*\alpha_j}(t, p) = -\frac{(c_i - p) + sd\left[B_j^i(t) + 2A_j^i(t)p\right]}{2g_i}, \ i = 3, 4, 5. \quad (14.18)$$

Functions $A_j^{1,2}(t)$, $B_j^{1,2}(t)$, $C_j^{1,2}(t)$, $A_j^i(t)$, $B_j^i(t)$, $C_j^i(t)$, $t \in [t_0 + j\Delta t, t_0 + j\Delta t + \overline{T}]$ are defined by the system of differential equations: $A_j^{1,2}(t_0 + j\Delta t + \overline{T})$, $B_j^{1,2}(t_0 + j\Delta t + \overline{T})$, $C_j^{1,2}(t_0 + j\Delta t + \overline{T})$:

$$\dot{A}_j^{1,2}(t) = A_j^{1,2}(t)(\rho + 2s) + \sum_{k\in\{1,2\}}\left(\frac{\alpha_k^j}{4g_k} + \frac{(A_j^{1,2}(t)d_js)^2}{\alpha_k^j g_k}\right) +$$

$$+ \sum_{k\in\{3,4,5\}}\frac{2A_j^k(t)A_j^{1,2}(t)d_j^2s^2}{g_k} - \sum_{k\in N}\frac{A_j^{1,2}(t)d_js}{g_k},$$

$$\dot{B}_j^{1,2}(t) = B_j^{1,2}(t)(\rho + s) - 2A_j^{1,2}(t)a_js - \sum_{k\in\{1,2\}}\left(\frac{\alpha_j^k c_k}{2g_k} - \frac{A_j^{1,2}(t)B_j^{1,2}(t)d_j^2s^2}{\alpha_k^j g_k}\right)$$

$$- \sum_{k\in\{3,4,5\}}\left(\frac{A_j^{1,2}(t)B_j^k(t)d_j^2s^2}{g_k} + \frac{A_j^k(t)B_j^{1,2}(t)d_j^2s^2}{g_k}\right) +$$

$$- \sum_{k\in N}\left(\frac{B_j^{1,2}(t)d_js}{2g_k} - \frac{A_j^{1,2}(t)c_kd_js}{g_k}\right)$$

$$\dot{C}_j^{1,2}(t) = C_j^{1,2}(t)\rho - B_j^{1,2}(t)a_js + \sum_{k\in\{1,2\}}\left(\frac{\alpha_j^k c_k^2}{4g_k} + \frac{(B_j^{1,2}(t)d_js)^2}{4\alpha_k^j g_k}\right) +$$

$$+ \sum_{k\in\{3,4,5\}}\frac{B_j^{1,2}(t)B_j^k(t)d_j^2 s^2}{2g_k} + \sum_{k\in N}\frac{B_j^{1,2}(t)c_k d_js}{2g_k},$$

$A_j^{1,2}(t_0 + j\Delta t + \overline{T}) = 0,\ B_j^{1,2}(t_0 + j\Delta t + \overline{T}) = 0,\ C_j^{1,2}(t_0 + j\Delta t + \overline{T}) = 0.$

$A_j^i(t_0 + j\Delta t + \overline{T}),\ B_j^i(t_0 + j\Delta t + \overline{T}),\ C_j^i(t_0 + j\Delta t + \overline{T}),$ for $i = 3,\dots,5$:

$$\dot{A}_j^i(t) = A_j^i(t)(\rho + 2s) + \sum_{k\in\{1,2\}}\left(\frac{1}{4g_i} + \frac{2A_j^i(t)A_j^{1,2}(t)d_j^2 s^2}{\alpha_k^j g_k}\right) +$$

$$+ \sum_{k\in\{3,4,5\}}\frac{2A_j^k(t)A_j^i(t)d_j^2 s^2}{g_i} - \sum_{k\in N}\frac{A_j^i(t)d_js}{g_k},$$

$$\dot{B}_j^i(t) = B_j^i(t)(\rho + s) - 2A_j^i(t)a_js - \frac{c_i}{2g_i} - \sum_{k\in N}\left(\frac{B_j^i(t)d_js}{2g_k} - \frac{A_j^i(t)c_k d_js}{g_k}\right)$$

$$+ \sum_{k\in\{1,2\}}\left(\frac{A_j^i(t)B_j^{1,2}(t)d_j^2 s^2}{\alpha_k^j g_k} + \frac{A_j^{1,2}(t)B_j^i(t)d_j^2 s^2}{\alpha_k^j g_k}\right) +$$

$$+ \sum_{k\in\{3,4,5\}}\left(\frac{A_j^i(t)B_j^k(t)d_j^2 s^2}{g_k} + \frac{A_j^k(t)B_j^i(t)d_j^2 s^2}{g_k}\right)$$

$$\dot{C}_j^i(t) = C_j^i(t)\rho + \frac{c_i^2}{4g_i} - B_j^i(t)a_js + \sum_{k\in\{1,2\}}\frac{B_j^i(t)B_j^{1,2}(t)d_j^2 s^2}{2\alpha_k^j g_k} +$$

$$+ \sum_{k\in\{3,4,5\}}\frac{B_j^i(t)B_j^k(t)d_j^2 s^2}{2g_k} + \sum_{k\in N}\frac{B_j^k(t)c_k d_js}{2g_k},$$

$A_j^i(t_0 + j\Delta t + \overline{T}) = 0,\ B_j^i(t_0 + j\Delta t + \overline{T}) = 0,\ C_j^i(t_0 + j\Delta t + \overline{T}) = 0.$

Substituting $q_{\alpha_j}^*(t)(t, p)$ (14.18) into (14.15) yields the dynamics of the feedback Nash equilibrium trajectory corresponding to the vector of weights for the players 1, 2:

$$\dot{p}(t) = s\left(a - d\sum_{i\in N}q_i^{*\alpha_j}(t, p) - p(t)\right),\ p(t_0 + j\Delta t) = p_{j,0}. \qquad (14.19)$$

We denote by $p_{\alpha_j}^*(t)$ the solution of system (14.19) corresponding to the fixed weights $\alpha_j = (\alpha_1^j, \alpha_2^j)$. Further we call $p_{\alpha_j}^*(t)$ the Pareto optimal trajectory.

14.3.3 Individual Players Payoffs Under Cooperation

In the previous section we defined the payoff of players 3, 4, 5, but we did not define the individual payoff of players 1, 2. The next step is to define the individual payoffs of the players under cooperation. This can be achieved by substituting the Pareto optimal trajectory $p^*_{\alpha_j}(t)$ (14.19) and strategies $q^*_{\alpha_j}(t, p)$ (14.18) into the formula of each player's payoff (14.3). The payoff of players 1, 2 in the truncated subgame $\tilde{\Gamma}_j(p_{j,0}, t_0 + j\Delta t, t_0 + j\Delta t + \overline{T})$ along the Pareto optimal trajectory $p^*_{\alpha_j}(t)$ with strategies $q^*_{\alpha_j}(t, p)$ involved is denoted by $W_1^{\alpha_j}(t, p)$ and $W_2^{\alpha_j}(t, p)$. The other and more preferable way is to apply the results from the paper [38], where a special form of Hamilton–Jacobi–Bellman equation is solved for each player, but with fixed strategies $q^*_{\alpha_j}(t, p)$ (14.18) (so no maximization in the right-hand side of the equation is involved). The size of the current paper being limited, we cannot present it here.

14.3.4 Subgame Consistent Solutions

In the paper [38] the notion of subgame consistency property of a cooperative solution in a game with non-transferable payoffs is defined as follows:

Definition 14.5 The payoff vector $W(t, p^*(t)) = (W_1(t, p^*(t)), \ldots, W_n(t, p^*(t)))$ along the trajectory $p^*(t)$ is subgame consistent in the game $\Gamma_c(p_0, T - t_0)$ with non-transferable payoffs if it satisfies the following conditions:

1. the payoff vector

$$W(t, p^*(t)) = (W_1(t, p^*(t)), \ldots, W_n(t, p^*(t))), \ t \in [t_0, T] \tag{14.20}$$

   should be Pareto optimal;
2. the payoff vector should satisfy the individual rationality property

$$W_i(t, p^*(t)) \geq V_i(t, p^*(t)), \ t \in [t_0, T], \ i = \overline{1, n}, \tag{14.21}$$

where $V_i(t, p^*(t))$ is an individual payoff of player $i = \overline{1, n}$.

Definition 14.6 Let us define the set $S^\tau_j = \bigcap_{t_0 + j\Delta t \leq t < \tau} S^j_t$ for $\tau \in [t_0 + j\Delta t, t_0 + (j+1)\Delta t)$, $j = 0, \ldots, l$, where S^j_t is a set of $\alpha_j = (\alpha^j_1, \alpha^j_2)$ that satisfy individual rationality at the moment of time t for $i = \overline{1, n}$:

$$W_i^{\alpha^j}(t, p^*_{\alpha^j}(t)) \geq V_i^j(t, p^*_{\alpha^j}(t)), \tag{14.22}$$

where $V_i(t, p^*(t))$ is the payoff of player $i \in N$ in the non-cooperative game with Nash equilibrium strategies, but calculated along the Pareto optimal trajectory $p^*_{\alpha_j}(t)$.

$S_j^{t_0+(j+1)\Delta t}$ is a set of $\alpha_j = (\alpha_1^j, \alpha_2^j)$ that satisfy individual rationality (14.22) throughout the time interval before the information about the game structure is updated, at the moment of time $t = t_0 + (j + 1)\Delta t$. Set $S_j^{t_0+(j+1)\Delta t}$ provides a corresponding set of subgame consistent solutions:

$$\left\{ W^{\alpha_j}(t, p^*_{\alpha_j}(t)) = (W_1^{\alpha_j}(t, p^*_{\alpha_j}(t)), \ldots, W_n^{\alpha_j}(t, p^*_{\alpha_j}(t))), \right.$$

$$\left. t \in [t_0 + j\Delta t, t_0 + (j + 1)\Delta t], \ \alpha_j \in S_j^{t_0+(j+1)\Delta t}, \ i = \overline{1, n}, \ j = 0, \ldots, l \right\}.$$

$$(14.23)$$

For a more general case in [38], where individual rationality property (14.22) holds on the whole time interval it was proved that:

Theorem 14.3 *A solution optimality principle under which the players agree to choose the same weights* $\alpha = (\alpha_1, \ldots, \alpha_n) \in \Lambda$ *in all the subgames* $\Gamma_c(p^*_\alpha(t), T - t), \ t \in [t_0, T]$ *such that*

$$W_i^\alpha(t, p^*_\alpha(t)) \geq V_i(t, p^*_\alpha(t)), \ t \in [t_0, T], \ i = \overline{1, n}. \tag{14.24}$$

yields a subgame consistent solution to the cooperative game $\Gamma_c(p_0, T - t_0)$.

14.3.5   Solution Concept

Suppose that at the beginning of each truncated subgame $\tilde{\Gamma}_j(p_{j,0}, t_0 + j\Delta t, t_0 + j\Delta t + \overline{T})$ weights $\alpha_j \in S_j^{t_0+(j+1)\Delta t}$ are chosen. Then each truncated subgame develops along the Pareto optimal trajectory $p^*_{\alpha_j}(t)$ and as a consequence, the whole game $\Gamma_c(p_0, T - t_0)$ develops along the following trajectory. If we denote $p^*_{j,0} = p^*_{\alpha_{j-1}}(t_0 + j\Delta t)$, then:

Definition 14.7 The conditionally Pareto optimal trajectory $\{\hat{p}^*(t)\}_{t=t_0}^T$ is a combination of Pareto optimal trajectories $p^*_{\alpha_j}(t)$ for each truncated subgame $\tilde{\Gamma}_j(p^*_{j,0}, t_0 + j\Delta t, t_0 + j\Delta t + \overline{T})$ and defined weights $\alpha_j \in S_j^{t_0+(j+1)\Delta t}$:

$$\{\hat{p}^*(t)\}_{t=t_0}^T = \begin{cases} p^*_{\alpha_0}(t), \ t \in [t_0, t_0 + \Delta t), \\ \ldots, \\ p^*_{\alpha_j}(t), \ t \in [t_0 + j\Delta t, t_0 + (j + 1)\Delta t), \\ \ldots, \\ p^*_{\alpha_l}(t), \ t \in [t_0 + l\Delta t, t_0 + (l + 1)\Delta t]. \end{cases} \tag{14.25}$$

Now let us define the solution for the whole game $\Gamma_c(p_0, T - t_0)$. Let $p_{j,1}^* = p_{\alpha_j}^*(t_0 + (j + 1)\Delta t)$:

Definition 14.8 The resulting solution for the game $\Gamma_c(p_0, T - t_0)$ with the Looking Forward Approach corresponding to the chosen weights $\alpha_j \in S_j^{t_0 + (j+1)\Delta t}$ for each truncated subgame $\tilde{\Gamma}_j(p_{j,0}, t_0 + j\Delta t, t_0 + j\Delta t + \overline{T})$, $j = 0, \ldots, l$ is defined in the following way, let $t \in [t_0 + j\Delta t, t_0 + (j + 1)\Delta t]$:

$$\hat{W}_i(t, \hat{p}^*(t)) = \sum_{m=j+1}^{l} \left[W_i^{\alpha_m}(t_0 + m\Delta t, p_{m,0}^*) - W_i^{\alpha_m}(t_0 + (m + 1)\Delta t, p_{m,1}^*) \right] +$$

$$+ \left[W_i^{\alpha_j}(t, p_{\alpha_j}^*(t)) - W_i^{\alpha_j}(t_0 + (j + 1)\Delta t, p_{j,1}^*) \right]. \quad (14.26)$$

Therefore the set of all possible solutions in the game $\Gamma_c(p_0, T - t_0)$ has the following form:

$$\left\{ \hat{W}(t, \hat{p}^*(t)) = (\hat{W}_1(t, \hat{p}^*(t)), \ldots, \hat{W}_n(t, \hat{p}^*(t))), \ t \in [t_0, T], \right.$$

$$\left. \alpha_j \in S_j^{t_0 + (j+1)\Delta t}, \ \forall j = 0, \ldots, l \right\}. \quad (14.27)$$

In a game with a moving information horizon, we call a solution conditionally Pareto optimal if it is constructed using the Pareto optimal solutions in each truncated subgame. In this sense, we call the resulting solution $\hat{W}(t, \hat{p}^*(t))$ conditionally Pareto optimal.

14.3.6 Subgame Consistency of Solution Concept

In [25] the subgame consistency property of the resulting solution (14.26) was proved for a case of full cooperation. A similar result can be obtained for a partial cooperation:

Theorem 14.4 The resulting solution $\hat{W}(t, \hat{p}^*(t))$ is subgame consistent in the game $\Gamma_c(\hat{p}^*(t), T - t)$ with the Looking Forward Approach if the following condition is satisfied, let $\forall t \in [t_0 + j\Delta t, t_0 + (j + 1)\Delta t]$, $j = 0, \ldots, l$, $i = \overline{1, n}$:

$$W_i^{\alpha_j}(t, p_{\alpha_j}^*(t)) - V_i^j(t, p_{\alpha_j}^*(t)) \geq$$

$$\geq W^{\alpha_j}(t_0 + (j + 1)\Delta t, p_{j,1}^*) - V_i^j(t_0 + (j + 1)\Delta t, p_{j,1}^*), \quad (14.28)$$

14.3.7   Historical Cooperative Agreement

In order to make a comprehensive comparison of theoretical results and the real situation, it is necessary to mention the complexity of such an agreement at countries scale. It affects not only oil market situations inside the countries but also its balance and all macroeconomic indicators. The agreement is of both economic and political nature. According to the agreement of November 30, 2016, the OPEC countries and 11 non-OPEC countries decided to reduce their production by 1.18–7.69% for different countries. Herewith, Iran, Libya, and Nigeria are allowed to boost their output. After the agreement was reached, oil prices did start to increase. However, by the end of May 2017, they had fallen to almost the same level. Countries-outsiders, notably the US shale producers, increased their output. Most of the countries inside the agreement succeeded in fulfilling it. After 6 months all these counties decided to prolong the agreement for the next year.

To model this cooperative agreement we construct linear type strategies for players 1 and 2. We use Nash equilibrium strategies in the non-cooperative differential game for the other players. The optimal strategies for players 3, 4, 5 are obtained in the same way as in Sect. 14.1 using the Hamilton–Jacobi–Bellman equation with the strategies (14.29) involved.

According to the statistical data, we suppose that the strategies of players $i = 1, 2$ under the cooperative agreement have the following form:

$$\bar{q}_1^*(t) = 3.5\% q_1^{avr}(t_{nov})\left(1 - 3.5\% t/T\right) \qquad (14.29)$$

$$\bar{q}_2^*(t) = 3.1\% q_2^{avr}(t_{nov})\left(1 - 3.1\% t/T\right), \qquad (14.30)$$

where $q_{avr}(t_{nov}) = (q_1^{avr}(t_{nov}), q_2^{avr}(t_{nov})) = (39.838, 18.771)$ is the average quantity of produced oil in t_{nov}, which is November 2016 and T corresponds to May 2017.

14.3.8   Numerical Simulation

The second game starts immediately after the summit and lasts from December 2016 until May 2017. As we did for the previous time interval, we calculate the average oil prices for each period based only on the two major trading classifications of Brent crude and light crude which are accessible on the Finam agency data source. We take the average price in December 2016 as the initial price $p_0 = 53, 73$ for the second game.

Date	Total world supply, MMBD	Brent, $ for barrel	Light, $ for barrel	Average price, $ for barrel
12.2016	97,324	55,008	52,449	53,728
01.2017	96,627	55,607	52,951	54,279
02.2017	96,412	56,122	53,661	54,892
03.2017	96,411	52,592	49,873	51,232
04.2017	97,075	53,841	51,216	52,529
05.2017	97,529	51,533	48,744	50,139

In order to test the constructed model with the Looking Forward Approach, we use the same players' parameters c_i and g_i as for the previous time interval and we suppose the parameter of demand a_j to be equal to 300 at each Δt-period. The players use the appraised demand with parameters a_j and b_j as the forecast for the next $\overline{T} = 3$ periods. Parameter d_j is obtained using the formula (14.12).

Date	a	d
01.2017	300	2.530
02.2017	300	2.543
03.2017	300	2.542
04.2017	300	2.580
05.2017	300	2.549

i	Producer	c	g
1	OPEC	3.169	0.7
2	Non-OPEC	17.333	0.7
3	US shale	20.238	0.7
4	US non-shale	18.182	0.7
5	Others	20.867	0.7

In Fig. 14.5 we can see the conditionally Pareto optimal trajectory $\hat{p}^*(t)$ (solid line) and the set of Pareto optimal trajectories $p^*_{\alpha_j}(t)$ (dashed lines) corresponding the weights $\alpha_j \in S_j^{t_0+(j+1)\Delta t}$ in each truncated subgame $\tilde{\Gamma}_j(p^*_{j,0}, t_0 + \Delta t, t_0 + \Delta t + \overline{T})$. In Fig. 14.6 the optimal strategies corresponding to the $\{\hat{p}^*_{\alpha_j}(t)\}_{t=t_0}^T$ and the historical quantities of oil production are presented for each group of the countries.

14.3.9 Analysis of Optimality of Corporate Agreement

In Fig. 14.7 we can see the individual payoff of player $i = 1$ under cooperation $W_1^j(t, p^*_{\alpha_j}(t))$ (solid lines) for all possible weights $\alpha_j \in S_j^{t_0+(j+1)\Delta t}$ and $\alpha_j^* \in S_j^{t_0+(j+1)\Delta t}$, $j = 3$. In Fig. 14.8 we can see the payoff of players $\hat{W}_i(t, \hat{p}^*(t))$ (solid lines), $i \in N$ and the payoff of players $\hat{V}_i(t, \hat{p}^*(t))$ (dashed lines), $i \in N$ corresponding to the Nash equilibrium strategies along the conditionally Pareto optimal trajectory $\hat{p}^*(t)$.

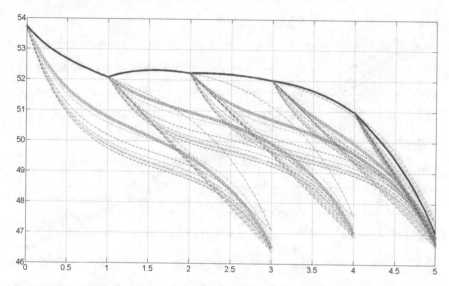

Fig. 14.5 Conditionally Pareto optimal trajectory $\hat{p}^*(t)$ (solid line), set of Pareto optimal trajectories $p^*_{\alpha_j}(t)$ (dashed lines) corresponding to weights $\alpha_j \in S_j^{t_0+(j+1)\Delta t}$ in each truncated subgame $\tilde{\Gamma}_j(p^*_{j,0}, t_0 + \Delta t, t_0 + \Delta t + \overline{T})$

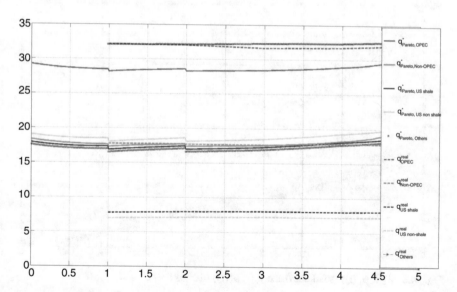

Fig. 14.6 Optimal strategies defined with the Looking Forward Approach (solid lines), and the corresponding historical quantities of oil production (dashed lines)

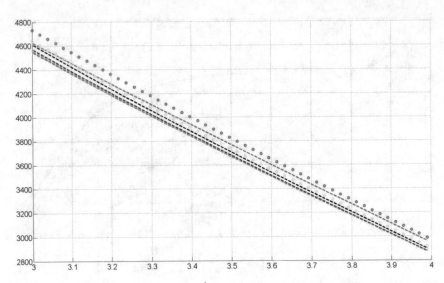

Fig. 14.7 The set of individual payoffs $W_1^j(t, p_{\alpha_j}^*(t))$ of player 1 for all possible weights $\alpha_j \in S_j^{t_0+(j+1)\Delta t}$ in the truncated subgame $\tilde{\Gamma}_j(p_{j,0}^*, t_0 + \Delta t, t_0 + \Delta t + \overline{T})$, $j = 3$

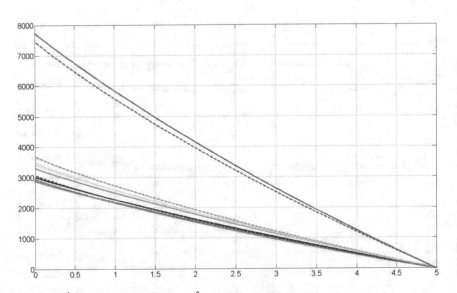

Fig. 14.8 $\hat{W}_i(t, \hat{p}^*(t))$ (solid lines) and $\hat{V}_i(t, \hat{p}^*(t))$ (dashed lines), $i \in N$

14.3.10 Comparison of Historical Cooperative Solution and Proposed Solution

This section presents the comparison of the historical cooperative agreement proposed in (14.29) with the strategies $\bar{q}^*(t)$ involved as well as the subgame consistent cooperative solution proposed in (14.26). In Fig. 14.9 we can see the conditionally Pareto optimal trajectory $\hat{p}^*_{\alpha_j}(t)$ corresponding to the chosen weights $\alpha^*_j \in S_j^{t_0+(j+1)\Delta t}$ in all the truncated subgames $\tilde{\Gamma}_j(p^*_{j,0}, t_0 + \Delta t, t_0 + \Delta t + \overline{T})$, $j = 0, \ldots, l$ trajectory $\bar{p}^*(t)$ corresponding to the historical cooperative agreement and the historical oil price dynamics.

In Fig. 14.10 the optimal strategies (14.8) corresponding to the $\hat{p}^*_{\alpha_j}(t)$, strategies $\bar{q}^*(t)$ (14.29) corresponding to the historical cooperative agreement are presented for each group of countries and the historical quantities of oil production for each group of countries. In Fig. 14.11 we can see the payoff of players $\hat{W}_i(t, \hat{p}^*(t))$ (solid lines), $i \in N$ along the trajectory $\hat{p}^*(t)$ and payoffs corresponding to the historical cooperative agreement (dashed lines).

On the whole, the decision to cooperate was justified for the countries under consideration. It can be seen by comparing the cooperative and non-cooperative solutions on the second time interval. In Fig. 14.12 we can see the conditionally Pareto optimal trajectory $\hat{p}^*_{\alpha_j}(t)$ (solid line) corresponding to the chosen weights

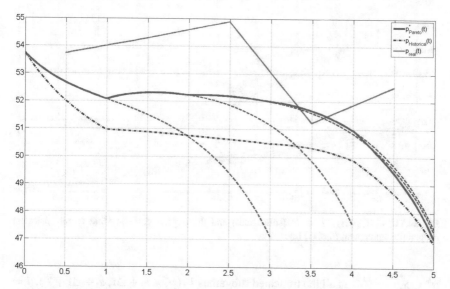

Fig. 14.9 The conditionally Pareto optimal trajectory $\hat{p}^*_{\alpha_j}(t)$ (thick solid line) corresponding to chosen weights $\alpha^*_j \in S_j^{t_0+(j+1)\Delta t}$, $j = 0, \ldots, l$, the trajectory $\bar{p}^*(t)$ (dashed line) corresponding to the historical cooperative agreement (14.29) and the historical oil price dynamics (thin solid line)

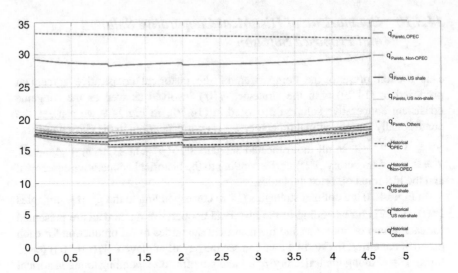

Fig. 14.10 The optimal strategies (14.8) defined with the Looking Forward Approach (solid lines), and the strategies (14.29) corresponding to the historical cooperative agreement (dashed lines)

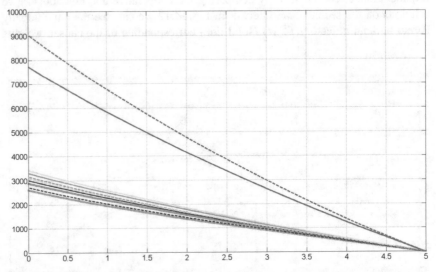

Fig. 14.11 $\hat{W}_i(t, \hat{p}^*(t))$, $i \in N$ (solid lines) and the payoffs corresponding to the historical cooperative agreement (dashed lines)

$\alpha_j^* \in S_j^{t_0+(j+1)\Delta t}$ in all the truncated subgames $\tilde{\Gamma}_j(p_{j,0}^*, t_0 + \Delta t, t_0 + \Delta t + \overline{T})$, $j = 0, \ldots, l$ and the conditionally non-cooperative trajectory of the oil price dynamics $\hat{p}_{NE}(t)$ corresponding to the Nash equilibrium (dashed line).

Nevertheless, the analysis of individual countries' profits confirms the profitability of the agreement only for the first player. In Fig. 14.13 we can see the payoff of

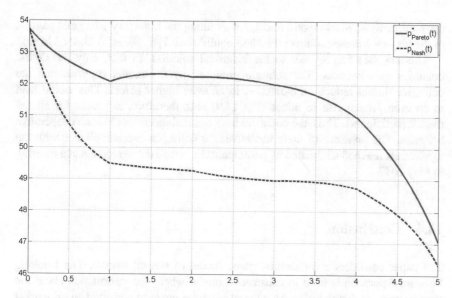

Fig. 14.12 The conditionally Pareto optimal trajectory $\hat{p}^*_{\alpha_j}(t)$ (solid line) and the conditionally non-cooperative trajectory of the oil price $\hat{p}_{NE}(t)$ (dashed line)

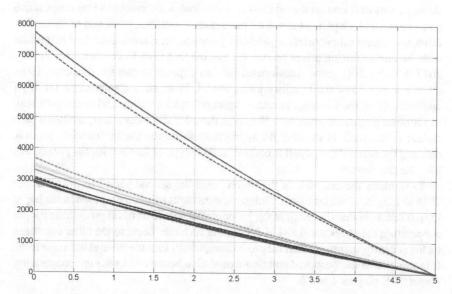

Fig. 14.13 $\hat{W}_i(t, \hat{p}^*(t))$, $i \in N$ (solid lines) and the payoffs of players corresponding to the Nash equilibrium $\hat{V}_i(t, \hat{p}_{NE}(t))$ (dashed lines)

the players $\hat{W}_i(t, \hat{p}^*(t))$ (solid lines), $i \in N$ along the trajectory $\hat{p}^*(t)$ and payoffs of the players corresponding to the Nash equilibrium $\hat{V}_i(t, \hat{p}_{NE}(t))$ (dashed lines).

This analysis corresponds to the historical solution. In fact, only the OPEC countries gain benefits. The solution proposed by the model suggests that the countries should reduce oil extraction to an even higher extent. This would lead to an even greater rise in prices (Fig. 14.9) and, therefore, to profits for all the market participants. Thus, the countries have rational incentives to such cooperative behavior. The presence of these incentives for collusion and compliance with the agreements reached stimulated its participants to extend the agreement at a meeting in May 2017.

14.4   Conclusion

The paper considers a differential game model of the oil market. The Looking Forward Approach is used to construct a model where the information about the process updates dynamically. An attempt has been made to construct an oil market model for two time periods with two different market conditions. Namely, we assume that from December 2015 to November 2016 a non-cooperative oligopoly structure was in action on the oil market. Therefore, we construct a non-cooperative game model and adapt it to the real oil price data. Thereafter, at the end of November 2016, the largest oil exporters signed an agreement for a reduction of oil extraction with the aim of raising prices on oil. We assume that from December 2016 to May 2017 the oligopoly game transformed into a cooperative one for this case. Here, we use a non-transferable utility game model. Numerical results show the high applicability of the Looking Forward Approach for modeling and simulating real-life conflict-controlled processes. However, there are still numerous questions which remain open, such as defining an appropriate value for the information horizon, defining the form of the payoff functions, and the type of forecast for the parameters used for the players.

To simulate the behavior of the players over the second period from December 2016 to May 2017 we construct a subgame consistent solution proposed in the paper [38]. It turns out that the result of the proportional reduction of oil production for the cooperating countries consisted in decreasing revenues because the other countries, on the contrary, increased their oil production, which kept the price at the same level. It is shown that the proposed solution would give better results for the cooperating players, i.e. players 1 and 2.

Acknowledgement The research of first author was supported by a grant from the Russian Science Foundation (Project No 18-71-00081).

References

1. Al-Sultan, A.: Alternative models for OPEC behavior. J. Energy Dev. **18**(2), 263–281 (1995)
2. Basar, T., Olsder, G.: Dynamic Noncooperative Game Theory. SIAM, Philadelphia (1999)
3. Berger, K., Hoel, M., Holden, S., Olsen, O.: The oil market as an oligopoly. Discussion paper. Cent. Bur. Stat. **32**, 1–19 (1988)
4. Bockem, S.: Cartel formation and oligopoly structure: a new assessment of the crude oil market. Appl. Econ. **36**(12), 1355–1369 (2004). https://doi.org/10.1080/0003684042000191093B
5. Cellini, R., Lambertini, L.: A differential oligopoly game with differentiated goods and sticky prices. Eur. J. Oper. Res. **176**, 1131–1144 (2007). https://doi.org/10.1016/j.ejor.2005.09.013
6. Dahl, C., Yucel, M.: Testing alternative hypotheses of oil producer behavior. Energy J. **12**(4), 117–138 (1991). https://doi.org/10.5547/ISSN0195-6574-EJ-Vol12-No4-8
7. Danielsen, A., Kim, S.: OPEC stability: an empirical assessment. Energy Econ. **10**(3), 174–184 (1998). https://doi.org/10.1016/0140-9883(88)90001-1
8. Goodwin, G., Seron, M., Dona, J.: Constrained Control and Estimation: An Optimisation Approach. Springer, London (2005)
9. Gromova, E., Petrosian, O.: Control of informational horizon for cooperative differential game of pollution control. In: 2016 International Conference Stability and Oscillations of Nonlinear Control Systems (Pyatnitskiy's Conference), pp. 1–4 (2016). https://doi.org/10.1109/STAB.2016.7541187
10. Gulen, S.: Is OPEC a cartel? Evidence from cointegration and causality tests. Energy J. **17**(2), 43–57 (1996)
11. Haurie, A.: A note on nonzero-sum differential games with bargaining solution. J. Optim. Theory Appl. **18**, 18–31 (1976). https://doi.org/10.1007/BF00933792
12. Isaacs, R.: Differential Games. Wiley, New York (1965)
13. Krasner, S.: The great oil Sheikdown. Foreign Policy **13**, 123–138 (1973). https://doi.org/10.2307/1147771
14. Kuchkarov, I., Petrosian, O.: On class of linear quadratic non-cooperative differential games with continuous updating. Lect. Notes Comput. Sci. **11548**, 635–650 (2019). https://doi.org/10.1007/978-3-030-22629-9_45
15. Kwon, W., Han, S.: Receding Horizon Control: Model Predictive Control for State Models. Springer, London (2005)
16. Ludkovski, M., Sircar, R.: Game theoretic models for energy production. In: Commodities, Energy and Environmental Finance. Fields Institute Communications, vol. 74, pp. 317–333. Springer, Berlin (2015). https://doi.org/10.1007/978-1-4939-2733-312
17. Molchanov, P.: A Statistical Analysis of OPEC Quota Violations. In: Economics, pp. 1–31. Duke University, Durham (2003)
18. Moran, T.: Managing an oligopoly of would-be sovereigns: the dynamics of joint control and selfcontrol in the international oil industry past, present, and future. Int. Organ. **41**(4), 575–607 (1987). https://doi.org/10.1017/S0020818300027612
19. Nash, J.: Equilibrium points in n-person games. Proc. Nat. Acad. Sci. **36**(1), 48–49 (1950). https://doi.org/10.1073/pnas.36.1.48
20. Nash, J.: Non-cooperative games. Ann. Math. **54**(2), 286–295 (1951). https://doi.org/10.2307/1969529
21. Petrosian, O.: Looking forward approach in cooperative differential games. Int. Game Theory Rev. **18**(2), 1–14 (2016). https://doi.org/10.1142/S0219198916400077
22. Petrosian, O.: Looking forward approach in cooperative differential games with infinite-horizon. Vestnik S.-Petersburg Univ. Ser. 10. Prikl. Mat. Inform. Prots. Upr. **4**, 18–30 (2016). https://doi.org/10.21638/11701/spbu10.2016.402
23. Petrosian, O., Barabanov, A.: Looking forward approach in cooperative differential games with uncertain-stochastic dynamics. J. Optim. Theory Appl. **172**(1), 328–347 (2017). https://doi.org/10.1007/s10957-016-1009-8

24. Petrosian, O., Kuchkarov, I.: About the looking forward approach in cooperative differential games with transferable utility. In: Frontiers of Dynamic Games: Game Theory and Management, St. Petersburg, 2018, pp. 175–208. Birkhäuser, Cham (2019). https://doi.org/10.1007/978-3-030-23699-110

25. Petrosian, O., Nastych, M., Volf, D.: Differential game of oil market with moving informational horizon and non-transferable utility. In: Constructive Nonsmooth Analysis and Related Topics (dedicated to the memory of V.F. Demyanov) (CNSA), 2017, pp. 1–4 (2017). https://doi.org/10.1109/CNSA.2017.7974002

26. Petrosian, O., Shi, L., Li, Y., Gao, H.: Moving information horizon approach for dynamic game models. Mathematics 7(12), 1239 (2019). https://doi.org/10.3390/math7121239

27. Petrosian, O., Tur, A.: Hamilton–Jacobi–Bellman equations for non-cooperative differential games with continuous updating. In: Mathematical Optimization Theory and Operations Research pp. 178–191 (2019). https://doi.org/10.1007/978-3-030-33394-2_14

28. Petrosyan, L.: Time-consistency of solutions in multi-player differential games. Vestnik of Leningrad State University. Series 1. Mathematics. Mechanics. Astronomy, vol. 4, pp. 46–52 (1977)

29. Petrosyan, L., Yeung, D.: A time-consistent solution formula for bargaining problem in differential games. J. Opt. Theory Appl. 16(4), 1–11 (2014). https://doi.org/10.1142/S0219198914500169

30. Plaut, S.: OPEC is not a cartel. Challenge 24(5), 18–24 (1981). https://doi.org/10.1080/05775132.1981.11470722

31. Polasky, S.: Do oil producers act as 'oil'igopolists? J. Environ. Econ. Manage. 23(3), 216–247 (1992). https://doi.org/10.1016/0095-0696(92)90002-E

32. Rawlings, J., Mayne, D.: Model Predictive Control: Theory and Design. Nob Hill Publishing, Madison (2009)

33. Salant, S.: Exhaustible resources and industrial structure: a Nash–Cournot approach to the world oil market. J. Polit. Econ. 84, 1079–1094 (1976). https://doi.org/10.1086/260497

34. Smith, J.: Inscrutable OPEC? Behavioral tests of the cartel hypothesis. Energy J. 26(1), 51–82 (2005). https://doi.org/10.2139/ssrn.353140

35. Wang, L.: Model Predictive Control System Design and Implementation Using MATLAB. Springer, London (2005)

36. Yeung, D., Petrosian, O.: Cooperative stochastic differential games with information adaptation. In: International Conference on Communication and Electronic Information Engineering (CEIE 2016) pp. 1–13 (2017). https://doi.org/10.2991/ceie-16.2017.47

37. Yeung, D., Petrosian, O.: Infinite horizon dynamic games: a new approach via information updating. Int. Game Theory Rev. 20(1), 1–23 (2017). https://doi.org/10.1142/S0219198917500268

38. Yeung, D., Petrosyan, L.: Subgame consistent solutions of a cooperative stochastic differential game with nontransferable payoffs. J. Optim. Theory Appl. 124(3), 701–724 (2005). https://doi.org/10.1007/s10957-004-1181-0

39. Youhanna, S.: A note on modelling OPEC behavior 1983–1989: a test of the cartel and competitive hypotheses. Am. Econ. 38(2), 78–84 (1994). https://doi.org/10.1177/056943459403800209

Chapter 15
Cooperative Decision Making in Cooperative Control Systems by Means of Game Theory

Simon Rothfuß, Jannik Steinkamp, Michael Flad, and Sören Hohmann

Abstract Current state cooperative control systems assisting the human in various applications, e.g. assisted driving, lack the ability of emancipated cooperative decision making. Due to certain situations in which this a crucial skill, the authors' objective is the automation design leading to cooperative control systems able to take part in human-machine negotiations.

In this work, we therefore introduce two games for modeling the cooperative decision making process. First, the event-based game is introduced in a complete information setting with focus on modeling the system dynamics the decision making process is based on, e.g. a vehicle. We also provide an analytical as well as a numerical solution approach to find parameterized strategies that are in a Nash equilibrium.

The second game is an enhanced war of attrition that models the negotiation process in an incomplete information setting. It includes a new cost structure that allows a better approximation of the costs in a realistic application of cooperative control systems. We proof that the resulting strategy leads to a Bayesian Nash equilibrium.

Keywords Cooperative decision making · Cooperative control systems · War of attrition

S. Rothfuß (✉) · J. Steinkamp · M. Flad · S. Hohmann
Institute of Control Systems (IRS) at Karlsruhe Institute of Technology (KIT), Karlsruhe, Germany
e-mail: simon.rothfuss@kit.edu

L. A. Petrosyan et al. (eds.), *Frontiers of Dynamic Games*,
Static & Dynamic Game Theory: Foundations & Applications,
https://doi.org/10.1007/978-3-030-51941-4_15

245

15.1 Introduction

In modern society humans are increasingly supported by automated assistance systems facilitating tasks in a wide range of applications. Examples are collaborative robots for lifting heavy equipment or driving assistance systems for lane keeping. All of these cooperative systems require a thorough automation design that takes into account the presence of and interaction with the human in order to ensure user acceptance and safety. One example of a design paradigm for such assistive systems can be found in Shared Control approaches, e.g. driving assistance systems based on game theory [9, 16]. Cooperative systems following this or other design approaches greatly contribute to reduced manual workload and increased human safety of today's society.

However, current state-of-the-art cooperative systems involving human-machine interaction lack the ability for cooperative decision making. An example are driving assistance systems like the lane keeping assistant. It supports the driver by warning and counter-steering if the vehicle gets too close to lane markings. However the decision to change or remain within the lane is made by the driver alone who is able to overrule the actions of the automation. Another example are the collision avoidance assistance systems that support the execution of the evasion maneuver after the driver decided on the concrete maneuver, i.e. evasion direction [8, 9].

Besides these parallel cooperative systems on action-level, there are conduct-by-wire concepts that enable sequential cooperation on a maneuver basis. The driver controls the car via maneuver commands and e.g. may use a touchpad instead of a steering wheel [3].

All these concepts exhibit the general approach which is the application of the master-slave principle with the human in the leading role who is supported by an assistance system. Nevertheless, this setting is not suitable for situations in which the automation has more or more reliable information about the cooperative task than the human. If the human, lacking important information, stays in the lead poor decisions may result inadvertently and vice versa. In these situations an emancipated discussion between the cooperating partners is needed. This has not been considered in literature so far.

As a consequence, our goal is the automation design of emancipated cooperative decision making systems capable to negotiate with humans. The interaction between automation and human may take place via some sort of dynamical system.

To support this objective we introduce an exemplary situation in which emancipated cooperative decision making is vital. This situation is an emergency evasion maneuver situation on a freeway, depicted in Fig. 15.1. The green vehicle is controlled by the human driver and an Advanced Driving Assistance System (ADAS) and is therefore called ego-vehicle in the following. The vehicle in front of the ego-vehicle performs an emergency braking maneuver. Assume the only way for the ego-vehicle to circumvent a crash is to change the lane to the right or left. This choice could be ambiguous or at least one of the driving agents (driver or ADAS) could prefer one of the two evasion options. In both cases, we assume

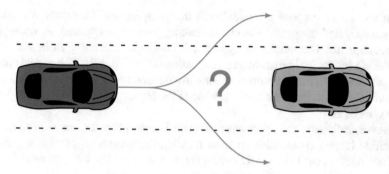

Fig. 15.1 Exemplary evasion maneuver with two options: left and right

that the driving agents need to agree on one option in order to successfully evade the obstacle. Assuming that both agents behave rationally, they will base their evaluation and ultimately their preference on the information they obtain about the situation. Therefore, the maneuver negotiation will strongly depend on the reliability of information the agents individually obtain. Also, the information sources of agents my differ. This difference in information status can easily appear as the following example demonstrates: the human is usually capable of recognizing and analyzing complex environments while the ADAS relies on fast, precise and reliable information about difference in velocity to the obstacle vehicle via radar sensors. Furthermore the system state, e.g. the position and motion of the vehicle, can have an influence on decision making. E.g. the closer the obstacle, the higher the pressure to agree on a decision becomes. In addition, it takes more effort to avoid a collision the smaller the distance between vehicle and obstacle becomes.

As a result both agents need to align their information status in order to reach a mutual decision. In critical situations like in the ADAS context haptic communication is a suitable communication channel due to its direct cognition and low latency [13]. However the haptic channel is not suited for the exchange of all relevant information. Instead one can only transmit intentions or other low-level information. This communicated information can then be included in further individual decision making. In the view of one agent the resulting negotiation process consists in the attempt to understand the opponent and to make the opponent understand oneself in order reach an agreement with limited communication resources.

Similar situations beyond the ADAS context can be found in robotics and other human-machine cooperation contexts, e.g. a worker and a robot have to cooperatively decide how to align a heavy work piece for future processing steps w.r.t. human context awareness and safety sensor information of the automation. Another example is the cooperative negotiation of an operator and a production management system to schedule tasks of a plant, combining the computational power of the automation with the experience of the human operator.

Since our objective is the design of a suitable automation for cooperative decision making while considering the introduced thoughts on cooperative decision making from above, we propose a model-based design approach. Therefore, a model of the emancipated cooperative decision making process is required. A widespread approach in the community designing automation to interact with humans is to establish a model for human-human interaction (cf. [9, 16, 17]). This model enables the automation to imitate human actions. An advantage of this approach is a more natural interaction from the point of view of the human, which leads to a high user acceptance [12].

Hence, following the approach to imitate human behavior, we require models capturing human cooperative decision making, i.e. human negotiation techniques and behavior. In our view the following features are essential for such models.

1. In a human negotiation, participants often have equal rights and follow an objective throughout the negotiation [19].
2. The model of the negotiation process should capture the dynamicity of a human-machine cooperation, i.e. the negotiation process itself (exchange of offers, deadlines, etc.) as well as the system dynamics the interaction is usually based on, e.g. in case of cooperative vehicle guidance the vehicle dynamics.
3. The negotiation model should be able to deal with incomplete information scenarios, respecting the low-level communication channel. In addition and in general, the specific human negotiation behavior has to be assumed to be unknown. However, some probabilistic assumptions on the behavior can be made.
4. The resulting negotiation strategy of such models should be deterministic w.r.t. the the available information in order to facilitate automation design.

To the knowledge of the authors, only the fields of negotiation theory and game theory offer models for decision making with the potential to adapt them to the features from above. Negotiation theory originates in the context of multi-agents systems [2] and was also successfully applied by Rothfuss et al. to model emancipated human-machine negotiation [17, 18]. In this paper we focus on the alternative models of game theory.

Like in negotiation theory, decision makers in game theory are usually modeled as emancipated and individual rational players. This already fulfills the first requirement from above. In the following we present some state-of-the-art models from game theory that are in or close to the scope of human-machine negotiation.

Looking at current models for assistance systems, differential games are often applied to model human-human or human-machine interaction on a trajectory basis. Na and Cole [9] and Flad et al. [16] provide such approaches in the context of driving assistance systems. In these cases, the assistance system supports the driver in the tracking of a given reference trajectory of the vehicle. To do this, the system also takes the system dynamics of the vehicle into account. In the context of cooperative decision making, the decision options would be various, conflicting reference trajectories. However, these approaches assume conflict-free references w.r.t. the given application. Therefore, they are not designed to resolve conflict situations, i.e. agreeing on a set of conflict-free reference trajectories, in an appropriate way.

In contrast to this, the class of coordination games explicitly focuses on making decisions how to coordinate players or their attributes. As an example, Zlotkin and Rosenschein [21] formulate the Postmen problem describing the strategic distribution of workload among players. In the incomplete information setting with a fixed interaction protocol, they propose an extended Zeuthen strategy [20] to achieve a Nash equilibrium. In our case however, we aim to model a decision on how to work on one cooperative task in contrast to distributing several tasks.

Another example of coordination games are revision games proposed by Calcagno et al. [5]. In these games players can revise their choice of action at decision times that are given by a exogenous Poison process until a deadline is reached. They apply a generalized form of backward induction to find a solution strategy. In the presented complete information game setting a trivial instantaneous solution, i.e. an agreement is found. Similar to this game setup, Caruana and Einav [6] introduce a game with switching costs that occur if players change their actions. In a two options complete information game, that can be described as the battle of sexes game, an instantaneous agreement is found. Hence, both aforementioned games are only able to find an instantaneous agreement solution. The models are not able to describe the process of negotiation which leads to the agreement. However, this process is essential considering negotiations with human participants as humans are not able to resolve conflicts instantaneously.

The war of attrition is a widely known coordination game introduced by Smith [15] for modeling conflicts among deer. The key aspect of the game with incomplete information setting is to achieve a price by outlasting the other players while facing increasing costs over time. A unique Bayesian Nash equilibrium can be found. This model has a wide range of application, e.g. in oligopoly theory [10], establishment of technical standards [7], auction theory [14] and strategic negotiation behavior [1, 4]. With respect to the required features of our application, the war of attrition model fulfills already all requirements except the consideration of an interaction system model. However, all of the coordination and bargaining games from above lack this interaction system model. Therefore, an adaptation of the war of attrition to our application seems the most promising.

In summary and to the knowledge of the authors there is no model available in literature that comprises all of the required features. To fill this void, our contribution is the introduction of a new game-theoretic model in the scope of human-machine negotiation as well as adapting an existing game to this scope. First, we propose a new event-based game definition for modeling cooperative decision making between human and machines with complete information in Sect. 15.2. It intentionally focuses on a complete information setting for a more descriptive and holistic nature. This is achieved by a detailed description of the negotiation and interaction process. Furthermore, a system dynamics model that is influenced by players during the negotiation process and its effects on the negotiation process is considered. It therefore explicitly aims at the second of the above introduced features.

Second, we revisit the war of attrition game model in Sect. 15.3 and adapt it to the desired scenario with incomplete information. The adaptation includes an extended cost function. Furthermore, we obtain a unique Bayesian Nash equilibrium strategy. Hence, the adapted war of attrition model focuses on the third (incomplete information) and fourth (deterministic solution) of the above mentioned requirements.

To demonstrate the two new games by providing examples, we introduce a specific problem definition that will be considered by the examples in the following sections.

Problem 15.1 The challenge is to provide a model of the negotiation process between human H and automation A in the scenario depicted in Fig. 15.2. The scenario setup is as follows:

- There are two decision makers, the human H and the automation A.
- The decision makers face a deadline D for decision making, i.e. the obstacle O with width w_{obs} at distance s_{obs}.
- The decision makers are in a vehicle V with constant speed v in a 2D-space (s, q) towards the deadline D. Without loss of generality the vehicle is assumed to be a point mass.
- The two decision options $left$ and $right$ describe the evasion direction of vehicle V around obstacle O.
- The decision makers shall take into account the vehicle motion when determining their strategies.

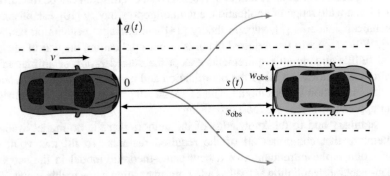

Fig. 15.2 Exemplary evasion scenario with obstacle at distance s_{obs}, obstacle width w_{obs} and potential maneuvers left and right

15.2   Event-Based Game

This section introduces an event-based game as a new way of modeling human-machine negotiation. First, we provide the general definition of the event-based game. Second, we propose the application of parameterized strategies and present two solution approaches for determining these strategies. The section concludes with a discussion of the event-based game as a modeling approach for human-machine negotiation.

15.2.1   Definition

In the following we introduce the general event-based game definition focusing on the consideration of system dynamics and the timing of a human negotiation process.

Definition 15.1 An event-based game Γ_{EB} is a 6-tuple $(\mathscr{P}, \mathscr{S}, \mathscr{E}, \mathscr{A}, \Sigma, \Pi)$. It has the following components:

- the set of players $\mathscr{P}$ with $|\mathscr{P}| = N$
- the set of actions for each player $\mathscr{A} = \mathscr{A}_1 \times \cdots \times \mathscr{A}_N$
- the set of possible events for each player $\mathscr{E} = \mathscr{E}_1 \times \cdots \times \mathscr{E}_N$
- the deterministic system dynamics $\mathscr{S}$ that typically transforms actions into events or triggers events according to internal system states. An example of such systems are hybrid discrete event systems based on a differential state space model:

$$e = \Phi(x, a) \qquad \text{(event trigger function)}$$

$$\dot{x} = f(x, u) \qquad \text{(differential state space equation)}$$

$$u = b(x, a, e) \qquad \text{(hybrid input function)}$$

with $e \in \mathscr{E}$, $a \in \mathscr{A}$, state $x \in \mathbb{R}^n$ with initial state x_0 and input $u \in \mathbb{R}^p$.
- the set of strategy sets for all players $\Sigma = \{\Sigma_i \mid i \in \mathscr{P}\}$. A strategy $\sigma \in \Sigma_i$ of player i is defined as a mapping of a sequence of event-time-tuples $\left((e, t)_k\right)_{k \in \mathbb{R}^+}$ to sequence of action-time-tuple $\left((a, t)_l\right)_{l \in \mathbb{R}^+}$:

$$\sigma : \left\{ \left((e, t)_k\right)_{k \in \mathbb{R}^+} \right\}_i \rightarrow \left\{ \left((a, t)_l\right)_{l \in \mathbb{R}^+} \right\}_i$$

with $e \in \mathscr{E}_i$, $a \in \mathscr{A}_i$ and $t \in \mathbb{R}^+$.
- the set of payoff functions $\Pi = \{\pi_i \mid i \in \mathscr{P}\}$, that assign a payoff to each strategy combination of all players and the occurring sequence of event-time-tuples:

$$\pi_i : \Sigma \times \mathscr{E}_i \times t \rightarrow \mathbb{R}.$$

Remark 15.1 The system $\mathscr{S}$ acts as a communication and interaction channel among players. Its event-based character suits the unpredictable timing of human communication.

Note that Definition 15.1 describes a complete information game. Therefore, players may determine their strategy according to their payoff function w.r.t. the strategies of all other players and their consequences.

Remark 15.2 The above definition of strategies may result in non-causal strategies. In an implementation causal strategies can be ensure by additional constraints on the time-sequences of events and actions of all players: all identical event-time sequences until a specific time t^* (indicated by $\{\cdot\}^*$) must lead to the identical action-time sequence.

$$\left\{\left((e,t)_k\right)_{k\in\mathbb{R}^+}\right\}_i \rightarrow \left\{\left((a,t)_l\right)_{l\in\mathbb{R}^+}\right\}_i, \ \forall i, \ t \leq t^*.$$

Given this game definition we provide in the following approaches to determine strategies for all players maximizing their individual payoff and hence find Nash equilibria of the game.

15.2.2   Solution with Parameterized Strategies

The payoffs gained by the individual players of an event-based game depend on the actions played by all players and the corresponding events of the discrete event system. Thus, if the strategy of each player i can be parameterized by a vector $\boldsymbol{\theta}_i$ and the given system is deterministic the payoffs can be expressed as a function Ψ of the profile of all parameters $\boldsymbol{\theta} = (\boldsymbol{\theta}_1, \ldots, \boldsymbol{\theta}_n)$:

$$\pi_i = \Psi_i(\boldsymbol{\theta}).$$

The best-response correspondence for player i is defined as the set of optimal strategies for which holds

$$\mathbf{BR}_i(\boldsymbol{\theta}_{\neg i}) = \arg\max_{\boldsymbol{\theta}_i} \pi_i(\boldsymbol{\theta}) \tag{15.1}$$

w.r.t. the fixed profile of strategies for all other players $\boldsymbol{\theta}_{\neg i}$.

Note that the best response is not necessarily unique.

A Nash equilibrium consists of strategies that are best responses to each other. Therefore the following lemma is stated.

Lemma 15.1 (Nash Equilibrium in Event-Based Games with Parameterized Strategies) *A profile of strategy vectors $\boldsymbol{\theta}^* = (\boldsymbol{\theta}_1^*, \ldots, \boldsymbol{\theta}_N^*)$ corresponds to a Nash equilibrium if*

$$\boldsymbol{\theta}_i^* \in \mathbf{BR}_i(\boldsymbol{\theta}_{\neg i}^*), \ \forall i \in \mathscr{P}.$$

We present two approaches to find equilibria in an event-based game. For an exemplary scenario with a low number of actions and events and a first order differential system we present an analytical function for all equilibria in Sect. 15.2.2.1. For larger setups iterative numerical solution methods can be applied. As an example we propose the Cournot dynamics methods in Sect. 15.2.2.2.

15.2.2.1 Example Yielding Analytical Solution

We model the example given in Problem 15.1 and Fig. 15.2 as an event-based game with two players ($\mathscr{P} = \{1, 2\}$). They can choose to steer to the left (L), to steer to the right (R) or not to steer at all (M). Thus, their respective action sets are $\mathscr{A}_1 = \mathscr{A}_2 = \{L, M, R\}$. Furthermore, it is assumed that both players $i \in \mathscr{P}$ are able to detect a collision as well as the Time to Collision (TTC) falling below a threshold τ_i. This leads to the event set for both players:

$$\mathscr{E}_1 = \{\text{Crash, TTC falls below } \tau_1\}, \tag{15.2a}$$

$$\mathscr{E}_2 = \{\text{Crash, TTC falls below } \tau_2\}. \tag{15.2b}$$

Events and action are coupled by the system $\mathscr{S}$ in (15.3)

$$s(t) = \int_0^t v_s \, dt \quad s(0) = 0, \ v_s = \text{const.} \tag{15.3a}$$

$$q(t) = \int_0^t v_q \, dt \quad q(0) = 0, \ v_q = \sum_{i \in \{1,2\}} \left(v_{q,i} \mathbb{1}_{\{a_i = L\}} - v_{q,i} \mathbb{1}_{\{a_i = R\}} \right) \tag{15.3b}$$

with the players' actions input, i.e. steering input, described by means of

$$\mathbb{1}_{\{a_i = a\}} = \begin{cases} 1 & \text{if } a_i = a \text{ at time } t, a \in \mathscr{A}_i \\ 0 & \text{if } a_i \neq a \text{ at time } t, a \in \mathscr{A}_i \end{cases} \tag{15.4}$$

It describes a vehicle as point mass with position (s, q) and velocity (v_s, v_q) cooperatively steered by the two players heading towards an obstacle. The obstacle is static and its position is fixed at $(s_{\text{obs}}, 0)$. It has width w_{obs}. Thus crossing the line between the points $(s_{\text{obs}}, -w_{\text{obs}}/2)$ and $(s_{\text{obs}}, w_{\text{obs}}/2)$ leads to a crash. The TTC is defined as

$$\text{TTC} = \begin{cases} s_{\text{obs}} - s/v_s & s \leq s_{\text{obs}} \wedge |q| \leq w_{\text{obs}}/2 \\ \infty & \text{else} \end{cases} \tag{15.5}$$

Each player's actions shall be restricted to use a trigger strategy of the following kind: Start with action M (no steering) and start to steer L or R after a time-threshold is met, causing a constant lateral velocity $v_{q,i}$. Thus, this strategy can be expressed in parametric form using the threshold $\tau_i = \theta_i$.

The payoff-functions in (15.6) penalize a collision as well as steering effort. The costs for steering are weighted individually with the factor r_i. Note that this factor could also be part of θ_i. However, we set it as a known constant value for reasons of clarity and comprehensibility.

$$\pi_i(\tau_i, \tau_{-i}) = -\mathbb{1}_{\text{Crash}} - r_i \cdot \mathbb{1}_{\text{TTC falls below } \tau_i} \tag{15.6a}$$

$$= \begin{cases} 0 & \neg\text{Crash} \wedge \neg\text{TTC falls below } \tau_i \\ -r_i & \neg\text{Crash} \wedge \text{TTC falls below } \tau_i \\ -1 & \text{Crash} \wedge \neg\text{TTC falls below } \tau_i \\ -1 - r_i & \text{Crash} \wedge \text{TTC falls below } \tau_i \end{cases} \tag{15.6b}$$

Incorporating the descriptions of the system and the events and simplifying leads to parameterized payoff functions. In the following the resulting payoff function of player one including the system dynamics is stated.

$$\pi_1(\theta_1, \theta_2) = \begin{cases} 0 & \theta_1 < \theta_2 \wedge \left| v_{q,2}(\theta_2^+ - \theta_1^+) \right| > w_{\text{obs}}/2 \\[2mm] -r_1 & \theta_1 > \theta_2 \wedge \left| v_{q,1}(\theta_1^+ - \theta_2^+) \right| > w_{\text{obs}}/2 \\[2mm] -r_1 & \begin{aligned} &\theta_1 < \theta_2 \wedge \left| v_{q,1}\theta_1^+ + v_{q,2}\theta_2^+ \right| > w_{\text{obs}}/2 \\ &\quad \wedge \left| v_{q,2}(\theta_2^+ - \theta_1^+) \right| \leq w_{\text{obs}}/2 \end{aligned} \\[2mm] -r_1 & \begin{aligned} &\theta_1 \geq \theta_2 \wedge \left| v_{q,1}\theta_1^+ + v_{q,2}\theta_2^+ \right| > w_{\text{obs}}/2 \\ &\quad \wedge \left| v_{q,1}(\theta_1^+ - \theta_2^+) \right| \leq w_{\text{obs}}/2 \end{aligned} \\[2mm] -1 & \theta_1 \leq 0 \wedge \left| v_{q,2}\theta_2^+ \right| \leq w_{\text{obs}}/2 \\[2mm] -1 - r_1 & \theta_1 > 0 \wedge \left| v_{q,1}\theta_1^+ \right| \leq w_{\text{obs}}/2 \wedge \theta_2 \leq 0 \\[2mm] -1 - r_1 & \theta_1 > 0 \wedge \left| v_{q,1}\theta_1^+ + v_{q,2}\theta_2^+ \right| \leq w_{\text{obs}}/2 \wedge \theta_2 > 0 \end{cases} \tag{15.7}$$

with θ_i^+ defined as: $\theta_i^+ = \max(\theta_i, 0)$.

Using the values $r_i = 0.1$ and $v_{q,i} = 1, i \in \{1, 2\}$ as an example leads to:

$$\pi_1(\theta_1, \theta_2) = \begin{cases} 0 & \theta_1 < \theta_2 \wedge \left|\theta_2^+ - \theta_1^+\right| > 1 \\ -0.1 & \theta_1 > \theta_2 \wedge \left|\theta_1^+ - \theta_2^+\right| > 1 \\ -0.1 & \theta_1 < \theta_2 \wedge \left|\theta_1^+ + \theta_2^+\right| > 1 \wedge \left|\theta_2^+ - \theta_1^+\right| \leq 1 \\ -0.1 & \theta_1 \geq \theta_2 \wedge \left|\theta_1^+ + \theta_2^+\right| > 1 \wedge \left|\theta_1^+ - \theta_2^+\right| \leq 1 \\ -1 & \theta_1 \leq 0 \wedge \left|\theta_2^+\right| \leq 1 \\ -1.1 & \theta_1 > 0 \wedge \theta_1 \leq 1 \wedge \theta_2 \leq 0 \\ -1.1 & \theta_1 > 0 \wedge \theta_1 + \theta_2 \leq 1 \wedge \theta_2 > 0 \end{cases} \tag{15.8}$$

Based on the parameterized payoff functions, the best-response correspondences can be obtained by optimization

$$BR_1(\theta_2) = \arg\max_{\theta_1} \pi_1(\theta_1, \theta_2) \tag{15.9}$$

$$BR_2(\theta_1) = \arg\max_{\theta_2} \pi_2(\theta_1, \theta_2). \tag{15.10}$$

Figure 15.3 shows the graphs of both best-response correspondences BR_1 and BR_2. Their intersection is the set of parameter profiles constituting Nash equilibria. This set is not a singleton.

Fig. 15.3 Nash equilibria of thresholds in the event-based game applied to the evasion maneuver scenario

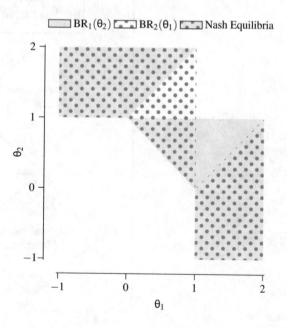

15.2.2.2 Iterative Numerical Solution Methods

If no analytical solution can be found, iterative numerical solution methods can be applied to find numerically a Nash equilibrium in an event-based game i.e. solve (15.9) and (15.10). Figure 15.4 shows the Cournot dynamics method which we propose to solve an event-based game in parameterized form with complete information. It is based on alternating best-responses of both players to the parameters of the respective other player, starting from some initial parameters. The best-responses in an event-based game can be found by numerical optimization based on a simulation of the system's behavior. If neither of the players changes his set of parameters in an iteration w.r.t. a small, chosen $\epsilon > 0$ a Nash equilibrium

Fig. 15.4 Best-response dynamics to numerically find a Nash equilibrium by bilaterally and iteratively determining the best response to the other player's strategy

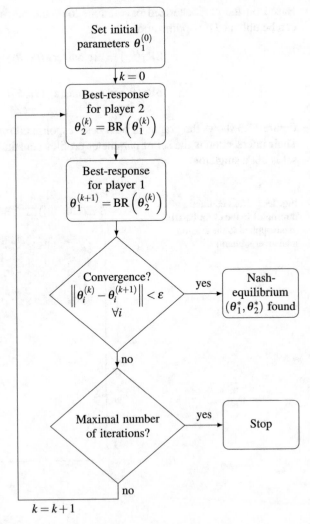

is found. The method is not guaranteed to converge. Therefore, in a practical implementation a stopping criterion based on a maximum number of iterations is included.

15.2.3 Discussion

The event-based game introduces the ability to jointly model the features of a human-human (and hence a human-machine) negotiation process and the system dynamics the interaction is based on. The individual negotiation participants are represented by individual players. The individual strategical thinking of human negotiators is enabled by individual payoff functions evaluating strategies and their consequences. The exchange of offers among the players is modeled via actions and events, allowing to model a selective and individual human perception and influence of the negotiation process. As humans trigger actions and are able to receive events at non-predefined times, the definition allows for events and actions to occur at arbitrary times. Last but not least, the event-based systems dynamics are included, connecting actions and events and allow for events driven by autonomous system behavior. This resembles the human ability to not only react to actions of the opponent but also to relevant system behavior.

In order to find equilibria, we provide an approach that parameterizes the strategy and payoff function. Solutions can be found in analytical form or by iterative numerical solution methods. A practical implementation of the approach in case of multiple equilibria might require extensions that determine which equilibrium to choose. In this situation one might consider to reformulate the event-based game with incomplete information via a distribution of several types of each player potentially leading to a unique equilibrium.

15.3 Game Model Based on War of Attrition

In this section we reconsider the war of attrition game due to its ability to model a concession process, i.e. a special yet important form of negotiation process, yielding a unique equilibrium. In addition to a unique equilibrium its incomplete information setting is another advantage compared to the event-based game of Sect. 15.2. The minor downside in comparison to the event-based game is the relaxation of the initial requirement of a detailed system model consideration towards disagreement costs which increase linearly over time.

In the following, we first introduce the basic principle of the war of attrition. This is followed by its adaptation to our negotiation setting, i.e. the integration of a better approximation of the system behavior, and the proof that the solution leads to a Bayesian Nash equilibrium. Last but not least we present a exemplary implementation of the enhanced model.

15.3.1 Basic Principle

In a war of attrition players compete for a price by trying to outlast the other players. The players' valuation v of the price is only known by themselves. Common knowledge is the probability density function $f(v)$ and the cumulative distribution function $F(v)$, respectively, of the valuation of the other players. All players face the same costs for waiting in disagreement. These costs increase linearly with slope 1. In this setup the optimal waiting-time $\tau(v)$ for outlasting the other players in the unique perfect Bayesian Nash equilibrium is given by the following equation [11, p. 216–219].

$$\tau(v) = \int_0^v \tilde{v} \frac{f(\tilde{v})}{1 - F(\tilde{v})} \, d\tilde{v}. \tag{15.11}$$

This waiting-time yields the highest expected payoff. It is also a special case of the result given in [4].

In the following the war of attrition is used to model the strategic interaction between two drivers cooperatively steering a car in an evasive steering maneuver. By the nature of the war of attrition only two maneuvers i.e. actions are available. It is therefore assumed that both players have individual preferences for evading the obstacle on the left and on the right, respectively. The absolute difference between these preferences is interpreted as the valuation of the price in a war of attrition that arises between the two players if they prefer different options and consequently steer in opposite directions until one of them gives in.

In the basic form the war of attrition is only able to approximate the interaction system behavior by cost function which is linearly increasing over time. In order to be able to model more complex interaction system behaviors we adapt the conventional war of attrition model in the following towards a solution of Problem 15.1 by enhancing the cost structure.

15.3.2 Adaptation

To incorporate the system's behavior the war of attrition model is extended by a non-linear time-dependent cost function $c(t)$. Consider the following assumptions on this cost function:

Assumption 15.1 $c(t) : \mathbb{R}^+ \to \mathbb{R}^+$ is a continuous, strictly increasing and therefore invertible function with its inverse function c^{-1}

$$c(t) \in \mathscr{C}^1, \tag{15.12}$$

$$c'(t) > 0 \ \forall t. \tag{15.13}$$

In addition, consider this assumption on the threshold function:

Assumption 15.2 $\tau_i(v) : \mathbb{R}^+ \to \mathbb{R}^+$ is differentiable with an integrable derivative and

$$\tau_i(0) = 0. \tag{15.14}$$

Based on these assumptions the following lemma on the threshold for maximizing the expected payoff is stated.

Lemma 15.2 *The threshold of player i in a war of attrition maximizing his expected payoff w. r. t. the density distribution of valuation f_v of player $\neg i$, a cost function $c(t)$ and the player i's valuation v_i is given by*

$$\tau_i(v_i) = c^{-1} \left(\int_0^v \tilde{v} \frac{f(\tilde{v})}{1 - F(\tilde{v})} \, d\tilde{v} \right) \tag{15.15}$$

w.r.t. Assumptions 15.1 and 15.2.

Proof Following the approach of Fudenberg and Tirole [11, pp. 216-219] we set up the objective function for maximizing the expected payoff (15.16) w.r.t. the threshold $\tau(v)$ of player i, the density distribution of thresholds f_τ of player $\neg i$, the cost function $c(t)$ and the player i's valuation v_i.

$$\pi_i = \int_0^{\tau_i(v_i)} (v_i - c(t)) f_\tau(t) \, dt. \tag{15.16}$$

With the derivative of π_i by $\tau_i(v_i)$ the necessary condition for the maximum is found:

$$v_i \cdot f_\tau(\tau_i(v_i)) - c'(\tau_i(v_i))(1 - F_\tau(\tau_i(v_i))) = 0. \tag{15.17}$$

The proof of sufficiency of condition (15.17) is analogous to Fudenberg and Tirole [11, pp. 217–218] and therefore it is not presented here.

According to the fundamental theorem of calculus and with the assumption $\tau_i(v)$ being a differentiable function with integrable derivative the density distribution and the cumulative distribution function of (15.17) can be transformed to

$$\frac{v_i \cdot f_v(v_i)}{\tau'(v_i)} - c'(\tau_i(v_i))(1 - F_v(v_i)) = 0. \tag{15.18}$$

The transformed condition (15.18) is rearranged in (15.19) to be integrated w. r. t. v_i and (15.14).

$$c'(\tau_i(v_i)) \tau'(v_i) = v_i \cdot f_v(v_i) / (1 - F_v(v_i)). \tag{15.19}$$

With the cost function $c(t)$ being continuous, strictly increasing and therefore invertible (Assumption 15.1: conditions (15.12) and (15.13)) the resulting threshold function is

$$\tau_i(v_i) = c^{-1} \left(\int_0^{v_i} \tilde{v} \frac{f(\tilde{v})}{1 - F(\tilde{v})} \, \mathrm{d}\tilde{v} \right). \tag{15.20}$$

Furthermore, this function fulfills Assumption 15.2 of $\tau_i(v)$ being a differentiable function with integrable derivative.   ■

As a result (15.15) yields the optimal threshold for conceding w.r.t. the expected payoff in the incomplete information setting. It considers the invertible cost function $c(t)$. This cost function might resemble an approximation of the interaction system behavior and allows for modeling soft negotiation deadlines.

15.3.3 Bayesian Nash Equilibrium

In the following we provide the proof that the result of Lemma 15.2 is a Bayesian Nash equilibrium.

Theorem 15.1 *The following symmetric strategy profile yields a sub-game-perfect Bayesian Nash equilibrium: Start to act towards your preferred option. Give in and start to act towards the other option if your threshold is reached and the other player has not given in. The threshold is calculated according to* (15.15).

Proof We have to show, that the proposed strategy is a best response to itself.

If both players prefer the same option, an agreement is reached immediately without costs.

If players prefer different options, both will realize the conflict and hence the war of attrition they are in. By following the above introduced symmetric strategy of waiting until their thresholds that individually and statistically maximize their payoff at all times, they find themselves in a sub-game-perfect Bayesian Nash equilibrium.

Therefore, this also applies for the overall game.   ■

15.3.4 Example

In the following, we provide an exemplary parameterization of the enhanced war of attrition and the resulting chart of thresholds for the scenario given in Problem 15.1.

15.3.4.1 Cost Function

The solely time-dependent cost function should be an approximation of the system model in the evasion maneuver. Hence, the steering effort and a collision have to be penalized. It is assumed that a quadratic term of time is a good approximation of penalizing the effort for evasive steering. This is motivated by the fact that later steering leads to overproportionally higher effort to get around the obstacle. Furthermore, it is exemplary assumed that the collision will occur after 4 s with a given v, s_{obs} and w_{obs}. Therefore, we apply a cubic component yielding high a penalty for waiting longer than 3 s and subsequently risking a collision.

The combined cost function is given by (15.21) and depicted in Fig. 15.5.

$$c(t) = t^2 + 100 \, (t - 3)^3 \, \mathbb{1}_{t>3} \text{ with } \mathbb{1}_{\{t>3\}} = \begin{cases} 1 & \text{if } t > 3, \\ 0 & \text{else.} \end{cases} \tag{15.21}$$

15.3.4.2 Density Function

As an example, we apply the following density function of the valuation v, i.e. the preference difference, of the opponent.

$$f(v) = \frac{4}{3} \left(\frac{1}{2} \frac{1}{20} \varphi \left(\frac{v}{20} \right) + \frac{1}{2} \frac{1}{10} \varphi \left(\frac{v - 50}{10} \right) \right) \mathbb{1}_{v \in [0, 100]}. \tag{15.22}$$

It is a normalized superposition of two Gaussian distributions with expected values 0 and 50 and standard deviation of 20 and 10 on the interval $v \in [0, 100]$ and is depicted in Fig. 15.6.

In an application scenario analogous to Problem 15.1, e.g. on a freeway, the preference differences v could be seen as the result of a criticality analysis of the two evasion options.

Fig. 15.5 Exemplary cost function $c(t)$

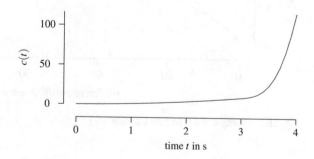

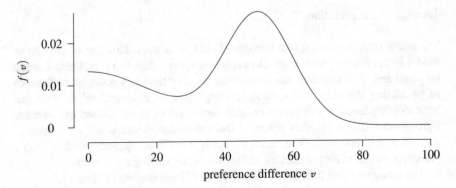

Fig. 15.6 Exemplary density function $f(v)$

15.3.4.3 Resulting Thresholds

With the exemplary cost (15.21) and density function (15.22) the resulting thresholds w.r.t. the own preference difference v can be calculated according to (15.15) of Lemma 15.2. For the range of $v \in [0, 100]$ the thresholds are depicted in Fig. 15.7. One can observe the continuous but nonlinear increase of thresholds with increasing preference v. This is due to the continuous nonlinear cost function. Furthermore a saturation behavior of the threshold around 4 s is visible. This is caused by the cost function heavily penalizing a collision at that time.

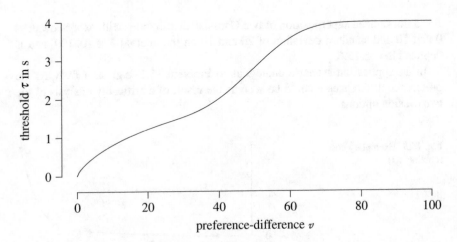

Fig. 15.7 Example of a threshold function $\tau(v)$

15.3.5   Discussion

The previous sections reveal the possibility that the war of attrition model can be adapted for modeling the human-automation negotiation process in the evasion maneuver scenario with two evasion maneuver options. The game yields one continuous time, easy to calculate solution for a threshold at which the player will change from his preferred option to the other one. It depends on costs resembling the system behavior and disagreement costs, the distribution of preference differences of the other player and the player's own preference difference. This solution is a unique Bayesian Nash equilibrium.

In application, the identification of the opponent's preference difference distribution is required as well as the cost function. Although, the cost function is only time-dependent and has to be common knowledge to all players, we argue that both aspects are valid assumptions in a practical implementation. I. e. deadline and system behavior are both aspects of negotiation scenario that usually can be observed by all players.

In future work, we will extend the war of attrition model towards more than two potential maneuver options towards a realistic application of the proposed model. Furthermore, we will investigate how to determine and estimate the cost function and preference difference distribution of the opponent, respectively.

15.4   Conclusion

In this work, we proposed two game concepts for describing the negotiation process in cooperative decision making with humans. These models form the basis for the objective to design an emancipated automation that is able to take part in cooperative decision making with a human.

Contrary to state-of-the-art models, the proposed modeling approaches are able to explicitly describe the process of cooperative decision making with human participants. Moreover, the behavior of the system through which participants interact with each other is also considered in the models. Accounting for this dynamical system is crucial for a meaningful solution in several applications. An example is given by the dynamics of a vehicle which are relevant for a safe human-machine cooperative guidance.

The newly introduced event-based game has the ability to provide a detailed description of the negotiation process and the dynamical system the automation and the human interact with. We proposed to utilize parameterized strategies to analytically or numerically calculate equilibria. The solution may be ambiguous and usually requires complete information. Hence, in a practical implementation additional identification methods may be required.

Furthermore, we enhanced the conventional war of attrition model by means of a time-dependent cost function. This allows the consideration of the approximated interaction system behavior, e.g. in form of negotiation constraints like deadlines. With the new cost function and the incomplete information setting we proved that the resulting solution leads to a unique Bayesian Nash equilibrium.

In summary, the introduced two models form a basis for the design of automated negotiation agents that are able to interact with humans on the decision level in an emancipated and intuitive way.

References

1. Abreu, D., Gul, F.: Bargaining and reputation. Econometrica **68**(1), 85–117 (2000) https://doi.org/10.1111/1468-0262.00094
2. Baarslag, T., Gerding, E.H., Aydogan, R., Schraefel, M.C.: Optimal negotiation decision functions in time-sensitive domains. In: 2015 IEEE/WIC/ACM International Conference on Web Intelligence and Intelligent Agent Technology (WI-IAT), pp. 190–197. IEEE, Piscataway (2015). https://doi.org/10.1109/WI-IAT.2015.161
3. Bruder, R., Didier, M.: Design of human-machine-interfaces for DAS. In: Winner, H., Hakuli, S., Lotz, F., Singer, C. (eds.) Handbook of Driver Assistance Systems: Basic Information, Components and Systems for Active Safety and Comfort, pp. 797–811. Springer International Publishing, Cham (2016). https://doi.org/10.1007/978-3-319-12352-3_34
4. Bulow, J., Klemperer, P.: The generalized war of attrition. Am. Econ. Rev. **89**(1), 175–189 (1999)
5. Calcagno, R., Kamada, Y., Lovo, S., Sugaya, T.: Asynchronicity and coordination in common and opposing interest games. Theor. Econ. **9**(2), 409–434 (2014). https://doi.org/10.3982/TE1202
6. Caruana, G., Einav, L: A theory of endogenous commitment. Rev. Econ. Stud. **75**(1), 99–116 (2008)
7. David, P., Monroe, H.: Standards development strategies under incomplete information. In: MERIT Research Memorandum. University of Limburg Maastricht, Maastricht (1994)
8. Della Penna, M., van Paassen, M.M., Abbink, D.A., Mulder, M., Mulder, M.: Reducing steering wheel stiffness is beneficial in supporting evasive maneuvers. In: 2010 IEEE International Conference on Systems, Man and Cybernetics, pp. 1628–1635. IEEE, Piscataway (2010) https://doi.org/10.1109/ICSMC.2010.5642388
9. Flad, M., Otten, J., Schwab, S., Hohmann, S.: Steering driver assistance system: a systematic cooperative shared control design approach. In: 2014 IEEE International Conference on Systems, Man, and Cybernetics (SMC), pp. 3585–3592. IEEE, Piscataway (2014). https://doi.org/10.1109/SMC.2014.6974486
10. Fudenberg, D., Tirole, J.: A theory of exit in duopoly. Econometrica **54**(4), 943–960 (1986). https://doi.org/10.2307/1912845
11. Fudenberg, D., Tirole, J.: Game Theory. MIT Press, Cambridge (1991)
12. Groten, R., Feth, D., Peer, A., Buss, M.: Shared decision making in a collaborative task with reciprocal haptic feedback – an efficiency-analysis. In: 2010 IEEE International Conference on Robotics and Automation, pp. 1834–1839. IEEE, Piscataway (2010). https://doi.org/10.1109/ROBOT.2010.5509906
13. Johns, M., Mok, B., Sirkin, D., Gowda, N., Smith, C., Talamonti, W., Ju, W.: Exploring shared control in automated driving. In: 2016 11th ACM/IEEE International Conference on Human-Robot Interaction (HRI), pp. 91–98. IEEE, Piscataway (2016). https://doi.org/10.1109/HRI.2016.7451738
14. Krishna, V., Morgan, J.: An analysis of the war of attrition and the all-pay auction. J. Econ. Theory **72**(2), 343–362 (1997). https://doi.org/10.1006/jeth.1996.2208

15. Maynard Smith, J.: The theory of games and the evolution of animal conflicts. J. Theor. Biol. **47**(1), 209–221 (1974). https://doi.org/10.1016/0022-5193(74)90110-6
16. Na, X., Cole, D.J.: Game-theoretic modeling of the steering interaction between a human driver and a vehicle collision avoidance controller. IEEE Trans. Human Mach. Syst. **45**(1), 25–38 (2015). https://doi.org/10.1109/THMS.2014.2363124
17. Rothfuss, S., Schmidt, R., Flad, M., Hohmann, S.: A concept for human-machine negotiation in advanced driving assistance systems. In: 2019 IEEE International Conference on Systems, Man and Cybernetics (SMC), pp. 3116–3123. IEEE, Piscataway (2019). https://doi.org/10.1109/SMC.2019.8914282
18. Rothfuss, S., Ayllon, C., Flad, M., Hohmann, S.: Adaptive negotiation model for humanma-chine interaction on decision level. In: 2020 IFAC Word Congress. IFAC, Geneva (2020)
19. Turnwald, A., Wollherr, D.: Human-like motion planning based on game theoretic decision making. Int. J. Soc. Robot. **11**(1), 151–170 (2019). https://doi.org/10.1007/s12369-018-0487-2
20. Zeuthen, F.: Problems of Monopoly and Economic Warfare. CRC Press, Boca Raton (2019)
21. Zlotkin, G., Rosenschein, J.S.: Negotiation and task sharing among autonomous agents in cooperative domains. In: Proceedings of the 11th International Joint Conference on Artificial Intelligence, , IJCAI'89, Vol. 2, pp. 912–917. Morgan Kaufmann Publishers Inc., San Francisco (1989)

Chapter 16
Social Inefficiency of Free Entry Under the Product Diversity

Alexander Sidorov

Abstract The paper addressed to a question whether the free entry of profit-seeking large firms (oligopolies) is advantageous for consumers, or the governmental restrictions to enter may have the positive effect on consumers' well-being. The negative welfare effect of excessive enter is well-known in case of homogeneous good, though there was hypothesis that consumers' love for variety in case of differentiated good may offset this effect. The main result of this paper is that this almost never happened. We study a general equilibrium model with imperfect Bertrand-type price competition. Firms assumed to have non-zero impact to market statistics, in particular, to consumer's income via distribution of non-zero profit across consumers-shareholders. It is proved that the governmental restrictions in certain bounds increases Social welfare under quite natural assumptions on utilities, which hold for most of the commonly used classes of utility functions, such as Quadratic, CARA, HARA, CES, etc.

Keywords Bertrand competition · Additive preferences · Ford effect · Excessive enter · Consumer's welfare

16.1 Introduction

The typical presumption of the most of economic theories is that free entry is desirable for social efficiency. As several articles have shown, however, when firms must incur fixed set-up costs upon entry, the number of firms entering a market need not equal the socially desirable number. Spence [13] and Dixit and Stiglitz [6], for example, demonstrate that in a monopolistically competitive market, free entry can result in too little entry relative to the social optimum. In more later work von Weizsäcker [14] and Perry [11] point to a tendency for excessive

A. Sidorov (✉)
Sobolev Institute of Mathematics, Novosibirsk, Russia

© The Editor(s) (if applicable) and The Author(s), under exclusive licence
to Springer Nature Switzerland AG 2020
L. A. Petrosyan et al. (eds.), *Frontiers of Dynamic Games*,
Static & Dynamic Game Theory: Foundations & Applications,
https://doi.org/10.1007/978-3-030-51941-4_16

267

entry in homogeneous product markets. Nevertheless, despite these findings, many economists continue to hold the presumption that free entry is desirable, in part, it seems, because the fundamental economic forces underlying these various entry biases remain somewhat mysterious. The empirical studies in broadcasting industry allow to draw the conclusion that the share of social losses due to excessive entry of radio stations is about 40%, see e.g., [4] and the more recent paper [5]. As for theoretical justification of this effect, the paper of Mankiw and Whinston, [8], consider the general model of oligopolistic competition between firms producing the homogeneous good. Authors formulated their assumptions in terms of equilibrium characteristics and the production cost function, which imply the entry excess of firms over the social optimum. These assumptions have a neat economic intuition and cover many well-known examples of the oligopolistic competition models, e.g., the linear Cournot oligopoly model, however, an assumption on homogeneity of good turns out to be crucial. In case of production diversity authors presented a counter-example with the opposite ranking of free-entry equilibrium number of firms and the social optimum. At the very end of Conclusion the following problem was formulated

> The introduction of product diversity, however, can reverse this bias toward excessive entry. Intuitively, a marginal entrant adds to variety, but does not capture the resulting gain in social surplus as profits. Hence, in heterogeneous product markets the direction of any entry bias is generally unclear, although efficient levels of entry remain an unlikely occurrence.

The purpose of this paper is to make the problem more clear. The goal is to provide a simple conditions, under which the number of entrants in a free-entry equilibrium is excessive or insufficient. Our analysis compares the number of firms that enter a market when there is free entry with the number that would be desired by a social planner who is unable to control the behavior of firms once they are in the market. That is, we consider the second-best problem of choosing the welfare-maximizing number of firms. We demonstrate that the crucial conditions for establishing the presence of an entry bias can be stated quite simply in terms of consumers' utility. In short, this paper shows that under the mild and natural assumption the free-entry number of firms is socially excessive. We also provide the sufficient condition for the opposite case and construct the corresponding example of utility function satisfying this condition.

16.2 The Model

Consider the one-sector economy with horizontally differentiated good and one production factor–labor. There is a continuum $[0, L]$ of identical consumers endowed with one unit of labor. The labor market is perfectly competitive and labor is chosen as the numéraire. There is a finite number $N \geq 2$ of "large" firms producing the varieties of some horizontally differentiated good and competing with prices. Each variety is produced by a single firm and each firm produces a single variety, thus the horizontally differentiated good may be represented as a finite-dimensional vector $\mathbf{x} = (x_1, \ldots, x_N) \in R_+^N$. The "large" size of firm implies that impact of single firm to market statistics is not negligible and should be strategically taken into account by other competitors. To operate, every firm needs a fixed requirement $f > 0$ and a marginal requirement $c > 0$ of labor, which may be normalized to 1 without loss of generality. Wage is also normalized to 1, thus the cost of producing q_i units of variety $i \in \{1, \ldots, N\}$ is equal to $f + 1 \cdot q_i$.

Consumers share the same additive preferences given by

$$U(\mathbf{x}) = \sum_{i=1}^{N} u(x_i), \tag{16.1}$$

where $u(\cdot)$ is thrice continuously differentiable, strictly increasing, strictly concave, and such that $u(0) = 0$. Following [15], we define the relative love for variety (RLV) as follows:

$$r_u(x) = -\frac{x u''(x)}{u'(x)},$$

which is strictly positive for all $x > 0$. Under the CES, we have $u(x) = x^\rho$ where ρ is a constant such that $0 < \rho \leq 1$, thus implying a constant RLV given by $1 - \rho$. The natural generalization of CES utility is the HARA function $u(x) = (x + \alpha)^\rho - \alpha^\rho$, $\alpha > 0$. Another example of additive preferences is provided by Behrens and Murata [1], who consider the CARA utility $u(x) = 1 - \exp(-\alpha x)$ where $\alpha > 0$ is the absolute love for variety; the RLV is now given by αx.

Very much like the Arrow-Pratt's relative measure of the risk aversion, the RLV measures the intensity of consumers' variety-seeking behavior. Following the paper [15], we suggest that the low-tier utility function $u(x)$ satisfies the following

Assumption 16.1

$$r_u(x) < 1, \quad r_{u'}(x) < 2 \tag{16.2}$$

for all x in some neighborhood of zero.

A consumer's income is equal to her wage plus her share in the total profits. Since we focus on symmetric equilibria, consumers must have the same income, which means that profits have to be uniformly distributed across consumers. In this case, a consumer's income y is given by

$$y = 1 + \frac{1}{L} \sum_{i=1}^{N} \Pi_i \geq 1, \tag{16.3}$$

where the profit made by the oligopoly selling amount q_i of variety $i \in \{1, \ldots, N\}$ at price p_i is given by

$$\Pi_i = (p_i - 1)q_i - f. \tag{16.4}$$

Evidently, the income level varies with firms' strategies p_i.

A consumer's budget constraint is given by

$$\sum_{i=1}^{N} p_i x_i = y. \tag{16.5}$$

The first-order condition for utility maximization yields

$$u'(x_k) = \lambda p_k, \tag{16.6}$$

where λ is the Lagrange multiplier

$$\lambda = \frac{\sum_{i=1}^{N} u'(x_i)x_i}{y} > 0, \tag{16.7}$$

which implies that the inverse demand may be represented in closed form

$$p_k = \frac{y u'(x_k)}{\sum_{i=1}^{N} u'(x_i)x_i} \tag{16.8}$$

for all varieties $k \in \{1, \ldots, N\}$.

Let $\mathbf{p} = (p_1, \ldots, p_N)$ be a price profile. Consumers' demand functions $x_i(\mathbf{p})$ are obtained by solving the system of Eqs. (16.8) where aggregate income of consumers y is now defined as follows:

$$y(\mathbf{p}) = 1 - Nf + \sum_{i=1}^{N} (p_i - 1)x_i(\mathbf{p}).$$

It follows from (16.7) that the marginal utility of income λ is a market aggregate that depends on the price profile $\mathbf{p}$. Indeed, the budget constraint

$$\sum_{j=1}^{N} p_j x_j(\mathbf{p}) = y(\mathbf{p})$$

implies that

$$\lambda(\mathbf{p}) = \frac{1}{y(\mathbf{p})} \sum_{i=1}^{N} x_i(\mathbf{p}) u'(x_i(\mathbf{p})). \tag{16.9}$$

Since $u'(x)$ is strictly decreasing, the demand function for variety i is thus given by

$$x_k(\mathbf{p}) = \xi(\lambda(\mathbf{p}) p_k), \tag{16.10}$$

where ξ is the inverse function to u'. Moreover, the i-th firm profit can be rewritten as follows:

$$\Pi_i(\mathbf{p}) = L(p_i - 1) x_i(\mathbf{p}) - f = L(p_i - 1)\xi(\lambda(\mathbf{p}) p_i) - f. \tag{16.11}$$

16.2.1 Market Equilibrium

The market equilibrium is defined by the following conditions:

(i) each consumer maximizes her utility (16.1) subject to her budget constraint (16.5),
(ii) each firm k maximizes its profit (16.4) with respect to p_k,
(iii) product market clearing:

$$Lx_k = q_k \qquad \text{for all } k \in \{1, \ldots, N\},$$

(iv) labor market clearing:

$$Nf + \sum_{i=1}^{N} q_i = L.$$

Market equilibrium is *symmetric* when $q_k = q_j$, $p_k = p_j$ for all $k \neq j$. Conditions (iii) and (iv) imply that

$$\bar{x} \equiv \frac{1}{N} - \frac{f}{L} \tag{16.12}$$

are the only candidate symmetric equilibrium demands for "oligopolistic" varieties.

This definition of equilibrium is similar to concepts used in [10] and [12] with exception of an assumption on the free entry until the zero-profit condition. The number of firms now is considered as an exogenous parameter.

16.2.2 First Order Condition for Bertrand Oligopoly Under the Ford Effect

As shown by (16.6) and (16.7), the income level influences firms' demands, whence their profits. As a result, firms must anticipate accurately what the total income will be. In addition, firms should be aware that they can manipulate the income level, whence their "true" demands, through their own strategies with the aim of maximizing profits. This feedback effect is known as the *Ford effect*.

In popular literature, this idea is usually attributed to Henry Ford, who raised wages at his auto plants to five dollars a day in January 1914. As specified in [2], the Ford effect may have different scopes of consumers income, which is sum of wage and a share of the distributed profits. The first specification (proposed in [9]) and used in [2]) is to suppose that firms take into account the effects of their decision on the total wage bill, but not on the distributed profits, which is still treated parametrically. This case may be referred as "Wage Ford effect" and it is exactly what Henry Ford meant. Another intermediate specification of The Ford effect is an opposite case to the previous one: firms take wage as given, but take into account the effects of their decisions on distributed profits. This case may be referred as "Profit Ford effect". Finally, the extreme case, Full Ford effect, assumes that firms take into account total effect of their decisions, both on wages and on profits. These two cases are studied in newly published paper [3]. In the presented research, we shall assume that wage is given. This includes the way proposed by O. Hart in [7], when the workers fix the nominal wage through their union. This assumption implies that only the Profit Ford effect is possible, moreover, firms maximize their profit anyway, thus being price-makers but not wage-makers, they have no additional powers at hand in comparison to No Ford case, with except the purely informational advantage—knowledge on consequences of their decisions. Nevertheless, as it was shown in [12], this advantage allows firms to get more market power, which justify the common wisdom "Knowledge is Power". The Ford effect assumption suggests actually that the large firms act as "sharks" rather than "dolphins", gathering the maximum market power.

The generalized *Bertrand equilibrium* is a vector $\mathbf{p}^*$ such that p_i^* maximizes $\Pi_i(p_i, \mathbf{p}_{-i}^*)$ for all $i \in \{1, \ldots, N\}$. Applying the first-order condition to (16.11) yields

$$\frac{p_i - 1}{p_i} = -\frac{\xi(\lambda p_i)}{\lambda p_i \xi'(\lambda p_i)\left(1 + \frac{p_i}{\lambda}\frac{\partial \lambda}{\partial p_i}\right)}, \qquad (16.13)$$

which involves $\partial \lambda / \partial p_i$ because λ depends on $\mathbf{p}$.

It was mentioned already that the "large" firms (oligopolies) have non-zero influence on market statistics, in particular, we can expect that $\partial \lambda / \partial p_k \neq 0$. By the standard interpretation, the Lagrange multiplier λ is a marginal utility of money, therefore "large" firms understand that the demand functions (16.10) must satisfy the budget constant as an identity. The consumer budget constraint, before symmetrization, can be rewritten as follows:

$$\sum_{i=1}^{N} p_i \xi(\lambda(\mathbf{p})p_i) = 1 - Nf + \sum_{i=1}^{N}(p_i - 1)\xi(\lambda(\mathbf{p})p_i),$$

which boils down to

$$\sum_{i=1}^{N} \xi(\lambda(\mathbf{p})p_i) = 1 - Nf. \tag{16.14}$$

Differentiating (16.14) with respect to p_k yields

$$\xi'(\lambda p_k)\lambda + \frac{\partial \lambda}{\partial p_k} \sum_{i=1}^{N} \xi'(\lambda p_i)p_i = 0$$

or, equivalently,

$$1 + \frac{p_k}{\lambda}\frac{\partial \lambda}{\partial p_k} = \frac{\sum_{i \neq k}^{N} \xi'(\lambda p_i)\lambda p_i}{\sum_{i=1}^{N} \xi'(\lambda p_i)\lambda p_i}. \tag{16.15}$$

Substituting (16.15) into (16.13) and symmetrizing the resulting expression we obtain

$$m(N) = r_u \left(\frac{1}{N} - \frac{f}{L}\right)\frac{N}{N-1}. \tag{16.16}$$

16.3 Consumers' Welfare Under the Free and Restricted Enter

Using (16.16) we can calculate the firm's profit at symmetric equilibrium. Indeed, the markup definition

$$m = \frac{p-c}{p} = \frac{p-1}{p}$$

implies

$$p - 1 = \frac{m}{1-m}.$$

Substituting (16.16) for m we obtain

$$\bar{\Pi} = L(p-1)\bar{x} - f = L \frac{r_u\left(\frac{1}{N} - \frac{f}{L}\right) - (N-1)\frac{f}{L}}{\left(1 - r_u\left(\frac{1}{N} - \frac{f}{L}\right)\right)N - 1}. \tag{16.17}$$

In what follows we shall use the notion $\varphi \equiv f/L$ to make formulas more compact. This allows us to determine Zero-Profit "number" of firms $\widehat{N}(\varphi)$ as root of equation $\bar{\Pi} = 0$, which is equivalent to equation

$$r_u\left(N^{-1} - \varphi\right) = (N-1)\varphi. \tag{16.18}$$

This number is typically non-integer; this is not a big problem, however, because this number only indicates that for all integers $N < \widehat{N}(\varphi)$ profit $\bar{\Pi} > 0$, while $N > \widehat{N}(\varphi)$ implies $\bar{\Pi} < 0$. The corresponding equilibrium consumption $x(\varphi) = \left(\widehat{N}(\varphi)\right)^{-1} - \varphi$.

Proposition 16.1 *For all sufficiently small φ there exist unique solution $\widehat{N}(\varphi)$ of Eq. (16.18). Moreover, for $\varphi \to 0$ we have $x(\varphi) \to 0$, $\widehat{N}(\varphi) \to \infty$.*

Proof This statement immediately follows from [12], Proposition 15.3. ■

Now consider the following Social Welfare function (actually, an indirect utility)

$$V(N) = Nu(\bar{x}) = N \cdot u\left(\frac{1}{N} - \frac{f}{L}\right), \tag{16.19}$$

with the firm's number as a variable. To save space we use the following notion $\varphi \equiv f/L$. The first order condition

$$V'(N) = u\left(\frac{1}{N} - \varphi\right) - \frac{1}{N}u'\left(\frac{1}{N} - \varphi\right) = 0$$

determines the Social optimum of firms' number[1] $N^*(\varphi)$. It is obvious that for CES utility with $u(x) = x^\rho$, which implies $r_u(x) = 1 - \rho$, the Social optimal number of firms is equal to

$$N^*(\varphi) = \frac{1 - \rho}{\varphi}.$$

---

[1] Of course, the actual number of firms is integer number, but this number indicates only that for $N < N + 1 \le N^*(\varphi)$ Social Welfare increases with number of firms $V(N) < V(N+1)$, while $N^*(\varphi) \le N < N + 1$ implies $V(N) > V(N+1)$.

On the other hand, the number of firms determined by zero-profit condition $\bar{\Pi}(N) = 0$ is equal to

$$\widehat{N}(\varphi) = \frac{1 - \rho}{\varphi} + 1 = N^*(\varphi) + 1.$$

This means that Social optimum is less than Free Entry number, though, the difference is not too large.

This result, i.e., inequality $\widehat{N}(\varphi) > N^*(\varphi)$, will be generalized to the wide class of utility functions. Moreover, we also present the counterexample with opposite ranking $\widehat{N}(\varphi) < N^*(\varphi)$.

Let

$$\varepsilon_u(x) \equiv \frac{x \cdot u'(x)}{u(x)}$$

be an elasticity of utility function $u(x)$, while

$$A(x) \equiv \frac{1 - r_u(x) + x}{2} + \sqrt{\left(\frac{1 - r_u(x) - x}{2}\right)^2 - x r_u(x)}. \tag{16.20}$$

Proposition 16.2 *For any $\varphi = f/L > 0$ the inequality $\widehat{N}(\varphi) > N^*(\varphi)$ holds if and only if $\varepsilon_u(x(\varphi)) > A(x(\varphi))$.*

Proof The Social welfare function is bell-shaped due to

$$V''(N) = \frac{1}{N^3} \cdot u''\left(\frac{1}{N} - \varphi\right) < 0,$$

therefore $\widehat{N}(\varphi) > N^*(\varphi)$ is equivalent to inequality $V'(\widehat{N}(\varphi)) < 0 = V'(N^*(\varphi))$. On the other hand, the inequality

$$V'(\widehat{N}(\varphi)) = u\left(\widehat{N}(\varphi)^{-1} - \varphi\right) - \widehat{N}(\varphi)^{-1} u'\left(\widehat{N}(\varphi)^{-1} - \varphi\right) = u(x(\varphi)) -$$

$$-(x(\varphi) + \varphi) \cdot u'(x(\varphi)) = u(x(\varphi)) \frac{x(\varphi) + \varphi}{x(\varphi)} \left(\frac{x(\varphi)}{x(\varphi) + \varphi} - \frac{x(\varphi) u'(x(\varphi))}{u(x(\varphi))}\right) < 0$$

holds if and only if,

$$\frac{x(\varphi)}{x(\varphi) + \varphi} - \varepsilon_u(x(\varphi)) < 0. \tag{16.21}$$

Note that $x(\varphi) = \widehat{N}(\varphi)^{-1} - \varphi$ is an implicit function, derived from Zero-profit condition

$$\bar{\Pi} = 0 \iff r_u\left(N^{-1} - \varphi\right) = (N-1)\varphi,$$

which generally cannot be represented in closed form. Its inverse function, $\varphi(x)$, however, has closed form solution. Indeed, Zero-profit condition may be rewritten in terms of $x = N^{-1} - \varphi$ and φ as follows

$$r_u(x) = (N-1)\varphi = \left(\frac{1}{x+\varphi} - 1\right)\varphi \iff \varphi^2 - (1 - r_u(x) - x)\varphi + xr_u(x) = 0$$

with obvious solution of corresponding quadratic equation

$$\varphi(x) = \frac{1 - r_u(x) - x}{2} - \sqrt{\left(\frac{1 - r_u(x) - x}{2}\right)^2 - xr_u(x)}. \qquad (16.22)$$

Note that the second solution of this equation,

$$\varphi_+(x) = \frac{1 - r_u(x) - x}{2} + \sqrt{\left(\frac{1 - r_u(x) - x}{2}\right)^2 - xr_u(x)},$$

is not admissible, because $\varphi_+(0) = 1 - r_u(0) \neq 0$, while $x(\varphi)$ converges to zero when $\varphi \to 0$. This implies that condition (16.21) is equivalent to

$$\frac{x}{x + \varphi(x)} < \varepsilon_u(x)$$

for $x > 0$, which after substitution of (16.22) and rearranging terms takes on the form $\varepsilon_u(x) > A(x)$ for $x = x(\varphi)$. *Vice versa*, the opposite ranking $\widehat{N}(\varphi) < N^*(\varphi)$ is equivalent to inequality $\varepsilon_u(x(\varphi)) < A(x(\varphi))$. ∎

In what follows we suggest that the following assumption holds.

Assumption 16.2 There exist finite limits of the following fractions:

$$\frac{u''(0)}{u'(0)} = \lim_{x \to 0} \frac{u''(x)}{u'(x)}, \quad \frac{u'''(0)}{u''(0)} = \lim_{x \to 0} \frac{u'''(x)}{u''(x)}. \qquad (16.23)$$

It is obvious that CES utility does not satisfy the Assumption 16.2, while HARA $u(x) = (x+\alpha)^p - \alpha^p$, CARA $u(x) = 1 - e^{-\alpha x}$ and Quadratic $u(x) = \alpha x - x^2/2$ utilities fit it well.

Lemma 16.1 *Let Assumptions 16.1 and 16.2 hold, then $r_u(0) = 0$, $\varepsilon_u(0) = 1$, $A(0) = 1$. Moreover, there exist the limit values*

$$r_u'(0) = -\frac{u''(0)}{u'(0)}, \ \varepsilon_u'(0) = \frac{1}{2}\frac{u''(0)}{u'(0)}, \ and \ A'(0) = \frac{u''(0)}{u'(0)}.$$

Proof Direct calculation shows that

$$r_u'(x) = -\frac{u''(x)}{u'(x)}\left(1 - \frac{xu''(x)}{u'(x)} + \frac{xu'''(x)}{u''(x)}\right),$$

which implies $r_u'(0) = -\frac{u''(0)}{u'(0)}$. Using the L'Hospital rule, we obtain

$$\lim_{x\to 0}\varepsilon_u(x) = \lim_{x\to 0}\frac{xu'(x)}{u(x)} = \lim_{x\to 0}\frac{u'(x)+xu''(x)}{u'(x)} = 1.$$

Moreover,

$$\varepsilon_u'(x) = \frac{u'(x)}{u(x)} - x\left(\frac{u'(x)}{u(x)}\right)^2 + \frac{xu''(x)}{u'(x)}\frac{u'(x)}{u(x)} = \varepsilon_u(x)\frac{1-r_u(x)-\varepsilon_u(x)}{x},$$

while $\lim_{x\to 0}(1 - r_u(x) - \varepsilon_u(x)) = 0$. Therefore, using the L'Hospital rule once again, we obtain

$$\lim_{x\to 0}\varepsilon'(x) = \lim_{x\to 0}\varepsilon_u(x)\frac{1-r_u(x)-\varepsilon_u(x)}{x} = \varepsilon_u(0)\left(-r_u'(0) - \lim_{x\to 0}\varepsilon_u'(x)\right),$$

which implies

$$\varepsilon_u'(0) = \frac{1}{2}\frac{u''(0)}{u'(0)}.$$

Calculating derivative

$$A'(x) = \frac{1}{2}\left(1 - r_u'(x)\right) - \frac{\frac{1-r_u(x)-x}{2}\left(1+r_u'(x)\right)+xr_u'(x)+r_u(x)}{2\sqrt{\left(\frac{1-r_u(x)-x}{2}\right)^2 - xr_u(x)}},$$

and substituting $x = 0$, we obtain

$$A'(0) = \frac{1}{2}\left(1 - r_u'(0)\right) - \frac{\frac{1-r_u(0)}{2}\left(1+r_u'(0)\right)+r_u(0)}{1-r_u(0)} = \frac{u''(0)}{u'(0)}.$$

∎

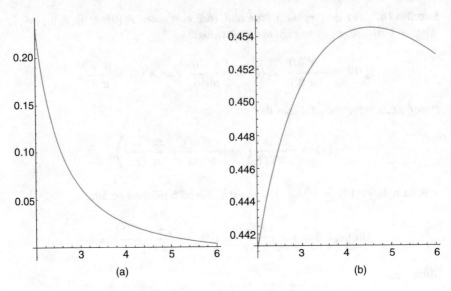

Fig. 16.1 (**a**) Firm's profit $\Pi(N)$. (**b**) Social welfare $V(N)$

Remark 16.1 It is easy to see that these assumptions are satisfied for all widely used non-CES "pro-competitive" classes of utility functions: HARA $u(x) = (x + \alpha)^\rho - \alpha^\rho$, $\alpha > 0$, CARA $u(x) = 1 - e^{-\alpha x}$, $\alpha > 0$, quadratic functions $u(x) = \alpha x - x^2/2$, $\alpha > 0$, as well as for any superposition of functions from these classes.

Theorem 16.1 *Let Assumptions 16.1 and 16.2 hold, then inequality (16.3) is satisfied for all sufficiently small $x > 0$, and thus $N^*(\varphi) < \widehat{N}(\varphi)$ holds for all sufficiently small φ.*

Proof Due to Lemma 16.1,

$$A'(0) - \varepsilon_u'(0) = \frac{u''(0)}{u'(0)} - \frac{u''(0)}{2u'(0)} = \frac{u''(0)}{2u'(0)} < 0,$$

which implies that $A(x) - \varepsilon_u(x) < 0$ for all $x > 0$ sufficiently small. ∎

To illustrate this result visually, let's consider the HARA utility $u(x) = \sqrt{x+1} - 1$ and $\varphi = f/L = 0.01$. Figure 16.1 shows that industry may accommodate with positive profit up to 6 firms, while the optimum number is approximately 4.

16.3.1 When Assumption 16.2 Does Not Hold

Consider two examples of utility function, that does not satisfy Assumption 16.2. These examples show that result may be ambiguous.

Case 16.1 Let $u(x) = x^\rho + \alpha x$ for $\alpha > 0$, then

$$\varepsilon_u(x) = 1 - \frac{1-\rho}{1+\alpha x^{1-\rho}}, \quad r_u(x) = \frac{\rho(1-\rho)}{\rho+\alpha x^{1-\rho}},$$

which implies $r_u(0) = 1 - \rho$, $\varepsilon_u(0) = \rho$, while

$$\varepsilon_u'(x) = \frac{\alpha(1-\rho)^2}{(1+\alpha x^{1-\rho})^2 \cdot x^\rho} \to +\infty, \quad r_u'(x) = -\frac{\alpha\rho(1-\rho)^2}{(\rho+\alpha x^{1-\rho})^2 \cdot x^\rho} \to -\infty,$$

when $x \to 0$. Differentiating the difference $A(x) - \varepsilon_u(x)$, we obtain

$$A'(x) - \varepsilon_u'(x) = \frac{1}{2} - \frac{\frac{1+r_u(x)-x}{2} - \frac{\alpha\rho(1-\rho)^2 \cdot x^{1-\rho}}{2(\rho+\alpha x^{1-\rho})^2}}{2\sqrt{\left(\frac{1-r_u(x)-x}{2}\right)^2 - xr_u(x)}} +$$

$$+\frac{\alpha(1-\rho)^2}{x^\rho}\left[\frac{\frac{\rho(1-r_u(x))}{2(\rho+\alpha x^{1-\rho})^2}}{2\sqrt{\left(\frac{1-r_u(x)-x}{2}\right)^2 - xr_u(x)}} + \frac{\rho}{2(\rho+\alpha x^{1-\rho})^2} - \frac{1}{(1+\alpha x^{1-\rho})^2}\right].$$

It is easy to see that

$$\frac{1}{2} - \frac{\frac{1+r_u(x)-x}{2} - \frac{\alpha\rho(1-\rho)^2 \cdot x^{1-\rho}}{2(\rho+\alpha x^{1-\rho})^2}}{2\sqrt{\left(\frac{1-r_u(x)-x}{2}\right)^2 - xr_u(x)}} \to -\frac{1-\rho}{\rho},$$

while term in brackets

$$\frac{\frac{\rho(1-r_u(x))}{2(\rho+\alpha x^{1-\rho})^2}}{2\sqrt{\left(\frac{1-r_u(x)-x}{2}\right)^2 - xr_u(x)}} + \frac{\rho}{2(\rho+\alpha x^{1-\rho})^2} - \frac{1}{(1+\alpha x^{1-\rho})^2} \to \frac{1-\rho}{\rho} > 0,$$

when $x \to 0$. This implies that

$$\lim_{x\to 0}(A'(x) - \varepsilon_u'(x)) = +\infty \Rightarrow A(x) > \varepsilon_u(x)$$

for all sufficiently small $x > 0$, or, equivalently, $N^*(\varphi) > \widehat{N}(\varphi)$ for all sufficiently small $\varphi = f/L$.

Case 16.2 Let $u(x) = x^\rho + \alpha x$ with $\alpha < 0$. This function satisfies $u'(x) > 0$ for all sufficiently small $x > 0$. Direct calculations show that

$$\varepsilon'_u(x) = \frac{\alpha(1-\rho)^2}{(1+\alpha x^{1-\rho})^2 \cdot x^\rho} \to -\infty, \quad r'_u(x) = -\frac{\alpha\rho(1-\rho)^2}{(\rho + \alpha x^{1-\rho})^2 \cdot x^\rho} \to +\infty,$$

and

$$\lim_{x \to 0} (A'(x) - \varepsilon'_u(x)) = -\infty \Rightarrow A(x) < \varepsilon_u(x)$$

for all sufficiently small $x > 0$, or, equivalently, $N^*(\varphi) < \widehat{N}(\varphi)$ for all sufficiently small $\varphi = f/L$.

Note that the case $\alpha = 0$, corresponding to the CES function, we obtain

$$\lim_{x \to 0} (A'(x) - \varepsilon'_u(x)) = -\frac{1-\rho}{\rho} < 0 \Rightarrow A(x) < \varepsilon_u(x)$$

with the same conclusion, which was proved directly at the very beginning of this Section.

16.4 Concluding Remarks

Economists have long believed that unencumbered entry is desirable for social efficiency. This view has persisted despite the illustration in several articles of the inefficiencies that can arise from free entry in the presence of fixed set-up costs. In this article we have attempted to elucidate the fundamental and intuitive forces that lie behind these entry biases. The previous papers with similar conclusions were based on assumption on the zero love for variety. Moreover, some papers, e.g., [8], suggested that in case of diversified goods the positive welfare effect of the love for variety may offset the negative effect of excessive enter. Our paper shows that generally this is not true–negative effect prevails for all known classes of utilities with non-decreasing love for variety. Nevertheless, the opposite example of insufficient enter was also built on the base of AHARA-utility with decreasing love for variety.

Acknowledgments I owe special thanks to C. d'Aspremont, J.-F. Thisse and M. Parenti for long hours of useful and fruitful discussions in CORE (Louvain-la-Neuve, Belgium) and in Higher School of Economics (St.-Petersburg Campus) on the matter of the Ford effect. The study was carried out within the framework of the state contract of the Sobolev Institute of Mathematics (project no. 0314-2019-0018). The work was supported in part by the Russian Foundation for Basic Research (project no. 18-010-00728).

References

1. Behrens, K., Murata, Y.: General equilibrium models of monopolistic competition: a new approach. J. Econ. Theory **136**, 776–787 (2007)
2. d'Aspremont, C., Dos Santos Ferreira, R., Gerard-Varet, L.: On monopolistic competition and involuntary unemployment. Q. J. Econ. **105**(4), 895–919 (1990)
3. d'Aspremont, C., Dos Santos Ferreira, R.: The Dixit–Stiglitz economy with a 'small group' of firms: a simple and robust equilibrium markup formula. Res. Econ. **71**(4), 729–739 (2017)
4. Berry, S.T., Waldfogel, J. Free entry and social inefficiency in radio broadcasting. RAND J. Econ. **30**(3), 397–420 (1999)
5. Berry, S.T., Eizenberg, A, and Waldfogel, J. Optimal product variety in radio markets. RAND J. Econ. **43**(3), 463–497 (2016)
6. Dixit, A.K., Stiglitz, J.E.: Monopolistic competition and optimum product diversity. Am. Econ. Rev. **67**, 297–308 (1977)
7. Hart, O.: Imperfect competition in general equilibrium: an overview of recent work. In: Arrow, K.J., Honkapohja, S. (eds.) Frontiers in Economics. Basil Blackwell, Oxford (1985)
8. Mankiw, N.G., Whinston, M.D.: Free entry and social inefficiency. RAND J. Econ. **17**(1), 48–58 (1986)
9. Marschak T., Selten R.: General equilibrium with price-making firms. Lecture Notes in Economics and Mathematical Systems. Springer, Berlin (1972)
10. Parenti, M., Sidorov, A.V., Thisse, J.-F., Zhelobodko, E.V.: Cournot, Bertrand or Chamberlin: toward a reconciliation. Int. J. Econ. Theory **13**(1), 29–45 (2017)
11. Perry, M.K.: Scale economies, imperfect competition, and public policy. J. Ind. Econ. **32**, 313–330 (1984)
12. Sidorov, A.V., Parenti, M., J.-F. Thisse: Bertrand meets Ford: benefits and losses. In: Petrosyan, L., Mazalov, V., Zenkevich, N. (eds.) Static and Dynamic Game Theory: Foundations and Applications, pp. 251–268. Birkhäuser, Basel (2018)
13. Spence, A.M.: Product selection, fixed costs, and monopolistic competition. Rev. Econ. Stud. **43**, 217–236 (1976)
14. von Weizsäcker, C.C.: A welfare analysis of barriers to entry. Bell J. Econ. **11**, 399–420 (1980)
15. Zhelobodko, E., Kokovin, S., Parenti M., Thisse, J.-F.: Monopolistic competition in general equilibrium: beyond the constant elasticity of substitution. Econometrica **80**, 2765–2784 (2012)

Chapter 17
An Alternative Pursuit Strategy in the Game of Obstacle Tag

Igor Shevchenko

Abstract According to the generalized Isaacs' approach, when solving a differential game, one has to fill the state space with trajectories on which the value function in some sense meets the main equation. Singular surfaces are manifolds where the value function or its derivatives are discontinuous. The obstacle tag game is a prototypical example which was used by R. Isaacs to illustrate some phenomena arising in differential games. The solution proposed by J.R. Isbell contains no singular surfaces. Afterward, several solutions with corner surfaces were described. J. Breakwell first constructed the field of optimal trajectories with a focal line and then with two switch envelops. A. Melikyan formed a field with two equivocal surfaces. In the paper, we consider the obstacle chase as an alternative pursuit game. In the part of the state space where the segment PE crosses the obstacle and alternatives are not consistent, as compared to Cases 7 and 8 of the Isbell's solution, the generated pursuit strategy with memory allows P to switch geodesic lines a finite number of times only on boundaries of the secondary domain, and thereby prevents sliding motions. Numerical simulations for particular states show that the guaranteed results for this strategy are quite close to the value functions for fixed alternatives and to those that constructed by J. Breakwell and A. Melikyan.

Keywords Switching between alternatives · Preventing sliding motions · Pursuit strategies with memory

I. Shevchenko (✉)
Pacific Branch of the Russian Federal Research Institute of Fisheries and Oceanography, Vladivostok, Russia

Far Eastern Federal University, Vladivostok, Russia
e-mail: igor.shevchenko@tinro-center.ru

L. A. Petrosyan et al. (eds.), *Frontiers of Dynamic Games*,
Static & Dynamic Game Theory: Foundations & Applications,
https://doi.org/10.1007/978-3-030-51941-4_17

17.1   Introduction

In the game of obstacle tag, let at $t = t_1$ the segment $P^{t_1} E^{t_1}$ cross the circular obstacle centered at C (see Fig. 17.1). Striving to catch E in minimal time, P may follow the geodesical line shortest at this instant. However, when E retreats on the continuation of this line, P may recognize that the other geodesic line becomes of equal length (when P, E, C are collinear at $t = t_2$) first, and then gets even shorter (at $t = t_3$). Switching the line, e.g., at $t = t_3$, P may reduce the initially evaluated chase time that equal to the length of the geodesic line at the current state divided by the speed difference. At the first International Symposium on the Theory and Applications of Differential games held in Amherst in 1969, R. Isaacs mentioned [1] that the ideas of his book [3] aren't suitable for analyzing the obstacle tag game in the described situation (see Problem 6.10.1). Different aspects of this game were studied by numerous authors (see, e.g., [1, 4–6] for the most relevant results and further references).

J.R. Isbell described a solution of this game without using any formalism [4]. He assumed that P moves along the geodesic line whereas E maintains collinearity of P, C, E (Case 7) or takes a secant line to the circle (Case 8) avoiding the situation shown in Fig. 17.1 by that ways. According to the generalized Isaacs' approach, singular surfaces is the main subject of zero-sum two-person differential games with full information. The state space has to be filled with the trajectories corresponding to the coupled optimal pursuit-evasion strategies. The value function is evaluated as the payoff for these trajectories. Singular surfaces are manifolds where the value function or its derivatives are discontinuous. However, there is no general theory of construction for singular surfaces. Commonly, a researcher needs to explore different known options for such surfaces [2]. The Isbell's solution is rather simple and includes no singular surfaces. It was revisited several times by J. Breakwell and his students with the use of a focal line first, and two switching envelops then [1, 5]. A. Melikyan and his students suggested that P and E have to move along straight lines in the attraction domains of the corner surfaces and their solution contains two equivocal surfaces [6].

A pursuit-evasion game is called alternative if it can be terminated by P at will on any of two given terminal manifolds, the payoff functionals of Boltza type on these manifolds differ only in their terminal parts (the integral part is common and equal to 1) and the optimal feedback strategies and the value functions are known [7–9].

Fig. 17.1 Switching preferable geodesic lines

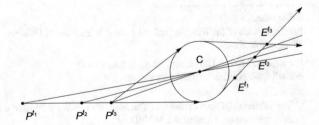

We consider the obstacle tag game as an alternative pursuit game. At every state where the segment PE crosses the obstacle, P has two alternatives, i.e., to follow the south or north geodesic line. For each of them, the guaranteed catch time is known, and P may choose those with lesser value. However, if a any state P chooses the shortest path to E, a sliding mode may arise on the manifold with collinear P, C, E, and the payoff is undefined there if the corresponding trajectories are defined as limits of Euler broken lines there.

First, we describe a setup of the game. Then, we analyze the structure of the game space in terms of relations of domination between alternatives, and their consistency. Finally, we describe a pursuit strategy with memory and evaluate the guaranteed result solving control optimization problems for E.

17.2   Setup

Let the obstacle be a circular hole of unit radius centered at $z_c = (x_c, y_c)$ in the plane. Let $z_p = (x_p, y_p)$ and $z_e = (x_e, y_e)$ be Cartesian coordinates of players, $\|z_p - z_c\| > 1$, $\|z_e - z_c\| > 1$. Let P and E have simple motions with speed 1 and β, $\beta < 1$, the players perfectly measure all coordinates and P strive to catch E in minimum time. We consider the game only for the initial states where the obstacle separates players. The game terminates at the first instant when P gets on the obstacle boundary or the line segment PE is tangential to the circle. P can follow the shortest geodesic line chosen at the initial state and guarantee that the time spent on E's point capture is less or equal to the initial distance between P and E along the geodesic line divided by $(1 - \beta)$. Our goal is to generate a pursuit strategy which allows P to choose geodesic lines if it would be advantageous to him, and evaluate corresponding guaranteed results for this strategy.

We will put the game into different reduced spaces of dimension three. Depending on the chosen state space, we will have different equations describing motions. The target set will be the set of states where P fixes his choice of the geodesic line when evaluates the guaranteed payoff finally.

Let the obstacle be centered at $z_c = (0, 1)$, and P be on the negative part of the x-axis. If $z_p = (-\rho_p, 0)$ then $\rho_p > 0$ is the distance from P to the obstacle along the tangential line. Let $z_e = (x_e, y_e)$ be Cartesian coordinates of E, $x_e, y_e \geq 0$, $\|z_e - z_c\| > 1$. Let $\alpha_p = \arctan \rho_p^{-1}$ and $y_e/(\rho_p + x_e) \leq \tan \alpha_p$. Then, the function V^s that evaluates the guaranteed time needed for point capture of E along the south geodesic line and its continuation may be described as

$$V^s(\rho_p, x_e, y_e) = \frac{\rho_p + \theta_e + \rho_e}{1 - \beta}, \qquad (17.1)$$

where $\theta_e = \alpha_e + \gamma_e$, $\alpha_e = \arctan \rho_e^{-1}$, $d_e = \sqrt{x_e^2 + (1 - y_e)^2}$, $\gamma_e = \arctan(y_e - 1)/x_e$, $\rho_e = \sqrt{d_e^2 - 1}$ (see Fig. 17.2).

Fig. 17.2 The first reduced space

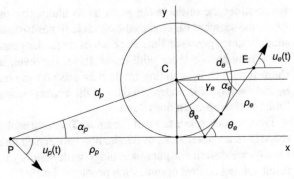

Fig. 17.3 The second reduced space

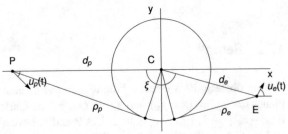

In the second reduced space [5], the obstacle is centered at the origin O. P lies on the negative part of the x-axis at $z_p = (-d_p, 0)$, $d_p \geq 1$, the distance from the origin to E equals $d_e \geq 1$ and ξ is the angle between OP and OE (see Fig. 17.3). The evaluation function may be represented as

$$V^s(d_p, \xi, d_e) = \frac{\sqrt{d_p^2 - 1} + \xi - \arccos d_p^{-1} - \arccos d_e^{-1} + \sqrt{d_e^2 - 1}}{1 - \beta}. \quad (17.2)$$

At the instant $t > 0$, let P and E be separated by the obstacle and move at angles $u_p(t)$ and $u_e(t)$ (see Fig. 17.4). Then, for $\arccos d_p^{-1} \leq \xi \leq \pi$, their motions may be described by the equations

$$\dot{d}_p(t) = \cos u_p(t),$$

$$\dot{\xi}(t) = \frac{\sin u_p(t)}{d_p(t)} + \beta \frac{\sin(u_e(t) + \pi - \xi(t))}{d_e(t)}, \quad (17.3)$$

$$\dot{d}_e(t) = \beta \cos(u_e(t) + \pi - \xi(t)).$$

Let $Z \subseteq R^3$ and M be the game space and terminal set, $U_p = \{u_p : \|u_p\| \leq 1\}$, $U_p = \{u_e : \|u_e\| \leq \beta\}$, $z(t) \in Z$, $u_p(t) \in U_p$, $u_e(t) \in U_e$ and

$$\dot{z}(t) = f(z(t), u_p(t), u_e(t)), \quad z(0) = z^0, \quad (17.4)$$

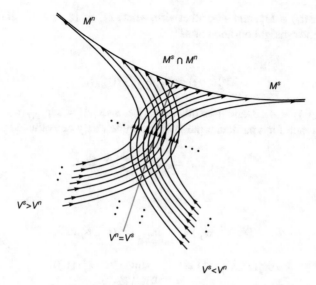

Fig. 17.4 A two-dimensional model of the state space: two fields of trajectories

be the equation that describes motions.

Strategies are rules that map available information into control values. We use equations like (17.4) to generate pencils of trajectories for given initial states and strategies, and then to evaluate the performance index for them. We consider only trajectories that are limits of Euler broken lines when diameters of time partitions tend to zero. This approach allows getting solutions that provide results close to guaranteed in numerical simulations of the game development.

Let $\Delta = \{t_0, t_1, \ldots, t_i, t_{i+1}, \ldots\}$ be a partition of the time axis R^+. For a given $z^0 \in Z$ and a chosen strategy $\mathscr{U}_p$ with available information I (e.g., $\mathscr{U}_p \div u_p$: $Z \to U_p$ for feedback strategies or $\mathscr{U}_p \div u_p$: $R^+ \times C^3_{[0,\infty)} \to U_p$ for memory strategies), let denote as $Z_p(z^0, \mathscr{U}_p, \Delta)$ the pencil of piecewise-constant solutions of the inclusion

$$\dot{z}(t) \in \text{co}\{f(z(t_i), u_p(t_i), u_e) : u_e \in U_e\}, \tag{17.5}$$

where $t \in [t_i, t_{i+1})$, $i \in N$, $t_0 = 0$, $t_i \to_{i\to\infty} \infty$, $u_p(t_i)$ generated by $\mathscr{U}_p$ with information available at the instant $t = t_i$. By this means $Z_p(z^0, \mathscr{U}_p, \Delta)$ contains continuous functions $z : R^+ \to Z$ for which there exists an absolutely continuous restriction onto $[0, \theta]$ for any $\theta > 0$ that meets (17.5) for almost all $t \in [0, \theta]$.

For the first (south) alternative, given z^0, $\mathscr{U}_p$, $M = M^s$, $\varepsilon > 0$, Δ and $z(\cdot) \in Z_p(z^0, \mathscr{U}_p, \Delta)$, let

$$\tau^s_\varepsilon(z(\cdot)) = \min\{t_i \in \Delta : z(t_i) \in M^s_\varepsilon\}, \tag{17.6}$$

if $\exists t_i \in \Delta : z(t_i) \in M_\varepsilon^s$, and $+\infty$ otherwise, where $M_\varepsilon^s = \{z : z \in Z, \min_{z' \in M^s} \|z - z'\| \le \varepsilon\}$ is the ε-neighbourhood of M^s.

Let

$$\mathscr{P}_\varepsilon^s(z(\cdot)) = \tau_\varepsilon^s + V^s(z(\tau_\varepsilon^s)), \tag{17.7}$$

if $\tau_\varepsilon^s = \tau_\varepsilon^s(z(\cdot)) < +\infty$, and $+\infty$ otherwise. Let also $|\Delta| = \sup_{i \in N}(t_{i+1} - t_i)$. The guaranteed result for a particular pursuit strategy $\mathscr{U}_p$ may be evaluated as

$$\mathscr{P}^s(z^0, \mathscr{U}_p) = \lim_{\varepsilon \to 0+} \mathscr{P}_\varepsilon^s(z^0, \mathscr{U}_p), \tag{17.8}$$

where

$$\mathscr{P}_\varepsilon^s(z^0, \mathscr{U}_p) = \lim_{|\Delta| \to +0} \mathscr{P}_\varepsilon^s(z^0, \mathscr{U}_p, \Delta), \tag{17.9}$$

$$\mathscr{P}_\varepsilon^s(z^0, \mathscr{U}_p, \Delta) = \sup_{z(\cdot) \in Z_p(z^0, \mathscr{U}_p, \Delta)} \mathscr{P}_\varepsilon^s(z(\cdot)).$$

For coupled pursuit and evasion strategies $\mathscr{U}_p$ and $\mathscr{U}_e$, the guaranteed result that defined according to the described scheme is denoted as $\mathscr{P}^s(z^0, \mathscr{U}_p, \mathscr{U}_e)$.

Similarly, we define the game with the second (south) alternative and the terminal set M^n, the guaranteed payoff $\mathscr{P}^n(z^0, \mathscr{U}_p)$ for $z^0 \in Z$ and $\mathscr{U}_p$, etc.

The game with free alternative is completed on $M = M^s \cup M^n$. For $z^0 \in Z$ and $\mathscr{U}_p$, if P fixes the preferable alternative, he guarantees the payoff

$$\mathscr{P}(z^0, \mathscr{U}_p) = \min(\mathscr{P}^s(z^0, \mathscr{U}_p), \mathscr{P}^n(z^0, \mathscr{U}_p)).$$

Our goal is to generate featured pursuit strategies for the game with free alternative and evaluate corresponding guaranteed payoffs.

17.3 Gradient Strategies

In the game with south alternative, define the (universal) gradient pursuit strategy for P that generates the control according to the following relation [10]

$$u_p^s(z) = \arg \min_{u_p \in U_p} \max_{u_e \in U_e} \frac{\partial V^s(z)}{\partial f(z, u_p, u_e)}, \quad z \in Z. \tag{17.10}$$

For $z = (d_p, \xi, d_e)$ and (17.2), (17.3), where $-\pi \le u_e, u_e \le \pi$, we have

$$\frac{\partial V^s(z)}{\partial f(z, u_p, u_e)} = \frac{\sin(u_p + \arccos d_p^{-1}) - \beta \sin(u_e - \xi + \arccos d_e^{-1})}{(1 - \beta)}. \tag{17.11}$$

Therefore, the guaranteed pursuit strategy corresponds to $u_p^s = -\arcsin d_p^{-1}$, i.e. P follows the south geodesic line at every state.

Moreover, the (universal) gradient evasion strategy that defined a similar way corresponds to $u_e^s = \arcsin d_e^{-1} - (\pi - \xi)$ when at every state E flees on the continuation of south geodesic line,

$$\min_{u_p^s} \max_{u_e^s} \frac{\partial V^s(z)}{\partial f(z, u_p, u_e)} = \frac{\partial V^s(z)}{\partial f(z, u_p^s, u_e^s)} = -1. \tag{17.12}$$

and

$$\min_{u_p^s} \frac{\partial V^s(z)}{\partial f(z, u_p, u_e)} = \frac{\partial V^s(z)}{\partial f(z, u_p^s, u_e)} < -1 \tag{17.13}$$

if $u_e \neq u_e^s$ [6, 11].

The same results are valid for the north alternative with the evaluation function V^n and controls u_p^n, u_e^n.

Therefore, in the game with a fixed alternative, at any state, the payoff V^s or V^n is guaranteed if P follows the respective geodesic line and E retreats on its continuation.

17.4   Decomposition of the State Space

Let us denote as $\mathscr{U}_p^0 \div u_p^0 : Z \rightarrow U_p$ the pursuit gradient strategy that at the state z generates the control $u_p^0(z^0, z) = u_p^s(z)$ (17.10) if the south geodesic lines is preferable for P at the state z^0 (or either of them if they are of equal length) and $u_p^n(z)$ otherwise. If P updates the target alternative at any current state, denote this universal strategy as $\mathscr{U}_p^Z \div u_p^t : Z \rightarrow U_p$. Also let us use similar notations $\mathscr{U}_e^0$ and $\mathscr{U}_e^Z$ for corresponding evasion strategies.

At any state, P has two alternative ways to chase E with known guaranteed results evaluated with V^s or V^n. Then the state space is filled with two families of ideal trajectories corresponding to the coupled geodesic pursuit-evasion strategies $\mathscr{U}_p^0$ and $\mathscr{U}_e^0$ (see Fig. 17.4). Starting at $z^0 \in Z$, P with $\mathscr{U}_p^0$ can guarantee the payoff equal to $\min(V^s(z^0), V^n(z^0))$.

Consider guaranteed results if P can switch between alternatives.

At the state $z^0 \in Z$, an alternative (south or north) is called consistent (stable) if it dominates the other one at the initial state and also at any state emerged when P and E move along the related geodesic line and its continuation. Let us divide the state space depending on the consistency of the relation of domination (see Fig. 17.5):

- Z^s and Z^n are subsets where the particular alternative (south or north) strictly dominates the other one and is consistent,
- $D^{s|n}$ disjoints Z^s and Z^n, and $V^s = V^n$ there,

Fig. 17.5 A two-dimensional
model of the state space: a
partition of the state space

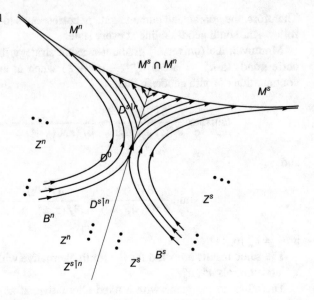

- $Z^{\bar{s}}$ and $Z^{\bar{n}}$ are subsets where the particular alternative strictly dominates other and is not consistent, $Z^{s\overline{|n}} = Z^{\bar{s}} \cup Z^{\bar{n}}$,
- $D^{s\overline{|n}}$ disjoints $Z^{\bar{s}}$ and $Z^{\bar{n}}$, and $V^s = V^n$ there,
- $D^0 \subset D^{s\overline{|n}}$ disjoints $D^{s|n}$ and $D^{s\overline{|n}}$,
- B^s disjoints Z^s and $Z^{\bar{s}}$, and when the players move along the south geodesic line and its continuation, the south alternative strictly dominates the north one, and there exists exactly one instant when the alternatives become equivalent,
- B^n is defined similar to B^s.

In the game of obstacle tag (see Fig. 17.6), for a given ρ_p, $Z^{s\overline{|n}}$ consists from two curvilinear triangles. They are joined along the half-line from P trough C. All D^0 for different ρ_p lie on this half-line

$$y = 1 + \sqrt{\beta}, \quad x \geq \sqrt{1 - \beta}. \tag{17.14}$$

Fig. 17.6 Decomposition of
the reduced space for a given
ρ_p

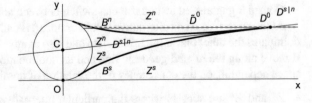

It is evident that $Z^{s\bar{|}n} = \varnothing$ for $\rho_p < \rho_p^*(\beta)$ where

$$\rho_p^*(\beta) = \sqrt{(1-\beta)/\beta}.$$

Therefore, actually the game may be terminated on the subsurface $\rho_p = \rho_p^*(\beta)$, and the state definitely leaves $Z^{s\bar{|}n}$ when P follows the geodesic line.

The dotted line in Fig. 17.6 shows the locus of E's terminal positions of the secondary domain $\tilde{D}^0$ for different ρ_p for some known solutions of the game(see, e.g., [5, 6]). It has the parametric representation

$$x_e = \qquad 1/\sqrt{1-(1-s)^2}, \qquad (17.15)$$

$$y_e = \beta/\sqrt{\beta^2 - (\beta-s)^2}, \ 0 < s \leq \beta. \qquad (17.16)$$

The half-line D^0 (see (17.14)) is a horizontal asymptote for it.

17.5 Alternative Pursuit Strategy with Memory

The strategy $\mathcal{U}_p^{\mathcal{Z}}$ is discontinuous on $D = D^{s|n} \cup D^{s\bar{|}n}$. When the state gets in the neighbourhood of $D^{s\bar{|}n}$, piecewise-constant solutions of the inclusion (17.5) with $\mathcal{U}_p^{\mathcal{Z}}$ stay there for some time. For P, it may lead to the switching control in sliding mode for which the payoff couldn't be evaluated with the use Euler broken lines.

For the initial state $z^0 \in Z^{s\bar{|}n} = Z^{\bar{s}} \cup D^{s\bar{|}n} \cup Z^{\bar{n}}$, let us allow P to remember the history of the game development and to update the target alternative no more than once on B^s or B^n when the state leaves $Z^{s\bar{|}n}$. Let us denote the corresponding strategy as $\mathcal{U}_p^{\mathcal{A}}$. The strategy $\mathcal{U}_p^{\mathcal{B}}$ prevents P from switching between alternative strategies in the neighbourhood of $D^{s\bar{|}n}$.

Let us evaluate the guaranteed result $\mathcal{P}^{s|n}$ (see (17.8)) when $z^0 \in Z^{s\bar{|}n}$, P applies $\mathcal{U}_p^{\mathcal{A}}(z^0, z)$ in $z \in Z^{s\bar{|}n}$ and $\mathcal{U}_p^{\mathcal{Z}}(z^0, z)$ for z the rest of the state space. In order to do that, we setup and solve optimization control problems for termination sets $D^{s\bar{|}n}$, B^s and B^n (see Fig. 17.7 for $z^0 \in Z^{\bar{s}}$). Thereafter, we assume that E moves at the angle ψ in a straight line within $Z^{s\bar{|}n}$ until the first instant when the state arrives on one of the terminal sets. The maximal of three corresponding estimations determines the guaranteed result.

On B^s and B^n, the guaranteed results correspond to the gradient strategies (see Sect. 17.3). From the boundaries, the state shifts on D^0 since B^s and B^n are themselves ideal trajectories for the coupled gradient strategies. Thus, in all cases, the state leaves the closure of $Z^{s\bar{|}n}$ through D^0 (see Fig. 17.7).

If the state under the E's control first gets on $D^{s\bar{|}n}$ and then on D^0 along $D^{s\bar{|}n}$ (see Fig. 17.7), the guaranteed result is described in [7]. It turns out that the preliminary

Fig. 17.7 A two-dimensional
model of the state space:
different options to leave $Z^{\bar{s}}$

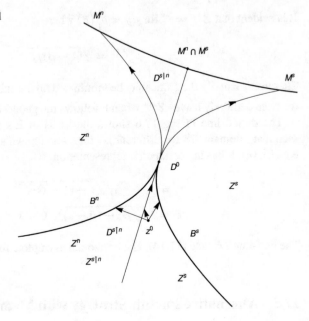

straight line is tangent to the curvilinear motion along $D^{s\bar{|}n}$ (the Isbell's Case 8 in
[4]) where at the state $(\rho_p, x_e, y_e) \in D^{s\bar{|}n}$ E chooses the angle [7]

$$\psi^{D^{s\bar{|}n}} = \arcsin \frac{(y_e - 1)}{\beta\sqrt{1+\rho_p^2}} + \arcsin \frac{1}{\sqrt{1+\rho_p^2}}.$$

It is important to note that the Isbell's solution for Case 7 [4] when the state shifts
on D^0 directly from within $Z^{s\bar{|}n}$ is just infeasible.

Thus far, for the initial state $z^0 \in Z^{s\bar{|}n}$ at any current state $z \in Z^{s\bar{|}n}$ with the
reduced coordinates (ρ_p, x_e, y_e), the angle ψ chosen by E determines the instants
$\tau^{s\bar{|}n}, \tau^{\bar{s}}, \tau^{\bar{n}}$ when the state arrives on $D^{s\bar{|}n}$, B^s, B^n, and the associated payoffs.
The maximal of them defines the guaranteed result $\mathscr{P}^{s\bar{|}n}$ for the described pursuit
strategy. As numerical simulations show, in the secondary optimization problem,
the preferable option for E always corresponds to the case when E from the initial
states $Z^{\bar{s}}$ shifts on $B^{\bar{n}}$, and on $B^{\bar{s}}$ from $Z^{\bar{n}}$. In this case, E takes the secant line
with minimal angle for which he gets from $Z^{\bar{s}}$ on $B^{\bar{n}}$ missing $B^{\bar{s}}$ or from $Z^{\bar{n}}$ on $B^{\bar{s}}$
missing $B^{\bar{n}}$.

Let $(d_p, \pi, 1)$ be the state vector in the second reduced space. An example of
the optimal evasion trajectory generated with the use of the described approach and
provided the guaranteed result is shown in Fig. 17.8 ($D^{s\bar{|}n}$, B^s, B^n, etc. are given for
$t = \tau^{D^0}$). Detail descriptions of solutions of the obstacle tag game mentioned in,
e.g., [1, 5, 6] are not available. However, it may be safely suggested that for the states
with E on the obstacle and collinear with P and C, the known the value function

Fig. 17.8 An example of optimal evasion in $Z^{\tilde{s}}$

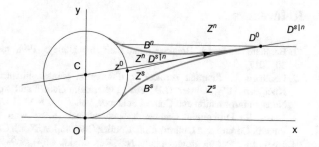

take the value defined on straight line motions of the players along the segment PE and its continuation. Then, the guaranteed result for them may be evaluated as $\tau^{\tilde{D}^0} + V^s(d_p - \tau^{\tilde{D}^0}, \pi, 1 + \beta\tau^{\tilde{D}^0})$ (see (17.2)) where $\tau^{\tilde{D}^0}$ is the first instant when E gets on $\tilde{D}^0$ (17.15) (see also Fig. 17.7). It's also worth noting that for $0.1 \leq \beta \leq 0.9$, $\rho_p > \rho_p^*(\beta)$, the maximum relative difference between these values and the guaranteed results for the described pursuit strategy with memory is about 1%.

17.6   Conclusion

In the situations shown in Fig. 17.1, evaluation of the guaranteed result as corresponding to the south geodesic line appears too pessimistic to P. On the other hand, if P chooses the shortest geodesic at any current state, this feedback strategy is discontinuous for collinear P, C, E. The resulting pencil of trajectories approximated by Euler broken lines doesn't include associated trajectory with P and E moving along the half-line. To form a pursuit strategy and to evaluate the guaranteed result, e.g., J. Breakwell and A. Melikyan described the fields of trajectories for coupled optimal strategies of the players with two switch envelops or equivocal lines [1, 5, 6]. The construction of such fields involves a cumbersome integration of characteristic equations for Hamilton-Jacobi-Isaacs equations.

We considered the obstacle chase game as an alternative pursuit game. The state space was divided into several parts depending on the domination and consistency features of alternatives at the initial state. In the parts corresponding to the situation similar to that shown in Fig. 17.1, the generated strategy with memory allows P to switch between alternatives only a finite number of times on their boundaries. The guaranteed results fit the evasion strategy whereby E takes a secant line to the obstacle until the state arrives on the boundary. Therefore, for the states in the special region, P uses a strategy that doesn't depend on the current position of E until the state reaches the boundary.

The approach can be modified to handle the games with convex obstacles of different shapes; see, e.g., [6]. However, decomposition of the state space will be asymmetrical and there will be different termination options for the alternatives when solving the secondary control problems for E.

References

1. Bernhard, P.: Isaacs, Breakwell, and their sons (June 6, 1998, revised April 15, 2015, March 30, 2017)
2. Bernhard, P.: Singular surfaces in differential games: an introduction. In: Hargedorn, P., Knobloch, H.W., Olsder, G.H. (eds.) Differential Games and Applications. Springer Lecture Notes in Information and Control Sciences, Vol. 3, pp. 1–33 (1977)
3. Isaacs, R.: Differential Games: A Mathematical Theory with Applications to Warfare and Pursuit, Control and Optimization. Courier Corporation, North Chelmsford (1999)
4. Isbell, J.R.: Pursuit around a hole. Nav. Res. Logist. **14**(4), 569–571 (1967)
5. Lewin, J.: Differential Games: Theory and Methods for Solving Game Problems with Singular Surfaces. Springer Science & Business Media, Berlin (2012)
6. Melikyan, A.: Generalized Characteristics of First Order PDEs: Applications in Optimal Control and Differential Games. Springer Science and Business Media, Berlin (2012)
7. Shevchenko, I.: Geometry of the Alternative Pursuit. FESU, Vladivostok (2003) (in Russian)
8. Shevchenko, I.: On reduction of alternative pursuit games. Adv. Dynam. Games Their Appl. **11**, 125–137 (2007)
9. Shevchenko, I.: Strategies for alternative pursuit games. Ann. ISDG **10**, 121–131 (2009)
10. Shevchenko, I., Stipanović, D.M.: A design of strategies in alternative pursuit games. In: Petrosyan, L.A., Zenkevich, N.A. (eds.) Contributions to Game Theory and Management, vol. IX, pp. 266–275. Saint Petersburg State University, St Petersburg (2016)
11. Subbotin, A.I.: Generalized Solutions of First Order PDEs: The Dynamical Optimization Perspective. Springer Science & Business Media, Berlin (2013)